Collisional Transport in Magnetized Plasmas

Collisional transport theory is of central importance to modern plasma physics. This book provides a self-contained treatment of the subject, starting from elementary concepts and developing the theory through to the research frontier.

Basic tools of kinetic plasma theory, such as the drift kinetic equation and the Coulomb collision operator are derived, and are then used to calculate classical and neoclassical transport occurring in high-temperature plasmas. Important phenomena such as neoclassical diffusion, bootstrap current, and plasma rotation are carefully explained.

Students, theoreticians and experimentalists in both fusion and space plasma physics will benefit from this book, which emerged from a graduate student level course taught at MIT.

PER HELANDER is a research scientist at the Culham Science Centre of the United Kingdom Atomic Energy Authority. After obtaining his doctorate from Chalmers University of Technology in Göteborg, Sweden, he held a post-doctoral position at Massachusetts Institute of Technology, where he remains a regular visitor. His scientific interests include most topics of theoretical plasma physics, in particular kinetic theory.

DIETER SIGMAR received his doctorate from the Technical University of Vienna where he is now a.o. Professor of Theoretical Physics. He has had a long career in the US Fusion Program, and is now retired from his position as head of the Plasma Theory Group at the Plasma Science and Fusion Center at Massachusetts Institute of Technology. He is a fellow of the American Physical Society and Corresponding Member of the Austrian Academy of Sciences.

CAMBRIDGE MONOGRAPHS ON PLASMA PHYSICS

General Editors: M. G. Haines, K. I. Hopcraft, I. H. Hutchinson, C. M. Surko and K. Schindler

Collisional Transport in Magnetized Plasmas

Per Helander
Culham Science Centre

Dieter J. Sigmar
Massachusetts Institute of Technology

CAMBRIDGE
UNIVERSITY PRESS

CAMBRIDGE UNIVERSITY PRESS
Cambridge, New York, Melbourne, Madrid, Cape Town, Singapore, São Paulo

Cambridge University Press
The Edinburgh Building, Cambridge CB2 2RU, UK

Published in the United States of America by Cambridge University Press, New York

www.cambridge.org
Information on this title: www.cambridge.org/9780521807982

First published 2002
This digitally printed first paperback version 2005

A catalogue record for this publication is available from the British Library

Library of Congress Cataloguing in Publication data

Helander, Per, 1967–
Collisional transport in magnetized plasmas / Per Helander, D. J. Sigmar
p.   cm. – (Cambridge monographs on plasma physics ; 4)
Includes bibliographical references and index.
ISBN 0 521 80798 0
1. Transport theory. 2. Plasma dynamics. 3. Magnetohydrodynamics.
I. Sigmar, D. J. II. Title. III. Series.

QC718.5.T7 H45   2002
530.4′.46–dc21   2001035677

ISBN-13 978-0-521-80798-2 hardback
ISBN-10 0-521-80798-0 hardback

ISBN-13 978-0-521-02098-5 paperback
ISBN-10 0-521-02098-0 paperback

To Åsa and Lisl

Errata

Misprints

Page	Position	Printed text	Should be
34	Eq. (3.32)	$\mathcal{L}(f_e)$	$\mathcal{L}(f_i)$
112	4 lines after first equation	$\partial_\alpha \overline{\overline{L}}/\partial\dot{\alpha} = 0$	$\partial\overline{\overline{L}}/\partial\dot{\alpha} = 0$
112	5 lines after first equation	$\partial_\beta \overline{\overline{L}}/\partial\dot{\beta} = Ze\alpha$	$\partial\overline{\overline{L}}/\partial\dot{\beta} = Ze\alpha$
121	3 lines after Eq. (7.8)	$(\mathbf{V}_a \cdot \nabla)\mathbf{V}_a$	$m_a n_a(\mathbf{V}_a \cdot \nabla)\mathbf{V}_a$
128	Eq. (7.18)	$\partial L/\dot{\varphi}$	$\partial L/\partial\dot{\varphi}$
128	Eq. (7.19)	$\partial\overline{L}/\dot{\overline{\varphi}}$	$\partial\overline{L}/\partial\dot{\overline{\varphi}}$
130	Last equation	$\frac{\Delta v_\parallel}{R\Omega_p}$	$\frac{\Delta v_\parallel}{\Omega_p}$
145	Last equation	$\frac{q^2}{\Omega_\varphi R r}$	$\frac{qv^2}{\Omega_\varphi R r}$
150	Figure caption, line 2	solid	dashed
150	Figure caption, line 3	dashed	solid
157	Eq. (8.26)	$\frac{5p_a}{2e_a B}$	$\frac{5p_a}{2e_a}$
159	Eq. (8.28)	$\langle RF_{a\perp\varphi}/e_a\rangle$	$-\langle RF_{a\perp\varphi}/e_a\rangle$
194	Second line of Eq. (11.22)	$\int_0^\infty \frac{m_a v^4}{T_a} f_{a0} dv$	$\int_0^\infty A(v)\frac{m_a v^4}{T_a} f_{a0} dv$
203	1st line after 1st equation	$\mathbf{E} \times \mathbf{B}$	$\mathbf{E} \times \mathbf{B}$ velocity
222	Line 2	$p = (2p_\perp^2 + p_\parallel)/3$	$p = (2p_\perp + p_\parallel)/3$
232	6th equation	$\xi\frac{h_a}{\partial\theta}$	$\xi\frac{\partial h_a}{\partial\theta}$
251	5th equation	$\frac{e\Phi_0}{T_i}$	$\frac{e\Phi_0}{T_e}$
253	3rd equation	$\frac{v_\perp^2}{2}\frac{B_p^2}{B^2}$	$\frac{u_\perp^2}{2}\frac{B_p^2}{B^2}$
253	Eq. (13.12)	$u_\parallel\nabla f_{a1}$	$u_\parallel\nabla_\parallel f_{a1}$
266	2nd equation	$\left.\frac{T'}{T}f_M\right](x)$	$\frac{T'}{T}f_M(x)\Big]$

Conceptual issues

- The discussion on p. 163 assumes that the loop voltage is a flux function, which is not always true. A more general calculation is straightforward.

- The calculation in Sec. 13.4 and the paper on which it relies are not self-consistent. The density associated with f_0 varies substantially over a squeezed gyro-orbit.

Contents

Preface

Transport theory occupies a position at the heart of plasma physics. Using only the laws of physics and mathematics, it aims to predict macroscopic properties of the complex and often mysterious form of matter we call plasma. In transport theory, quantities such as electrical and thermal conductivity are calculated from first principles of classical mechanics and electromagnetism. This book has emerged from a graduate student course on this subject taught at MIT over the last two decades, and from many interactions with colleagues and students. Its purpose is to develop a path leading the committed student from elementary concepts of plasma physics to central results of classical and neoclassical transport theory – topics which are of everyday use in fusion and laboratory plasma research, and also find fruitful application in space physics and astrophysics.

In the preface to his landmark treatise *The Theory of Sound* (1877), Lord Rayleigh noted that the most valuable contributions to his subject were to be found only in scattered periodicals, published in various parts of the world, that were practically inaccessible to those who did not happen to live in the neighbourhood of large public libraries. It may be argued that something similar is true for plasma transport theory today. For instance, the best known exposition of classical transport theory (Braginskii, 1965) has been out of print for decades and seems to have 'disappeared' from most libraries, so that today it mostly exists in the form of photocopied photocopies. At the other end of the literature spectrum, recent advances in neoclassical theory can only be found in original research papers. Thanks to the internet, these are much more accessible than they were in Lord Rayleigh's days, but can be difficult to read for a newcomer to the field.

This book is an attempt to bridge this gap between basic textbooks and research literature. Our aim has been to provide a self-contained presentation of the fundamental theoretical concepts and mathematical techniques of kinetic theory, and to use these tools to calculate the most important transport properties of highly ionized plasmas. We only treat collisional transport and do not discuss so-called anomalous transport originating from plasma turbulence, which is a subject that fully deserves to be discussed in its own right.

Our reader is assumed to be familiar with the standard mathematical methods of physics, including vector analysis, ordinary differential equations, elements of analytical mechanics and electrodynamics, but no more than a basic grasp of the principles of plasma physics is required. As a text on this introductory level, the presentation is not meant to be comprehensive. Readers who want a more complete exposition of collisional transport theory are referred to the reviews by Braginskii (1965, if found) and Hinton (1983) for classical transport, to Hinton and Hazeltine (1976), and Hirshman and Sigmar (1981) for neoclassical transport, and to the two volumes by Balescu (1988) for both subjects. At the end of each chapter in this book, we provide a guide to more specific references in the literature, chosen from the wider body of original historical works.

Acknowledgments

This book is the natural outgrowth of decades of research and teaching, and thus owes much of its existence to the influence and sustained interest of a number of mentors as well as many of our peers in plasma physics. In 1971, Prof. B. Coppi (MIT) introduced me to a manuscript which soon after appeared as the seminal Rosenbluth, Hazeltine and Hinton paper in *Physics of Fluids* (1972). The same year, P. Rutherford (PPPL) extended an invitation to work with him on the bootstrap and seed current on axis problem. Following B. Kadomtsev, J. Clarke (ORNL) alerted us to the importance of including multiple impurity ion species in neoclassical theory. This led to impurity transport as a doctoral thesis topic of Stephen P. Hirshman, whose extraordinary analytic strength, creativity and hard work culminated in our review paper in *Nuclear Fusion* (1981). At MIT, it was my great pleasure to work with Chi Tien Hsu and Peter Catto on up/down asymmetry in strongly rotating plasmas and banana-regime transport of alpha particles. The formative influence of several plasma experimentalists in tokamak transport is acknowledged as well, most notably R. Petrasso, K. Wenzel, S. Scott, and G. Fussmann who kept me honest (and humble) as a theorist. Finally, my very special appreciation goes to Prof. O. Hittmair at the Technical University in Vienna whose unceasing interest and encouragement in my research, teaching and writing activities has accompanied me throughout my career.

Dieter J. Sigmar, Cambridge, Massachusetts, USA

My own introduction to kinetic plasma theory came through my teachers, mentors and friends Dan Anderson and Mietek Lisak at Chalmers University, and the beauty of this subject was further revealed to me by Peter

Catto at Massachusetts Institute of Technology. In writing this book, I am grateful to my colleagues at Culham Science Centre for their generous advice and continuous interest, which makes working in this highly cooperative branch of science so enjoyable. It is a pleasure to express our special thanks to Jack Connor, Ian Hutchinson and Olga Siddons for carefully reading the manuscript and suggesting several valuable improvements.

Per Helander, Abingdon, UK

1

Introduction

The behaviour of an ionized gas, or a plasma, is strongly affected by the presence of a magnetic field. Indeed, many types of plasmas in space and in the laboratory are *confined* by magnetic fields. This is, for instance, true of planetary magnetospheres and of magnetic fusion experiments. Collisional plasma transport theory is the study of particle, momentum, and energy transport in a plasma due to collisions. Given Coulomb collisions and particle orbits in a magnetic field **B**, one aims to find out how, and at what rate, charged particles are displaced perpendicularly to **B**, thus leading to transport. At other times, the focus is on parallel transport. As the particles are essentially free to move along the field in the absence of collisions, the problem is then to determine to what extent collisions prevent them from doing so.

To answer questions about perpendicular transport we shall derive transport equations such as

$$\frac{\partial n}{\partial t} + \nabla \cdot \mathbf{\Gamma} = S_{\text{particle}},$$

$$\frac{\partial}{\partial t} \frac{3nT}{2} + \nabla \cdot \mathbf{q} = S_{\text{heat}}, \tag{1.1}$$

where n and T are the number density (particles per volume) and temperature, $\mathbf{\Gamma}$ and $\mathbf{q}$ are the particle and heat fluxes, and the right-hand sides contain sources of particles and heat. To understand the structure of these equations, let us integrate them over a volume V, bounded by a surface ∂V. The first equation then becomes

$$\frac{d}{dt} \int_V n \, dV + \int_{\partial V} \mathbf{\Gamma} \cdot \mathbf{n} \, dS = \int_V S_{\text{particle}} \, dV,$$

where we have used Gauss's law to convert the second term to a surface integral. The volume element is denoted by dV, the surface element by

1

dS, and the unit vector perpendicular to the surface (pointing outward) by **n**. The first term evidently describes the variation of the total number of particles in V, and the second term the outflow of particles from V. The sum of these terms must equal the total number of particles added to V on the right-hand side. Equation (1.1) is similar, but deals with the energy density $3nT/2$ rather than the particle density n.

The task of transport theory is essentially to calculate the particle and heat fluxes Γ and **q**. Typically, they are related to the gradients ∇n and ∇T according to transport laws such as

$$\Gamma = -D\nabla_\perp n,$$

$$\mathbf{q} = -\kappa\nabla_\perp T,$$

where $\nabla_\perp$ denotes the gradient perpendicular to the magnetic field. D is called the diffusion coefficient and κ the heat conductivity. Collisional transport is caused by a random walk (explained in more detail later), and the diffusion coefficient scales as

$$D \sim \nu(\Delta x)^2,$$

where ν is the frequency at which the steps of random walk are taken and Δx the step size. The former is of the order of the collision frequency and the latter is related to the Larmor radius. For instance, for diffusion of electrons caused by electron–ion collisions,

$$D \sim \nu_{ei}\rho_e^2,$$

and for ion heat transport driven by ion–ion collisions,

$$\kappa \propto \nu_{ii}\rho_i^2.$$

Here ν_{ei} and ν_{ii} are the electron–ion and ion–ion collision frequencies, and ρ_e and ρ_i the electron and ion gyroradii,

$$\rho_a = v_{Ta}/|\Omega_a|, \quad a = e, i,$$

where

$$v_{Ta} = \sqrt{2T_a/m_a}, \quad a = e, i,$$

are the thermal velocities, and

$$\Omega_a = e_a B/m_a$$

the cyclotron (or gyro-) frequencies.

In the present chapter, we give heuristic, random-walk derivations of the fluxes Γ and **q**. The rest of the book is devoted to the rigorous derivation of transport laws and transport equations from first principles.

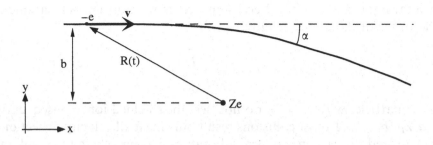

Fig. 1.1. An electron–ion collision.

1.1 Collision frequencies

Since Coulomb collisions are of fundamental importance for our topic, let us first develop the elementary scalings of the frequency v_{ei} for electron–ion collisions. From kinetic gas theory we would expect

$$v_{ei} \sim n_i \sigma v_{Te},$$

where $\sigma = \pi b^2$ is the cross section for collisions with impact parameter b. Because of the long range of the Coulomb force, it is not immediately clear how to estimate σ. To this end, consider an electron with charge $e_e = -e$ and velocity $\mathbf{v}$ travelling past an ion with charge $e_i = Ze$. The impact parameter b is assumed to be so large that the direction of $\mathbf{v}$ only changes by a very small angle α, as in Fig. 1.1.

As a result of the Coulomb force $e_i e_e / 4\pi\epsilon_0 r^2(t)$, where $r(t) \simeq (b^2 + v^2 t^2)^{1/2}$ is the electron–ion distance at time t, the electron acquires momentum in the vertical (y) direction,

$$m_e \Delta v_y = \int_{-\infty}^{\infty} \frac{e_i e}{4\pi\epsilon_0 r^2(t)} \frac{b}{r(t)} \, dt$$

$$= \frac{e_i e}{4\pi\epsilon_0} \int_{-\infty}^{\infty} \frac{b \, dt}{(b^2 + v^2 t^2)^{3/2}} = \frac{e_i e}{2\pi\epsilon_0 b v}.$$

If $v = v_{Te}$, the angle α becomes

$$\alpha = \frac{\Delta v_y}{v} = \frac{e_i e}{2\pi\epsilon_0 b m_e v_{Te}^2} = \frac{b_{min}}{b}, \tag{1.2}$$

where $b_{min} \equiv e_i e / 2\pi\epsilon_0 m_e v_{Te}^2$. As anticipated, the deflection angle α becomes small if the impact parameter b is large; if $b \gg b_{min}$ then $\alpha \ll 1$ and it is justified to approximate the actual electron orbit (a hyperbola, see Fig. 1.1)

with a straight line. A typical collision occurs with an impact parameter
of the order of the Debye length,

$$\lambda_D = \sqrt{\frac{\epsilon_0 T}{ne^2}},$$

because particles with $b \gg \lambda_D$ do not 'see' the shielded ion. Consequently,
if $\Lambda \equiv \lambda_D/b_{min} \gg 1$ most collisions result only in small velocity deflections.
This is indeed the case in a typical laboratory plasma, where the Coulomb
logarithm can be shown to be

$$\ln \Lambda = 18.4 - 1.15 \,^{10}\log \frac{n}{\mathrm{m}^{-3}} + 2.3 \,^{10}\log \frac{T_e}{\mathrm{eV}},$$

and typically is $\ln \Lambda = 10 - 20$. It is crucial for our analysis that this
number is much larger than unity. Only then do most collisions result
in small-angle deflections of the velocity vector, and only then are most
of the results in this book accurate. Note that $\Lambda \sim n\lambda_D^3$ is the average
number of particles inside the Debye sphere.

The velocity vector **v** undergoes a random walk with very small steps
because of the grazing nature of the Coulomb collisions. This results in
a diffusion process in velocity space with a diffusion coefficient related to
the collision frequency v_{ei}. The latter is defined so that $1/v_{ei}$ is the time
after which enough small-angle deflections have accumulated to cause a
significant (say, 90-degree) scattering of **v**. This follows from the general
estimate for a diffusion coefficient $v(\Delta x)^2$ (to be derived in a slightly
different context in the next section), where the step size Δx is of the order
of one radian. (Order unity factors are not important here.) We can use
the same formula for estimating v_{ei} itself if we bear in mind that collisions
with different step sizes $\Delta x = \alpha = \Delta v_y/v$ occur with different frequencies
v. In order to obtain the total diffusion coefficient v_{ei}, we need to sum over
all these,

$$v_{ei} \sim \int \alpha^2 \, dv.$$

Here, dv is the frequency of collisions resulting in deflections between α
and $\alpha + d\alpha$, which can be estimated from Eq. (1.2),

$$dv = n_i v \, d\sigma \sim n_i v_{Te} b \, db.$$

Hence

$$v_{ei} \sim \int_{b_{min}}^{\lambda_D} \left(\frac{b_{min}}{b}\right)^2 n_i v_{Te} b \, db = \left(\frac{e_i e}{2\pi\epsilon_0 m_e}\right)^2 \frac{n_i \ln \Lambda}{v_{Te}^3}, \tag{1.3}$$

where we have used Eq. (1.2), and introduced the cutoffs b_{min} and λ_D in
the integral. This is justified since collisions with the impact parameter

$b = b_{\min}$ give a deflection of the order of the maximum possible, and collisions with impact parameters much larger than λ_D do not give any deflection at all because of the Debye shielding. Equation (1.3) also justifies the assumption $\alpha \ll 1$ since the main contribution to the integral comes from the region $b \gg b_{\min}$ where the deflection angle is small.

Note that the collisional cross section $\sigma = \nu_{ei}/n_i \nu_{Te}$ depends strongly on electron speed, $\sigma \propto \nu_{Te}^{-4}$, and that the collision frequency decreases rapidly when ν_{Te} becomes large. The reason for this is that the deflection angle α is very small for a fast electron, see Eq. (1.2). Such an electron spends such a short time in the interaction zone $r < \lambda_D$ that its velocity is not much affected.

In accordance with the estimate (1.3), we now define the electron collision time $\tau_{ei} \sim \nu_{ei}^{-1}$ for later use,

$$\tau_{ei} \equiv \frac{12\pi^{3/2}}{2^{1/2}} \frac{m_e^{1/2} T_e^{3/2} \epsilon_0^2}{n_i Z^2 e^4 \ln \Lambda}. \tag{1.4}$$

In practical units,

$$\tau_{ei} = 3.44 \cdot 10^{11} \, \text{s} \, \frac{1 \, \text{m}^{-3}}{n_e} \left(\frac{T_e}{1 \, \text{eV}} \right)^{3/2} \frac{1}{Z_i \ln \Lambda},$$

where we have noted that $Z n_i = n_e$ in a quasineutral plasma. The ion–ion collision time is longer by a factor of $(m_i/m_e)^{1/2}/Z^2$,

$$\tau_{ii} \equiv \frac{12\pi^{3/2}}{2^{1/2}} \frac{m_i^{1/2} T_i^{3/2} \epsilon_0^2}{n_i Z^4 e^4 \ln \Lambda}, \tag{1.5}$$

because the thermal velocity, and hence the number of encounters, is smaller by a factor $(m_e/m_i)^{1/2}$. Sometimes in plasma physics it is conventional to define the ion collision time as a factor $\sqrt{2}$ longer than our τ_{ii}. For instance, Braginskii (1965) follows this practice. We shall denote this collision time by $\tau_i = \sqrt{2}\tau_{ii}$.

Finally, the ion–electron collision frequency is yet smaller,

$$\frac{1}{\tau_{ie}} \equiv \frac{m_e}{m_i \tau_{ei}},$$

since it takes a factor m_i/m_e times more collisions to change the velocity of an ion than it does to do so for the much lighter electrons. A much more detailed discussion of collision statistics will be given in Chapter 3.

1.2 Random-walk estimates for classical transport in a straight magnetic field

Throughout this book, we assume that the plasma is magnetized in the sense that

$$\nu_{ei} \ll \Omega_e, \qquad \nu_{ii} \ll \Omega_i.$$

This means that the magnetic field, rather than collisions, determines the motion of plasma particles on short time scales. We also assume that all plasma parameters vary slowly on the length scale of the Larmor radius, e.g., for the density,

$$L_n \equiv \left(\frac{1}{n_e} \frac{dn_e}{dx} \right)^{-1} \gg \rho_i.$$

These orderings will ensure that collisional transport phenomena relax the gradients slowly compared with the collisional time scale.

Particle diffusion across B

Consider a plasma embedded in a straight and constant magnetic field $\mathbf{B}$. Each plasma particle is subject to the Lorentz force

$$m_a \dot{\mathbf{v}} = e_a \mathbf{v} \times \mathbf{B}. \tag{1.6}$$

Since this force is perpendicular to $\mathbf{B}$, the parallel velocity of the particle, $v_\parallel = \mathbf{v} \cdot \mathbf{b}$, is constant. Here $\mathbf{b} \equiv \mathbf{B}/B$ is the unit vector along the magnetic field. Since the Lorentz force is also perpendicular to $\mathbf{v}$, it performs no work on the particle, and the kinetic energy is conserved,

$$\frac{d}{dt} \left(\frac{m_a v^2}{2} \right) = 0,$$

as also follows from multiplying (1.6) by $\mathbf{v}$. Thus, since both $v_\parallel$ and $v^2 = v_\perp^2 + v_\parallel^2$ are constant, it follows that the perpendicular speed also remains constant. The trajectory is a helix around the magnetic field, with the guiding centre

$$\mathbf{R} = \mathbf{r} - \frac{1}{\Omega_a} \mathbf{b} \times \mathbf{v},$$

where $\mathbf{r}$ is the position of the particle, and $\Omega_a = e_a B / m_a$ again is the gyrofrequency. The guiding centre moves purely along the field because

$$\dot{\mathbf{R}} = \mathbf{v} - \frac{m_a}{e_a B} \mathbf{b} \times \dot{\mathbf{v}} = v_\parallel \mathbf{b},$$

and the fact that the particle orbit is a helix follows from the circumstance that the Larmor radius

$$\rho_a = |\mathbf{r} - \mathbf{R}| = v_\perp / \Omega_a$$

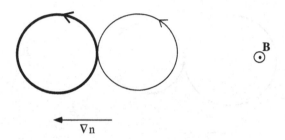

Fig. 1.2. Schematic diagram of a random-walk process in a magnetic field with a density gradient. There is a greater density of particles on the left Larmor orbit, bold.

is constant. This is the basis of magnetic confinement. The particle is essentially constrained (within a Larmor radius) to follow a magnetic field line, and cannot move across the field.

However, a change $\Delta\mathbf{v}$ in the velocity vector as a result of a collision leads to a shift in the guiding-centre position,

$$\Delta\mathbf{R} = -\frac{1}{\Omega}\mathbf{b} \times \Delta\mathbf{v}, \tag{1.7}$$

and hence to particle transport across the field.

Let us estimate the flux of particles across the field due to collisions when there is a density gradient in the negative x-direction, see Fig. 1.2 showing Larmor orbits of two adjacent electrons. Small deflections in $\mathbf{v}$ (as a result of grazing collisions) accumulate to large ones in the time τ_{ei} and lead to a step in x of the order of $\Delta x = \rho_e$. This gives rise to a flux in the x-direction

$$\Gamma_x = \tfrac{1}{2}(n_x - n_{x+\Delta x})V_x,$$

where V_x is the random-walk velocity of the electrons, and the factor $\tfrac{1}{2}$ represents an equal probability of a random jump to the left or to the right. Since $V_x = \Delta x/\tau_{ei} \sim \rho_e \nu_{ei}$ and

$$n(x + \Delta x) \simeq n(x) + \frac{\partial n}{\partial x}\Delta x,$$

the flux becomes

$$\Gamma_x^e \sim -D_\perp \frac{\partial n}{\partial x},$$

where the diffusion coefficient for transport across the magnetic field is

$$D_\perp \sim \frac{\nu_{ei}\rho_e^2}{2}.$$

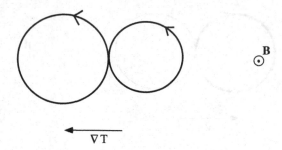

Fig. 1.3. Schematic diagram of classical heat diffusion. The Larmor radius to the left is larger since the temperature is higher there.

For ions colliding with electrons, the corresponding flux is equally large,

$$\Gamma^i_x \sim -\frac{\nu_{ie}\rho_i^2}{2}\frac{\partial n}{\partial x} \sim -\frac{\nu_{ei}\rho_e^2}{2}\frac{\partial n}{\partial x},$$

since the larger step size $\rho_i = (m_i/m_e)^{1/2}\rho_e$ is balanced by a smaller collision frequency $\nu_{ie} = (m_e/m_i)\nu_{ei}$. More fundamentally, collisional diffusion is 'ambipolar', i.e., $\Gamma_i = \Gamma_e$, in a steady-state plasma as a consequence of momentum conservation in Coulomb collisions, i.e., the ion–electron and electron–ion friction forces exactly balance each other, as will be discussed in Chapter 5.

Collisions with particles of the same species do not contribute to either the electron or ion flux, since when two identical particles (1 and 2) collide, the average guiding-centre position does not move. This follows from Eq. (1.7),

$$\Delta\mathbf{R}_1 + \Delta\mathbf{R}_2 = -\frac{1}{\Omega}\mathbf{b} \times (\Delta\mathbf{v}_1 + \Delta\mathbf{v}_2) = 0,$$

where $\Delta\mathbf{v}_1 + \Delta\mathbf{v}_2 = 0$ because of momentum conservation.

Heat diffusion across B

A similar random-walk argument applies to the transport of energy. Assume that there is a temperature gradient $\partial T/\partial x \neq 0$, but no density gradient, $\partial n/\partial x = 0$, see Fig. 1.3. When a particle makes a collisional step, it takes its energy with it. The net energy transport, or heat flux, $\mathbf{q} \sim n(mv_T^2/2)\mathbf{V}$, becomes

$$q_x = \frac{1}{2}\left[\left(n\frac{mv_T^2}{2}\right)_x - \left(n\frac{mv_T^2}{2}\right)_{x+\Delta x}\right]V_x,$$

with $V_x = v\Delta x, \Delta x \simeq \rho$, and $mv_T^2/2 = T$. Thus

$$q_x \sim -\frac{\rho^2 v n}{2} \frac{\partial T}{\partial x}$$

for each particle species. For heat diffusion, collisions among identical particles do lead to transport, and since $\rho_i = (m_i/m_e)^{1/2}\rho_e$ the biggest effect comes from the ions. Thus, the dominant heat flux is that of the ions, q_i. It is mainly effected through ion–ion collisions, and we have

$$q \simeq q_i = -\kappa_\perp^i \frac{\partial T}{\partial x}, \qquad \kappa_\perp^i \sim \frac{n_i \rho_i^2}{2\tau_{ii}} \equiv n_i \chi_i, \tag{1.8}$$

where χ_i is the ion heat diffusivity.

Heat diffusion along B

Transport parallel to the magnetic field is very fast since particles are free to move along the field as long as they do not collide. If the parallel length scale of temperature variation is longer than the mean-free path $\lambda = v_T/v$, a random-walk estimate for the parallel heat transport is possible. The particles can be imagined to undergo a random walk along the field, with a step size of the order of λ. Let z denote the coordinate along a field line, and suppose

$$\frac{1}{T} \frac{\partial T}{\partial z} \ll \frac{1}{\lambda}.$$

Then the temperature varies only slightly from one step to the next, and a random-walk estimate is possible for the heat flux $q_z = -\kappa_\parallel \partial T/\partial z$, as in the previous subsection. This gives

$$\kappa_\parallel \sim nv\lambda^2 \sim nv_T^2/v \sim nT\tau/m, \tag{1.9}$$

with $\tau = 1/v$ the collision time. It follows that the parallel heat conduction is dominated by the electrons rather than by the ions. Note that the thermal conductivity (1.9) is strongly dependent on temperature, $\kappa_\parallel^e \propto T_e^{5/2}$, but is independent of density because $\tau \propto n^{-1}$. Heat conduction is much larger in the parallel direction than in the perpendicular direction, $\kappa_\parallel/\kappa_\perp \sim (\Omega\tau)^2$.

Also note that $\kappa_\parallel$ is proportional to the collision time τ while $\kappa_\perp$ is inversely proportional to τ. In other words, increasing the collision frequency leads to increasing $\kappa_\perp$ and decreasing $\kappa_\parallel$. Perpendicular diffusion takes place because of collisions, while parallel transport is inhibited by them.

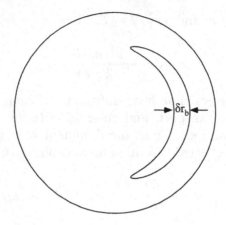

Fig. 1.4.   Banana orbit in a tokamak.

1.3 Random-walk estimate for ion heat diffusion in a tokamak

Here we outline a heuristic estimate of so-called neoclassical heat transport
(as opposed to 'classical' transport just considered), to be followed by a
rigorous treatment in Chapter 11. The curved magnetic field of a tokamak
decreases like $B \propto 1/R$ with major radius R. As a result, particles with
velocity vectors nearly perpendicular to the magnetic field, $|v_\parallel|/v_\perp \lesssim \epsilon^{1/2}$
are trapped in the relatively weak magnetic field on the outside of the
torus. Here $\epsilon \equiv r/R \ll 1$ is the inverse aspect ratio of the torus, with r the
minor radius. As shown in Chapter 7, the gyrocentres of trapped particles
trace out so-called 'banana' orbits with a width

$$\delta r_b \sim \rho_p \sqrt{\epsilon},$$

where $\rho_p = v_T/\Omega_p$ is the gyroradius in the poloidal magnetic field B_p, $\Omega_p =
eB_p/m$, see Fig. 1.4. B_p is usually much weaker than the total magnetic
field B, so the banana width δr_b is larger than the gyroradius, indicating
a large random-walk step size for these particles. Energy transport is
dominated by ions (again because of their larger step size) and is larger
than the classical result (1.8),

$$\chi_i^{\text{ban}} = f_t (\Delta x)^2 v_{\text{eff}}.$$

Here, the trapped fraction of the particles $f_t \simeq (2\epsilon)^{1/2}$ is small, but the step
size $\Delta x \sim \delta r_b$ and the 'effective' collision frequency v_{eff} are both relatively
large. v_{eff} is the frequency of collisions causing a trapped ion to take a
step of order δr_b. This only requires scattering by an angle of order $\epsilon^{1/2}$,
and because of the diffusive nature of Coulomb collisions

$$v_{\text{eff}}(\sqrt{\epsilon})^2 = v_{ii}.$$

Hence we obtain the ion banana thermal conductivity,

$$\chi_i^{\text{ban}} \sim \sqrt{2\epsilon}\rho_{pi}^2\epsilon(\nu_{ii}/\epsilon) = \sqrt{2\epsilon}\rho_{pi}^2\nu_{ii}, \qquad (1.10)$$

which is larger than the corresponding quantity for electrons by a factor of $(m_i/m_e)^{1/2}$, and larger than the classical result (1.8) by

$$\frac{\chi_i^{\text{ban}}}{\chi_i^{\text{class}}} = \sqrt{2\epsilon}\left(\frac{B}{B_p}\right)^2 \sim \frac{1}{2}\cdot 100$$

in a typical tokamak.

While this large, so-called 'neoclassical' enhancement of the classical heat conductivity is an essential consequence of the toroidal curvature of the confining magnetic field, it is often (but not always) overwhelmed by 'anomalous' transport due to fluctuating electromagnetic fields caused by plasma turbulence.

1.4 Random-walk estimate of Bohm diffusion

Because the collisional diffusion of electrons is so weak across the magnetic field, their transport is usually dominated by diffusion caused by turbulent fluctuations in the plasma. We now make a random-walk estimate of this so-called anomalous diffusion in the spirit of the previous section.

Electrostatic turbulence gives rise to a fluctuating electric field $\mathbf{E} = -\nabla\Phi$, which affects the motion of electrons. $\mathbf{E}$ is practically constant on the small length and time scales of electron gyration around $\mathbf{B}$. The gyration is not much affected itself, but superimposed on it is a drift

$$\mathbf{v}_E = \frac{\mathbf{E}\times\mathbf{B}}{B^2}$$

across the magnetic field. This follows from solving for the velocity $\mathbf{v}$ in the equation of motion

$$m_e\dot{\mathbf{v}} = -e(\mathbf{E} + \mathbf{v}\times\mathbf{B}),$$

by taking the vector product with $\mathbf{B}$,

$$\mathbf{v}_\perp = \frac{\mathbf{E}\times\mathbf{B}}{B^2} + \frac{m_e\dot{\mathbf{v}}\times\mathbf{B}}{eB^2}.$$

Taking a time average over the time gyration scale Ω_e^{-1} eliminates the last term.

Thus, the turbulent electric field produces a random drift velocity of electron guiding centres. The random-walk estimate for the electron particle diffusion coefficient perpendicular to $\mathbf{B}$ is

$$D_e = \frac{(\Delta x)^2}{\tau},$$

where τ is the time per step in the random walk,

$$\frac{\Delta x}{\tau} = v_E = -\frac{\nabla \Phi}{B} \sim \frac{\Phi}{\Delta x B}$$

is the velocity, and we have taken the step size $\Delta x \sim \lambda$ to be of the order of the perpendicular turbulent wavelength. Let the saturation amplitude be $e\Phi/T = k_0 < 1$. The exact calculation of k_0 is a difficult nonlinear problem; the traditional estimate is $k_0 = 1/16$, giving

$$\boxed{D_e^{\text{Bohm}} = \frac{T}{16eB}.}$$

For instance with $T_e = 1\,\text{keV}$ and $B = 1\,\text{T}$, we obtain

$$D_e^{\text{Bohm}} = 62.5\,\text{m}^2/\text{s}.$$

Note that Bohm diffusion is usually much faster than classical in a magnetized plasma,

$$D_e^{\text{Bohm}}/D_e^{\text{class}} \sim k_0 \Omega_e \tau_e.$$

Further reading

The dynamics of Coulomb collisions is explained in great detail by Trubnikov (1965), Sivukhin (1966) and Dendy (1990). Random-walk estimates of transport can be found throughout the literature, e.g., in Braginskii (1965), Goldston and Rutherford (1995), and Hazeltine and Waelbroeck (1998). The overall problem of plasma transport theory is posed lucidly in the Preface of the book by Balescu (1988).

Exercises

1. Above which temperature does the Bohm diffusivity exceed the

   (a) classical
   (b) neoclassical

   heat diffusivity across a magnetic field of 1 T? Assume $n = 10^{20}\,\text{m}^{-3}$, $\epsilon = 0.1$, $B/B_p = 10$.

   Solution: Rough figures are $D^{\text{Bohm}} > \chi_i^{\text{class}}$ when $T > 6\,\text{eV}$, and $D^{\text{Bohm}} > \chi_i^{\text{ban}}$ when $T > 50\,\text{eV}$. In a typical tokamak such cold plasmas tend not to be in the 'banana regime', as will be explained in Chapter 8. The actual collisional ion thermal conductivity is then somewhere between χ_i^{class} and χ_i^{ban}.

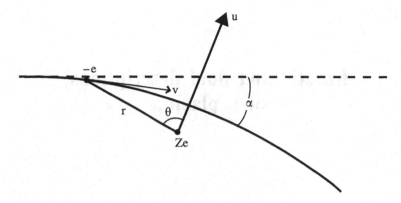

Fig. 1.5. An electron–ion collison. The transferred momentum is in the direction *u*.

2. Referring to Figs. 1.1 and 1.5, derive the Rutherford formula

$$\tan\frac{\alpha}{2} = \frac{Ze^2}{4\pi\epsilon_0 m_e v^2 b},$$

relating the deflection angle α in an electron–ion collision to the impact parameter b. Hint: Note that angular momentum is conserved because electrostatic Coulomb interaction is spherically symmetric, so that $r^2\dot\theta = $ const $= bv$.

Solution: In Fig. 1.5, the momentum transferred in the u-direction is equal to

$$\Delta p_u = 2m_e v \cos\theta_0 = \int_{-\infty}^{\infty} F(r)\cos\theta \, dt,$$

where θ_0 is the angle θ long before the collision, and $F(r) = Ze^2/4\pi\epsilon_0 r^2$ is the Coulomb force. Hence

$$2m_e v \cos\theta_0 - \int_{-\theta_0}^{\theta_0} \frac{Ze^2 \cos\theta}{4\pi\epsilon_0 r^2} \frac{d\theta}{\dot\theta} = \frac{Ze^2 \sin\theta_0}{2\pi\epsilon_0 vb}.$$

Rutherford's formula now follows after noting that $\alpha = \pi - 2\theta_0$ so that $\tan\alpha/2 = \cot\theta_0$.

2

Kinetic and fluid descriptions
of a plasma

2.1 The kinetic equation

A plasma is made up of many particles (electrons and different kinds of ions), each of which is characterized by its position $\mathbf{r}$ and velocity $\mathbf{v}$. Thus, each particle can be represented as a point in six-dimensional phase space $(\mathbf{r}, \mathbf{v})$. The distribution function $f_a(\mathbf{r}, \mathbf{v}, t)$ of a particle species 'a' is defined as the number of particles of that species per unit volume in phase space near the point $\mathbf{z} = (\mathbf{r}, \mathbf{v})$ at the time t. In other words,

$$f_a(\mathbf{r}, \mathbf{v}, t)\, d^3r\, d^3v$$

is the number of particles in the volume element $d^3r\, d^3v$ surrounding the point $(\mathbf{r}, \mathbf{v})$. Hence, if we integrate f_a over velocity space, we obtain the number density of particles in real space,

$$\int f_a(\mathbf{r}, \mathbf{v}, t)\, d^3v = n_a(\mathbf{r}, t).$$

In a plasma, each particle moves according to the equations of motion

$$\dot{\mathbf{r}} = \mathbf{v},$$

$$\dot{\mathbf{v}} = \frac{e_a}{m_a}\,(\mathbf{E} + \mathbf{v} \times \mathbf{B}),$$

where e_a is the electric charge of the particle in question and m_a its mass. The 'flow velocity' of particles in phase space can thus be represented by the six-dimensional vector $\dot{\mathbf{z}} = (\dot{\mathbf{r}}, \dot{\mathbf{v}})$. Since the distribution function $f(\mathbf{z}, t)$ is the number density of particles in phase space, it obeys the conservation equation

$$\frac{\partial f_a}{\partial t} + \frac{\partial}{\partial \mathbf{z}}\,(\dot{\mathbf{z}} f_a) = 0,$$

14

where the second term is the six-dimensional divergence of the phase-space particle flux $\dot{z}f_a$. More explicitly, we have the so-called Vlasov equation

$$\frac{\partial f_a}{\partial t} + \mathbf{v} \cdot \nabla f_a + \frac{e_a}{m_a}(\mathbf{E} + \mathbf{v} \times \mathbf{B}) \cdot \frac{\partial f_a}{\partial \mathbf{v}} = 0,$$

where we have used

$$\frac{\partial}{\partial \mathbf{v}} \cdot (\mathbf{E} + \mathbf{v} \times \mathbf{B}) = 0.$$

Here $\mathbf{E}$ and $\mathbf{B}$ are the total electric and magnetic fields. Close to each charged particle in the plasma they are, of course, very large as the fields from that particular particle dominate over the macroscopic, large-scale fields. In other words, $\mathbf{E}$ and $\mathbf{B}$ fluctuate strongly on short length scales comparable to the Debye length, and these fluctuations are caused by the discreteness of the particles. However, in a macroscopic description of the plasma, we can ignore small-scale fluctuations, and instead regard $\mathbf{E}$ and $\mathbf{B}$ as large-scale fields. In the following, we therefore understand $\mathbf{E}$ and $\mathbf{B}$ to be the average of the actual electric and magnetic fields over many Debye lengths. The effects of short-range electromagnetic fluctuations, or collisions, on the distribution function are instead lumped into what we call the collision operator,

$$C_a(f_a) = \left.\frac{\partial f_a}{\partial t}\right|_{\text{collisions}}.$$

This object is relegated to the right-hand side of the kinetic equation,

$$\frac{\partial f_a}{\partial t} + \mathbf{v} \cdot \nabla f_a + \frac{e_a}{m_a}(\mathbf{E} + \mathbf{v} \times \mathbf{B}) \cdot \frac{\partial f_a}{\partial \mathbf{v}} = C_a(f_a), \qquad (2.1)$$

where $\mathbf{E}$ and $\mathbf{B}$ now and henceforth denote the average fields. For Coulomb collisions we shall see that C_a is a Fokker–Planck operator; the corresponding operator for a gas is the Boltzmann operator. The kinetic equation (2.1) is often called the Fokker–Planck equation. If collisions are ignored, it becomes the Vlasov equation.

The collision operator

$$C_a = \sum_a C_{ab}(f_a, f_b) \qquad (2.2)$$

is a sum of contributions from collisions with each particle species 'b', including $b = a$, and we expect it to satisfy certain conservation rules. For instance, since collisions change only the velocity of the interacting particles, and not their spatial location, the density should not be affected by collisions,

$$\left.\frac{\partial n_a}{\partial t}\right|_{\text{collisions}} = \int C_{ab}(f_a)\, d^3v = 0. \qquad (2.3)$$

Similarly, the conservation laws of momentum and energy require that the collision operator satisfy

$$\int m_a \mathbf{v} C_{ab}(f_a) \, d^3v = -\int m_b \mathbf{v} C_{ba}(f_b) \, d^3v, \tag{2.4}$$

$$\int \frac{m_a v^2}{2} C_{ab}(f_a) \, d^3v = -\int \frac{m_b v^2}{2} C_{ba}(f_b) \, d^3v, \tag{2.5}$$

so that the force species a exerts on species b is equal and opposite to that which b exerts on a, and that no net energy is produced by collisions. Taking $b = a$ in these equations gives

$$\int m_a \mathbf{v} C_{aa}(f_a) \, d^3v = 0, \tag{2.6}$$

$$\int \frac{m_a v^2}{2} C_{aa}(f_a) \, d^3v = 0. \tag{2.7}$$

When we derive the collision operator in the next chapter, we shall see that it indeed meets these requirements.

Another important property of the collision operator is that it drives the distribution function toward local thermodynamic equilibrium, i.e., toward a (shifted) Maxwellian

$$f_{Ma}(\mathbf{r}, t) = n_a(\mathbf{r}, t) \left(\frac{m_a}{2\pi T_a(\mathbf{r}, t)} \right)^{3/2} \exp \left\{ \frac{m_a [\mathbf{v} - \mathbf{V}_a(\mathbf{r}, t)]^2}{2 T_a(\mathbf{r}, t)} \right\}, \tag{2.8}$$

where the density is n_a, the mean velocity is $\mathbf{V}_a$, and the temperature is T_a. The collision operator vanishes if, and only if, all species have the same mean velocity and temperature. In one species (with only 'like-particle' collisions), $C_{aa}(f_{Ma}) = 0$ for arbitrary mean velocity $\mathbf{V}_a$, manifesting Galilean invariance of the collision operator.

2.2 Fluid equations

Most useful information from the distribution function f_a is contained in its first few moments, i.e., integrals over velocity space of f_a multiplied by different functions of $\mathbf{v}$. The 'zeroth' moment is the density

$$n_a(\mathbf{r}, t) = \int f_a(\mathbf{r}, \mathbf{v}, t) \, d^3v,$$

and to form higher moments it is convenient to denote the average over the particle distribution by

$$\langle A \rangle_f \equiv \frac{1}{n_a} \int A f_a \, d^3v.$$

The macroscopic fluid velocity is now defined by

$$\mathbf{V}_a(\mathbf{r}, t) = \langle \mathbf{v} \rangle_f \,,$$

and is equal to the average velocity of all particles of species a in a certain point in space. The velocity of a particular particle differs from this average velocity by $\mathbf{v}'_a \equiv \mathbf{v} - \mathbf{V}_a$. The temperature T_a is defined so that $3T_a/2$ represents the average kinetic energy associated with these random velocities,

$$\tfrac{3}{2} T_a(\mathbf{x}, t) = \left\langle \frac{m_a \mathbf{v}'^2_a}{2} \right\rangle_f .$$

It should be noted that this definition of temperature has nothing to do with whether or not the plasma is in local thermodynamic equilibrium. If we take the $m_a v^2/2$ moment of the distribution function,

$$\frac{m_a n_a \langle v^2 \rangle_f}{2} = \frac{m_a n_a V_a^2}{2} + \frac{3 n_a T_a}{2},$$

we find that the total energy is the sum of the kinetic energy associated with the mean flow $\mathbf{V}$ and the thermal energy.

Fluid equations are formed by taking similar moments of the kinetic equation (2.1),

$$\frac{\partial n}{\partial t} + \nabla \cdot (n\mathbf{V}) = 0, \tag{2.9}$$

$$\frac{\partial mn\mathbf{V}}{\partial t} + \nabla \cdot \mathbf{\Pi} = ne(\mathbf{E} + \mathbf{V} \times \mathbf{B}) + \int m\mathbf{v}C(f)\,d^3v, \tag{2.10}$$

$$\frac{\partial}{\partial t}\left(\frac{3nT}{2} + \frac{mnV^2}{2}\right) + \nabla \cdot \mathbf{Q} = en\mathbf{E} \cdot \mathbf{V} + \int \frac{mv^2}{2}C(f)\,d^3v, \tag{2.11}$$

where we dropped the species index and introduced the following notation for the flux of momentum and energy,

$$\Pi_{jk} \equiv \langle mn v_j v_k \rangle_f \,,$$

$$\mathbf{Q} \equiv \frac{mn \langle v^2 \mathbf{v} \rangle_f}{2},$$

and used the symbolic tensor notation $(\nabla \cdot \mathbf{\Pi})_j = \partial \Pi_{jk}/\partial x_k$. Here and throughout this book, summation over repeated indices is understood. The equations (2.9)–(2.11) represent the conservation of particles, momentum, and energy, respectively, and they all have the structure of conservation laws. The first term on the left-hand side is the time derivative of the

conserved quantity in question, and the second term the divergence of
the flux of this quantity. The right-hand side of the continuity equation
(2.9) is zero because particles are conserved within each species, see (2.3).
The right-hand sides of the momentum (2.10) and energy (2.11) equations
contain the rate of change of momentum and energy due to the influence of
electric and magnetic fields, and the collisional transfer of these quantities
to other species,

$$\int m\mathbf{v}C(f)\,d^3v = \mathbf{R},$$

$$\int \frac{mv^2}{2}C(f)\,d^3v = Q + \mathbf{R}\cdot\mathbf{V},$$

where

$$Q \equiv \int \frac{mv'^2}{2}C(f)\,d^3v.$$

Thus, $\mathbf{R}$ is the total force exerted on the particle species in question as a
result of collisions with the other species in the plasma, and Q is the rate
of the corresponding thermal energy transfer. The total energy transfer
is equal to the sum of Q and the work $\mathbf{R}\cdot\mathbf{V}$ performed by the force $\mathbf{R}$.
As the effects of collisions with different species are additive in the sense
expressed by Eq. (2.2), the exchange of momentum and energy can be
written as sums,

$$\mathbf{R}_a = \sum_b \mathbf{R}_{ab},$$

$$Q_a = \sum_b Q_{ab},$$

where each term $\mathbf{R}_{ab}$ or Q_{ab} represents the interaction between species a
and b.

The pressure p, the viscosity tensor π_{jk}, and the heat flux $\mathbf{q}$ of each
species are defined as

$$p \equiv nm\left\langle v'^2\right\rangle_f /3 = nT,$$

$$\pi_{jk} \equiv mn\left\langle v'_j v'_k\right\rangle_f - p\delta_{jk}, \tag{2.12}$$

$$\mathbf{q} \equiv n\left\langle \frac{mv'^2}{2}\mathbf{v}'\right\rangle_f,$$

so that the fluxes of momentum and energy appearing in (2.10) and (2.11)
are

$$\Pi_{jk} = p\delta_{jk} + \pi_{jk} + mnV_jV_k, \tag{2.13}$$

$$Q_j = q_j + \frac{5pV_j}{2} + \pi_{jk}V_k + \frac{mnV^2}{2}V_j. \tag{2.14}$$

This shows that the energy flux **Q** is composed of a conductive heat flux **q**, a convective flux $5p\mathbf{V}/2$, viscous transport of energy, and convection of kinetic energy.

Using these definitions, the conservation laws (2.9)–(2.11) can be manipulated (see the exercises at the end of this chapter) to yield fluid moment equations for density, flow velocity, and temperature,

$$\frac{dn_a}{dt}\bigg|_a + n_a \nabla \cdot \mathbf{V}_a = 0, \tag{2.15}$$

$$m_a n_a \frac{d\mathbf{V}_a}{dt}\bigg|_a = -\nabla p_a - \nabla \cdot \boldsymbol{\pi}_a + e_a n_a (\mathbf{E} + \mathbf{V}_a \times \mathbf{B}) + \mathbf{R}_a, \tag{2.16}$$

$$\frac{3}{2} n_a \frac{dT_a}{dt}\bigg|_a + p \nabla \cdot \mathbf{V}_a = -\nabla \cdot \mathbf{q}_a - \boldsymbol{\pi}_a : \nabla \mathbf{V}_a + Q_a. \tag{2.17}$$

Here $\boldsymbol{\pi} : \nabla \mathbf{V} = \pi_{jk} \partial V_k / \partial x_j$ and

$$\frac{d}{dt}\bigg|_a \equiv \left(\frac{\partial}{\partial t} + \mathbf{V}_a \cdot \nabla \right)$$

is the so-called convective derivative, which is the time derivative in the frame moving at the fluid velocity $\mathbf{V}_a$. In other words, it is the rate of change experienced by the fluid element itself. For instance, the acceleration of a fluid element is $d\mathbf{V}_a/dt|_a$ and is equal to the total force acting upon it divided by the mass.

The continuity equation (2.15) shows that the fluid is incompressible if $\nabla \cdot \mathbf{V} = 0$ since then $dn/dt = 0$. The momentum equation (2.16) is similar to the Navier–Stokes equation for a liquid, with the addition of the Lorentz force and the inter-species friction force $\mathbf{R}_a$. The divergence of the flow velocity shows up again in the second term of the energy equation (2.17), where it describes heating by adiabatic compression. If the plasma is compressed, this term causes heating at the rate $d\ln T/dt = -(2/3)\nabla \cdot \mathbf{V}$. The first term on the right of the energy equation represents conductive transport of heat, and the second term describes viscous heating.

It is useful to note that the evolution equation (2.17) for T_a is also an equation for entropy production. If the continuity equation (2.15) is used to eliminate the velocity by $\nabla \cdot \mathbf{V}_a = -d\ln n_a/dt$, the equation can be written as

$$p \frac{ds_a}{dt}\bigg|_a = -\nabla \cdot \mathbf{q}_a - \boldsymbol{\pi}_a : \nabla \mathbf{V}_a + Q_a, \tag{2.18}$$

where $s_a \equiv \frac{3}{2} \ln(p_a/n_a^{5/3})$ is the entropy per particle within an unimportant constant. In other words, $n_a s_a$ is the entropy density (per unit volume) of species *a*. The terms on the right-hand side represent the production of entropy by collisional mechanisms, i.e., by heat conduction, heat generated by viscosity, and heat exchanged with other species.

The objective of fluid transport theory is to determine the space-time evolution of n_a, $\mathbf{V}_a$, and T_a, for all plasma species – in a fusion plasma typically electrons, hydrogen ions and heavier impurity ions. Each moment equation (2.15)–(2.17) couples to the one higher up in the hierarchy. The evolution equation for n_a contains $\mathbf{V}_a$; the evolution equation for $\mathbf{V}_a$ contains π_a and p_a; and the evolution equation for T_a (which equals p_a/n_a) contains the higher moment $\mathbf{q}$. As they stand, these equations do not form a closed system. We need constitutive relations for π, $\mathbf{R}$, $\mathbf{q}$, and Q, linking them back to the fundamental macroscopic quantities n, $\mathbf{V}$, and T. This is the task of kinetic theory, which, when completed, gives a closed system of fluid equations, whose solutions describe the transport evolution of n, $\mathbf{V}$, and T in space and time. We shall return to this issue in Chapter 4.

Further reading

Braginskii (1965) gives a wonderfully clear and comprehensive derivation of the fluid equations from the kinetic equation. Naturally, this topic is covered in most textbooks of plasma physics and kinetic theory, including Lifshitz and Pitaevskii (1981), Balescu (1988), and Hazeltine and Waelbroeck (1998). It is sometimes useful to derive more complicated fluid equations by taking higher velocity moments of the kinetic equation than we have done. This approach, which was pioneered by Grad, has been described in the context of plasma physics by Wang and Callen (1992).

Exercise

1. Derive the momentum equation (2.16) and the energy equation (2.17) from the conservation laws (2.9)–(2.11).

   Solution: Use the continuity equation (2.9) to write

   $$\partial_k(mnV_jV_k) = mnV_k\partial_kV_j + V_j\partial_k(mnV_k) = mnV_k\partial_kV_j - mV_j\partial n/\partial t,$$

   and substitute this in (2.10) to derive the momentum equation, using (2.13). For the energy equation, use

   $$\frac{\partial}{\partial t}\left(\frac{mnV^2}{2}\right) + \nabla\cdot\left(\frac{mnV^2}{2}\mathbf{V}\right)$$

   $$= mn\mathbf{V}\cdot\frac{\partial\mathbf{V}}{\partial t} + \frac{mV^2}{2}\left(\frac{\partial n}{\partial t} + \nabla\cdot(n\mathbf{V})\right) + mn\mathbf{V}\cdot\nabla\left(\frac{V^2}{2}\right)$$

   $$= mn\mathbf{V}\cdot\left(\frac{\partial\mathbf{V}}{\partial t} + (\mathbf{V}\cdot\nabla)\mathbf{V}\right) = \mathbf{V}\cdot(-\nabla p - \nabla\cdot\pi + ne\mathbf{E} + \mathbf{R})$$

and (2.14) to write (2.11) as

$$\frac{\partial}{\partial t}\left(\frac{3nT}{2}\right) + \nabla\cdot\left(\frac{5p\mathbf{V}}{2}\right) - \mathbf{V}\cdot\nabla p + \nabla\cdot\mathbf{q} + \boldsymbol{\pi}:\nabla\mathbf{V} + (ne\mathbf{E} + \mathbf{R})\cdot\mathbf{V} = Q,$$

where the first three terms can be rewritten as

$$\frac{3n}{2}\left(\frac{\partial}{\partial t} + \mathbf{V}\cdot\nabla\right)T + \frac{3T}{2}\frac{\partial n}{\partial t} + n\mathbf{V}\cdot\nabla T + \frac{5T}{2}\nabla\cdot(n\mathbf{V}) - \mathbf{V}\cdot\nabla p$$

$$= \frac{3n}{2}\frac{dT}{dt} + p\nabla\cdot\mathbf{V}.$$

Combining these equations gives (2.17).

3

The collision operator

As mentioned in Chapter 1, in a plasma with large Coulomb logarithm, $\ln \Lambda \gg 1$, most Coulomb collisions cause only small deflections of the velocity vector of a particle. This is very different from the situation in an ordinary gas, where the molecules bounce off each other like colliding billiard balls, see Fig. 3.1.

3.1 Derivation of the Fokker–Planck operator

General Fokker–Planck operator

The fact that the velocity of a plasma particle changes only gradually makes it possible to describe collisions by a so-called Fokker–Planck operator, which is essentially a diffusion operator in velocity space, and which is simpler than the more general Boltzmann operator used in kinetic gas theory.

To see how this is done, let us first consider a plasma particle species in one dimension with the distribution function $f(x, v, t)$. If we let $F(v, \Delta v)$ be the probability that the velocity of a particle changes from v to $v + \Delta v$ as a result of collisions in the time Δt, the distribution function obeys

$$f(v, t + \Delta t) = \int f(v - \Delta v, t) F(v - \Delta v, \Delta v) \, d\Delta v,$$

where we have suppressed the x-dependence for notational convenience. The meaning of this relation is clear: if a particle had the velocity $v - \Delta v$ at the time t, then the probability of it having the velocity v at the slightly later time $t + \Delta t$ is equal to $F(v - \Delta v, \Delta v)$. The density of particles with this history is thus $f(v - \Delta v, t) F(v - \Delta v, \Delta v)$. Summing over all Δv gives the total density of particles with velocity v at time $t + \Delta t$.

In a plasma most Coulomb collisions cause only a small change in the velocity of a particle. We therefore expect $F(v - \Delta v, \Delta v)$ to be highly

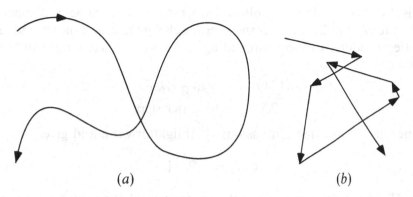

(a) (b)

Fig. 3.1. Typical trajectories of particles in a plasma (a), and a gas (b). The former resembles the path of a person lost in a desert at night (Arzimovich, 1965).

peaked around $\Delta v = 0$ in the second argument if Δt is small. (Note that this is not true in an ordinary gas.) It is therefore appropriate to treat Δv as a small quantity and to expand $f(v - \Delta v, t)$ and $F(v - \Delta v, \Delta v)$ in the first argument,

$$f(v, t + \Delta t) = \int \left[f(v, t)F(v, \Delta v) - \Delta v \frac{\partial f(v, t)F(v, \Delta v)}{\partial v} \right.$$

$$\left. + \frac{(\Delta v)^2}{2} \frac{\partial^2 f(v, t)F(v, \Delta v)}{\partial v^2} - \cdots \right] d\Delta v.$$

Because the sum of all probabilities of velocity changes is unity, we must have

$$\int F(v, \Delta v) d\Delta v = 1, \tag{3.1}$$

for all v. Using the notation

$$\langle \Delta v \rangle \equiv \int F(v, \Delta v) \Delta v \, d\Delta v,$$

$$\left\langle (\Delta v)^2 \right\rangle \equiv \int F(v, \Delta v)(\Delta v)^2 \, d\Delta v,$$

to denote the expectation values of Δv and $(\Delta v)^2$, we can write the rate of change in the distribution function due to collisions as

$$C(f) = \frac{\partial f(v, t)}{\partial t} \bigg|_{\text{collisions}} = \lim_{\Delta t \to 0} \frac{f(v, t + \Delta t) - f(v, t)}{\Delta t}$$

$$= -\frac{\partial}{\partial v} \left(\frac{\langle \Delta v \rangle}{\Delta t} f \right) + \frac{\partial^2}{\partial v^2} \left(\frac{\left\langle (\Delta v)^2 \right\rangle}{2 \Delta t} f \right) - \cdots. \tag{3.2}$$

This is the Fokker–Planck collision operator in one velocity dimension. The first term in (3.2) contains the average change in v, and the second term describes random, diffusive spreading in velocity space, with a diffusion coefficient

$$\frac{\langle (\Delta v)^2 \rangle}{2\Delta t} \sim \frac{(\text{step size})^2}{\text{time per step}}.$$

Generalizing to three dimensions is straightforward, and gives

$$C(f) = -\nabla_v \cdot \mathbf{j}, \tag{3.3}$$

where ∇_v is the divergence in velocity space and the vector $\mathbf{j}$ represents the flux in this space,

$$j_k \equiv \frac{\langle \Delta v_k \rangle}{\Delta t} f - \frac{\partial}{\partial v_l} \left(\frac{\langle \Delta v_k \Delta v_l \rangle}{2\Delta t} f \right). \tag{3.4}$$

The diffusion coefficient is now replaced by a tensor, $\langle \Delta v_k \Delta v_l \rangle / 2\Delta t$, reflecting the fact that the diffusion is not necessarily isotropic in velocity space.

Derivatives higher than second order have been neglected here. As we shall see later, these higher-order terms in the expansion are smaller by a factor $1/\ln \Lambda$. They describe the effect of close collisions resulting in large deflections of the velocity vector, and may be neglected in the Fokker–Planck approximation we are considering, provided $\ln \Lambda \gg 1$. This requirement is satisfied in most laboratory plasmas, and in many plasmas of astrophysical interest, but not for example in the solar plasma. At any rate, the accuracy of Fokker–Planck transport theory cannot be expected to be better than $1/\ln \Lambda$.

The first term in (3.4) describes the average force felt by a plasma particle due to collisions. It typically has the character of a drag force, tending to slow down the particle. In equilibrium, this pull towards the origin in velocity space, $\mathbf{v} = 0$, is balanced by diffusive spreading represented by the second term in (3.4). As we shall see, the result is a Maxwellian distribution function (2.8).

A plasma usually consists of several species (electrons and ions of various isotopes and charge states), and each distribution function f_a is affected by collisions with all the other species in an additive way, see (2.2). In other words, we can represent the velocity-space flux of particles of species a as a sum

$$\mathbf{j}^a = \sum_b \mathbf{j}^{ab}$$

of contributions from collisions with each species b (including a itself).

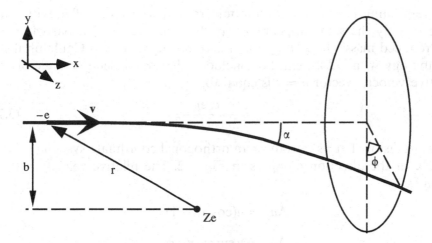

Fig. 3.2. Collision dynamics in the rest frame of particle b.

Collision dynamics

To calculate $\mathbf{j}^{ab}$ it is useful to recall the discussion of collision dynamics from Section 1.1. It was shown that in a collision between a light (a) and a heavy (b) particle, the velocity vector of the former is deflected by a small angle

$$\alpha = \frac{e_a e_b}{2\pi\epsilon_0 r m_a v_a^2} \ll 1, \tag{3.5}$$

where r is the impact parameter, cf. (1.2), see Fig. 3.2.

Now let the relative masses of the colliding particles be arbitrary. Writing the Lagrangian

$$L = \frac{m_a \dot{\mathbf{x}}_a^2}{2} + \frac{m_b \dot{\mathbf{x}}_b^2}{2} - \frac{e_a e_b}{4\pi\epsilon_0 |\mathbf{x}_a - \mathbf{x}_b|}$$

in the coordinates of the centre of mass and the relative position,

$$\mathbf{R} \equiv \frac{m_a \mathbf{x}_a + m_b \mathbf{x}_b}{m_a + m_b}, \qquad \mathbf{r} \equiv \mathbf{x}_a - \mathbf{x}_b,$$

gives

$$L = \frac{(m_a + m_b)\dot{\mathbf{R}}^2}{2} + \frac{1}{2}\frac{m_a m_b \dot{\mathbf{r}}^2}{m_a + m_b} - \frac{e_a e_b}{4\pi\epsilon_0 r}. \tag{3.6}$$

Since L is independent of $\mathbf{R}$, it follows from the equation of motion,

$$\frac{d}{dt}\left(\frac{\partial L}{\partial \dot{\mathbf{R}}}\right) = \frac{\partial L}{\partial \mathbf{R}},$$

that $(m_a + m_b)\ddot{\mathbf{R}} = 0$, so that the centre of mass moves at constant speed, $\dot{\mathbf{R}} = $ const. Thus, the first term in (3.6) is merely an additive constant. The two remaining terms describe motion around the centre of gravity, and have exactly the same appearance as the Lagrangian of a particle with the 'reduced mass' $m_* \equiv m_a m_b / (m_a + m_b)$ moving in a fixed Coulomb field. By analogy with (3.5) it can be concluded that the deflection angle of the relative velocity vector $\mathbf{u} = \dot{\mathbf{r}}$ is equal to

$$\alpha_* = \frac{e_a e_b}{2\pi\epsilon_0 r m_* u^2} \tag{3.7}$$

in the collision. Let us introduce an orthogonal coordinate system (x, y, z), with x in the direction of $\mathbf{v}_a$, as in Fig. 3.2. The relative velocity $\mathbf{u}$ then changes by

$$\Delta u_x = u(\cos\alpha_* - 1),$$

$$\Delta u_y = u\sin\alpha_* \cos\phi,$$

$$\Delta u_z = u\sin\alpha_* \sin\phi,$$

because of the collision. Here ϕ is the angle of the perpendicular component of the deflection vector to the xy-plane, see Fig. 3.2.

Using

$$\mathbf{x}_a = \mathbf{R} + \frac{m_b}{m_a + m_b}\mathbf{r} \quad \Rightarrow \quad \Delta\mathbf{v}_a = \frac{m_b}{m_a + m_b}\Delta\mathbf{u},$$

it is now straightforward to calculate the change $\Delta\mathbf{v}_a$ in the velocity vector of particle a. In a collision with an impact parameter r, $\mathbf{v}_a$ changes by an amount

$$\Delta v_x = \frac{m_b(\cos\alpha_* - 1)}{m_a + m_b}u \simeq -\left(1 + \frac{m_a}{m_b}\right)\left(\frac{e_a e_b}{2\pi\epsilon_0 m_a}\right)^2 \frac{1}{2r^2 u^3}, \tag{3.8}$$

in the x-direction, and

$$\Delta v_y = \frac{m_b \sin\alpha_* \cos\phi}{m_a + m_b}u = \frac{e_a e_b}{2\pi\epsilon_0 m_a}\frac{\cos\phi}{ur}, \tag{3.9}$$

$$\Delta v_z = \frac{m_b \sin\alpha_* \sin\phi}{m_a + m_b}u = \frac{e_a e_b}{2\pi\epsilon_0 m_a}\frac{\sin\phi}{ur}, \tag{3.10}$$

in the y- and z-directions. Here we have used (3.7) and made the approximations $\cos\alpha_* - 1 \simeq -\alpha_*^2/2$, and $\sin\alpha_* \simeq \alpha_*$ since the deflection angle α_* is small in most collisions.

We have now calculated the effect $\Delta\mathbf{v}_a$ of a collision on the velocity vector of one of the colliding particles. Note that relativistic effects were ignored, so that the present analysis is valid as long as the temperature is much smaller than the rest mass. For electron collisions this means that we are limited to temperatures much lower than 500 keV, which is sufficient for most laboratory and space plasmas.

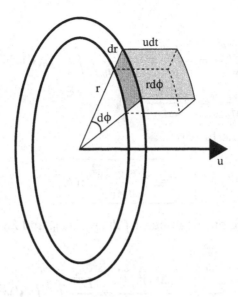

Fig. 3.3. The cross section defined by impact parameters in the interval $[r, r + dr]$ and angles in the interval $[\phi, \phi + d\phi]$ is equal to $d\sigma = r\,dr\,d\phi$. The volume traced out by this cross section in time dt by motion along the relative velocity vector $\mathbf{u} = \mathbf{v} - \mathbf{v}'$ is equal to $dV = u\,dt\,d\sigma$, and the number of collisions with these parameters is proportional to the number of particles in this volume.

Collision statistics

Having established what happens in a single collision, we now consider the cumulative effect of many such events. First we need to count the number of collisions between a given particle a and particles of species b. The number of such collisions taking place in time Δt with impact parameters between r and $r + dr$, and angles between ϕ and $\phi + d\phi$ is

$$\Delta t\, d\phi\, r\, dr \int f_b(\mathbf{v}')u\, d^3v', \qquad (3.11)$$

where, again, $\mathbf{u} \equiv \mathbf{v} - \mathbf{v}'$ is the relative velocity of the colliding particles, see Fig. 3.3. Here $d\sigma = r\,dr\,d\phi$ is the area spanned by dr and $d\phi$, and $u\,dt$ is the distance that particle a travels relative to particle b in time dt. Multiplying the product of these numbers, which is the volume $dV = r\,dr\,d\phi\,u\,dt$, by the phase-space density $f_b(\mathbf{v}')$ and integrating over the velocity $\mathbf{v}'$ of b-particles gives the frequency of collisions under consideration.

Next, by multiplying Eqs. (3.8) and (3.9) with the number of collisions (3.11) and integrating over r and ϕ, it follows that the average changes in the velocity vector of particle a as a result of collisions with particles b

are

$$\frac{\langle\Delta v_x\rangle^{ab}}{\Delta t} = -\frac{L^{ab}}{4\pi}\left(1+\frac{m_a}{m_b}\right)\int\frac{1}{u^2}f_b(\mathbf{v}')d^3v',$$

$$\frac{\langle\Delta v_y\rangle^{ab}}{\Delta t} = \frac{\langle\Delta v_z\rangle^{ab}}{\Delta t} = 0,$$

where the logarithmic factor

$$\boxed{L^{ab} \equiv \left(\frac{e_a e_b}{m_a\epsilon_0}\right)^2\ln\Lambda} \tag{3.12}$$

appears after cutting off the integral $\int dr/r$ at $r_{\min}$ and λ_D as in Section 1.1. Similarly,

$$\frac{\left\langle(\Delta v_y)^2\right\rangle^{ab}}{\Delta t} = \frac{\left\langle(\Delta v_z)^2\right\rangle^{ab}}{\Delta t} = \frac{L^{ab}}{4\pi}\int\frac{1}{u}f_b(\mathbf{v}')d^3v'.$$

On the other hand, $\left\langle(\Delta v_x)^2\right\rangle^{ab}/\Delta t$ is smaller by a factor of $1/\ln\Lambda$ and is thus negligible since it does not involve a divergent r-integral that needs to be cut off. For the same reason, all higher order terms $\langle\Delta v_j\Delta v_k\Delta v_l\rangle/\Delta t$ etc. in the Fokker–Planck expansion (3.2) are small.

We have now calculated the expectation values $\langle\Delta v_k\rangle$ and $\langle\Delta v_k v_l\rangle$ that we need for the collision operator (3.4). However, these results are expressed in a coordinate system aligned with the velocity vector of one of the colliding particles. We need to make a transformation to the 'laboratory system'. In an arbitrary orthogonal coordinate system with unit vectors $\mathbf{e}_k$ we have

$$\frac{\langle\Delta v_k\rangle^{ab}}{\Delta t} = \frac{\langle\mathbf{e}_k\cdot\hat{\mathbf{x}}\Delta v_x\rangle^{ab}}{\Delta t} = -\frac{L^{ab}}{4\pi}\left(1+\frac{m_a}{m_b}\right)\int\frac{u_k}{u^3}f_b(\mathbf{v}')d^3v', \tag{3.13}$$

and

$$\frac{\langle\Delta v_k\Delta v_l\rangle^{ab}}{\Delta t} = \frac{\langle\mathbf{e}_k\cdot(\hat{\mathbf{y}}\Delta v_y+\hat{\mathbf{z}}\Delta v_z)\,\mathbf{e}_l\cdot(\hat{\mathbf{y}}\Delta v_y+\hat{\mathbf{z}}\Delta v_z)\rangle^{ab}}{\Delta t}$$

$$= \left\langle(\mathbf{e}_k\cdot\hat{\mathbf{y}})(\mathbf{e}_l\cdot\hat{\mathbf{y}})\,(\Delta v_y)^2+(\mathbf{e}_k\cdot\hat{\mathbf{z}})(\mathbf{e}_l\cdot\hat{\mathbf{z}})\,(\Delta v_z)^2\right\rangle^{ab}/\Delta t$$

$$= \left\langle[\mathbf{e}_k\cdot\mathbf{e}_l-(\mathbf{e}_k\cdot\hat{\mathbf{x}})\,(\mathbf{e}_l\cdot\hat{\mathbf{x}})]\,(\Delta v_y)^2\right\rangle^{ab}/\Delta t$$

$$= \frac{\left\langle(\delta_{kl}-u_k u_l/u^2)(\Delta v_y)^2\right\rangle^{ab}}{\Delta t} = \frac{L^{ab}}{4\pi}\int U_{kl}f_b(\mathbf{v}')d^3v', \tag{3.14}$$

where we have introduced the tensor

$$U_{kl} \equiv \frac{u^2 \delta_{kl} - u_k u_l}{u^3}.$$

Collision operator

Having completed our analysis of the collision dynamics, we are now in a position to write down the collision operator (3.3), (3.4). For this purpose, it is useful to introduce so-called 'Rosenbluth potentials' (Rosenbluth, McDonald and Judd, 1957; Trubnikov, 1965)

$$\varphi_b(\mathbf{v}) \equiv -\frac{1}{4\pi} \int \frac{1}{u} f_b(\mathbf{v}') d^3 v', \tag{3.15}$$

$$\psi_b(\mathbf{v}) \equiv -\frac{1}{8\pi} \int u f_b(\mathbf{v}') d^3 v'. \tag{3.16}$$

Since

$$\frac{\partial u}{\partial v_k} = \frac{\partial}{\partial v_k} \sqrt{\sum_l (v_l - v_l')^2} = \frac{u_k}{u}$$

and

$$\frac{\partial^2 u}{\partial v_k \partial v_l} = U_{kl},$$

the expectation values (3.13) and (3.14) can then be written as

$$A_k^{ab} \equiv -\frac{\langle \Delta v_k \rangle^{ab}}{\Delta t} = \left(1 + \frac{m_a}{m_b}\right) L^{ab} \frac{\partial \varphi_b}{\partial v_k}, \tag{3.17}$$

$$D_{kl}^{ab} \equiv \frac{\langle \Delta v_k \Delta v_l \rangle^{ab}}{2\Delta t} = -L^{ab} \frac{\partial^2 \psi_b}{\partial v_k \partial v_l}. \tag{3.18}$$

Note that the average force felt by a particle of species a by collisions with particles of species b is equal to $-m_a A_k^{ab}$, while D_{kl}^{ab} can be interpreted as a diffusion tensor in velocity space. The Fokker–Planck collision operator (3.3) now becomes

$$\boxed{C_{ab}(f_a, f_b) = \frac{\partial}{\partial v_k} \left[A_k^{ab} f_a + \frac{\partial}{\partial v_l} \left(D_{kl}^{ab} f_a \right) \right].} \tag{3.19}$$

We can rewrite this result slightly by noting that if we let ∇_v^2 denote the Laplace operator in velocity space,

$$\nabla_v^2 = \sum_j \frac{\partial^2}{\partial v_j^2},$$

then $\nabla_v^2 u = U_{kk} = 2/u$, so that

$$\nabla_v^2 \psi_b = \varphi_b. \tag{3.20}$$

Therefore, A_k^{ab} and D_{kl}^{ab} are related to each other by the 'Einstein relation'

$$A_k^{ab} = -\left(1 + \frac{m_a}{m_b}\right) \frac{\partial D_{kl}^{ab}}{\partial v_l},$$

and we can write the collision operator with Rosenbluth potentials as

$$C_{ab}(f_a, f_b) = \ln \Lambda \left(\frac{e_a e_b}{m_a \epsilon_0}\right)^2 \frac{\partial}{\partial v_k} \left(\frac{m_a}{m_b} \frac{\partial \varphi_b}{\partial v_k} f_a - \frac{\partial^2 \psi_b}{\partial v_k \partial v_l} \frac{\partial f_a}{\partial v_l}\right). \tag{3.21}$$

Finally, we can rewrite the collision operator in yet another, and more symmetric, way by inserting the Rosenbluth potentials (3.15), (3.16) in (3.21), and using $\partial U_{kl}/\partial v_l = 2u_k/u$, to obtain

$$C_{ab}(f_a, f_b) = -\frac{e_a^2 e_b^2 \ln \Lambda}{8\pi \epsilon_0^2 m_a} \frac{\partial}{\partial v_k} \int U_{kl} \left[\frac{f_a(\mathbf{v})}{m_b} \frac{\partial f_b(\mathbf{v}')}{\partial v_l'} - \frac{f_b(\mathbf{v}')}{m_a} \frac{\partial f_a(\mathbf{v})}{\partial v_l}\right] d^3v',$$
$$\tag{3.22}$$

which is the expression originally found by Landau (1936) in the first derivation of the Coulomb collision operator.

By direct substitution, is not difficult to verify that the collision operator, whether written in the form (3.19), (3.21) or (3.22), satisfies the conservation laws (2.3)–(2.5). It is also clear that it is Galilean invariant, i.e., it is not affected by a transformation to a moving frame, $\mathbf{v} \to \mathbf{v} - \mathbf{V}$. This follows immediately from the observation that the velocity $\mathbf{v}$ does not appear explicitly but only as the relative velocity $\mathbf{u} = \mathbf{v} - \mathbf{v}'$ or as the derivative $\partial/\partial \mathbf{v}$. Both these objects are unaffected by a Galilean transformation. Finally we note $C_{ab}(f_a, f_b) = 0$ if f_a and f_b are Maxwellian (2.8) with the same temperature. This is most easily seen from Landau's form of the collision operator (3.22), where

$$U_{kl} \left[\frac{f_a(\mathbf{v})}{m_b} \frac{\partial f_b(\mathbf{v}')}{\partial v_l'} - \frac{f_b(\mathbf{v}')}{m_a} \frac{\partial f_a(\mathbf{v})}{\partial v_l}\right] = \frac{f_a f_b}{T} U_{kl}(v_l - v_l') = 0$$

for Maxwellian distribution functions with equal temperatures $T_a = T_b = T$.

3.2 Electron–ion and ion–impurity collisions

The general Coulomb collision operator can be simplified considerably whenever the colliding particles move at very disparate speeds. Consider,

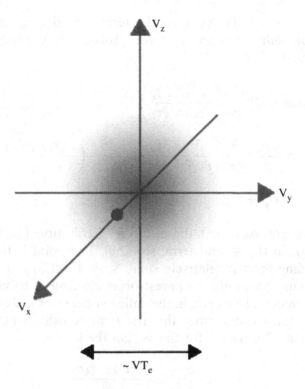

Fig. 3.4. The mean ion flow velocity $\mathbf{V}_i$ is usually much smaller than the electron thermal speed v_{Te}, which characterizes the thermal spread of electron velocities.

for instance, electrons colliding with ions moving with an average velocity $\mathbf{V}_i$. The characteristic velocity spread around $\mathbf{V}_i$ in the ion distribution function is of the order of the ion thermal speed, which is small in comparison with the electron thermal speed,

$$v_{Ti} = \sqrt{2T_i/m_i} \ll v_{Te},$$

unless the ion and electron temperatures are widely different, see Fig. 3.4. When calculating the Rosenbluth potentials (3.15) and (3.16), the ion distribution in velocity space, as seen by the electrons, may thus be approximated by a delta function,

$$f_i(\mathbf{v}) \simeq n_i \delta(\mathbf{v} - \mathbf{V}_i),$$

where normally $V_i \ll v_{Te}$, so that

$$\varphi_i \simeq -\frac{n_i}{4\pi} \frac{1}{|\mathbf{v} - \mathbf{V}_i|} \simeq -\frac{n_i}{4\pi v}\left(1 + \frac{\mathbf{v} \cdot \mathbf{V}_i}{v^2}\right),$$

$$\psi_i \simeq -\frac{n_i}{8\pi}|\mathbf{v} - \mathbf{V}_i| \simeq -\frac{n_i}{4\pi}v\left(1 - \frac{\mathbf{v} \cdot \mathbf{V}_i}{v^2}\right), \tag{3.23}$$

It follows from Eq. (3.21) that only the term related to ψ_i is important for electron–ion collisions since the term involving φ_i is multiplied by the small factor m_e/m_i. Thus

$$C_{ei}(f_e) \simeq -L^{ei} \frac{\partial}{\partial v_k} \left(\frac{\partial^2 \psi_i}{\partial v_k \partial v_l} \frac{\partial f_e}{\partial v_l} \right)$$

$$\simeq \frac{n_i L^{ei}}{8\pi} \frac{\partial}{\partial v_k} \left[\frac{\partial^2 v}{\partial v_k \partial v_l} \frac{\partial f_e}{\partial v_l} - \frac{\partial^2}{\partial v_k \partial v_l} \left(\frac{v_m V_{im}}{v} \right) \frac{\partial f_{Me}}{\partial v_l} \right]$$

$$= C_{ei}^0 + C_{ei}^1, \qquad (3.24)$$

where we have approximated the electron distribution function with a Maxwellian f_{Me} in the second term, see (2.8). The point is that in (3.23) the corresponding term is relatively small since $V_i \ll v_{Te}$. It is therefore appropriate to use here only the lowest-order electron distribution, which is usually Maxwellian. However, in the collision operator (3.24) both terms may be of the same order, since the first term vanishes when operating on a Maxwellian. The reason for this is that the tensor

$$W_{kl} = \frac{\partial^2 v}{\partial v_k \partial v_l} = \frac{v^2 \delta_{kl} - v_k v_l}{v^3} \qquad (3.25)$$

is orthogonal to the vector v_l, and therefore also to $\partial f_{Me}/\partial v = -(m_e v / T_e) f_{Me}$. This circumstance is also very useful in simplifying the first term C_{ei}^0 of the collision operator (3.24). Since it is a divergence of a vector,

$$C_{ei}^0 = \frac{n_i L^{ei}}{8\pi} \frac{\partial}{\partial v_k} \left(W_{kl} \frac{\partial f_e}{\partial v_l} \right) = \frac{n_i L^{ei}}{8\pi} \frac{\partial}{\partial \mathbf{v}} \cdot \left[\frac{1}{v} \frac{\partial f_e}{\partial v} - \frac{\mathbf{v}}{v^3} \left(\mathbf{v} \cdot \frac{\partial f_e}{\partial \mathbf{v}} \right) \right], \quad (3.26)$$

we need only include the components of that vector (i.e., the expression within the square brackets) that are perpendicular to $\mathbf{v}$. The underlying physical reason is that because of the small mass ratio between electrons and ions, electron–ion collisions do not change the magnitude v of the electron velocity vector, only its direction. Thus, the second term in C_{ei}^0 drops out entirely, and only the components of the first term in the θ- and φ-directions matter, where we use (v, θ, φ) for spherical coordinates in velocity space. Employing the formulas for the gradient and the divergence in spherical coordinates, we arrive at the expression

$$C_{ei}^0 = \frac{n_i L^{ei}}{8\pi v^3} \left[\frac{1}{\sin \theta} \frac{\partial}{\partial \theta} \left(\sin \theta \frac{\partial f_e}{\partial \theta} \right) + \frac{1}{\sin^2 \theta} \frac{\partial^2 f_e}{\partial \varphi^2} \right] \qquad (3.27)$$

for the first part of the collision operator (3.24). To evaluate the second part, C_{ei}^1, we use

$$\frac{\partial^2}{\partial v_k \partial v_l}\left(\frac{v_m}{v}\right)\frac{\partial f_{Me}}{\partial v_l} = \frac{v^2 \delta_{km} - v_k v_m}{v^3}\frac{m_e}{T_e}f_{Me} = W_{km}\frac{m_e}{T_e}f_{Me}$$

and

$$\frac{\partial W_{km}}{\partial v_k} = \frac{\partial^3 v}{\partial v_k \partial v_k \partial v_m} = -\frac{2v_m}{v^3}$$

to obtain

$$C_{ei}^1 = -\frac{n_i L^{ei}}{8\pi}\frac{m_e V_{im}f_{Me}}{T_e}\frac{\partial W_{km}}{\partial v_k} = \frac{n_i L^{ei}}{4\pi}\frac{m_e}{T_e}\frac{\mathbf{V}_i \cdot \mathbf{v}}{v^3}f_{Me}.$$

Thus, the total electron–ion collision operator becomes

$$C_{ei}(f_e) \simeq v_{ei}(v)\left(\mathscr{L}(f_e) + \frac{m_e \mathbf{v} \cdot \mathbf{V}_i}{T_e}f_{Me}\right), \qquad (3.28)$$

where

$$\mathscr{L}(f_e) \equiv \frac{1}{2}\left[\frac{1}{\sin\theta}\frac{\partial}{\partial\theta}\left(\sin\theta\frac{\partial f_e}{\partial\theta}\right) + \frac{1}{\sin^2\theta}\frac{\partial^2 f_e}{\partial\varphi^2}\right], \qquad (3.29)$$

is the so-called Lorentz scattering operator, and

$$v_{ei}(v) \equiv \frac{n_i L^{ei}}{4\pi v^3} = \frac{n_i e_i^2 e^2}{4\pi m_e^2 \epsilon_0^2 v^3}\ln\Lambda \qquad (3.30)$$

is a velocity-dependent electron–ion collision frequency. Note that it decreases with increasing velocity as v^{-3}, and can be written in terms of the electron–ion collision time (1.4) as

$$v_{ei}(v) = \frac{3\pi^{1/2}}{4\tau_{ei}}\left(\frac{v_{Te}}{v}\right)^3. \qquad (3.31)$$

The first part of the collision operator, $C_{ei}^0 = v_{ei}(v)\mathscr{L}(f_e)$, describes electrons colliding with infinitely heavy, stationary ($\mathbf{V}_i = 0$), ions (like ping-pong balls colliding with bowling balls, but with collisions such that the velocity vector only changes gradually). The magnitude of the electron velocity v is conserved in such collisions, and only its direction changes. This is why no v-derivatives appear in the final collision operator (3.28). Note that $2\mathscr{L}$ is simply the θ- and φ-parts of a Laplacian ∇_v^2, and thus describes diffusion on the surface of a sphere $v = $ constant. This operator is intrinsically spherically symmetric, reflecting the fact that the Coulomb collision operator has no directional preference. In later chapters, we often use this operator in cases where the azimuthal angle φ (frequently the gyroangle) is ignorable. Then

$$\mathcal{L} = \frac{1}{2}\frac{\partial}{\partial \xi}\left(1 - \xi^2\right)\frac{\partial}{\partial \xi},$$

with $\xi = \cos\theta$ the cosine of the pitch angle, describes 'pitch-angle scattering'. In the large-mass-ratio limit we are considering, electron–ion collisions do not drive the electron distribution function toward a Maxwellian. All they do is to make the electron distribution function isotropic in the ion rest frame.

It is useful to note that the electron–ion collision operator does not depend on the ion mass. Only the charge of the ions matters. This simplifies the treatment of electron–ion collisions in a plasma where several different ion species are present. For instance, if all these ions are stationary the total electron–ion collision operator is

$$C_{ei}\left(f_e\right) = \sum_j C_{ej}\left(f_e\right) = Z_{\text{eff}}\frac{3\pi^{1/2}}{4\tau_{ee}}\left(\frac{v_{Te}}{v}\right)^3 \mathcal{L}\left(f_e\right),$$

where the sum is taken over all ion species j, and where Z_{eff} is the *effective ion charge* defined by

$$Z_{\text{eff}} \equiv \frac{1}{n_e}\sum_j n_j Z_j^2 = \frac{\sum_j n_j Z_j^2}{\sum_j n_j Z_j}.$$

In other words, the electrons only 'feel' the presence of a single ion species, with an effective charge Z_{eff}.

To arrive at the simplification (3.28) of the electron–ion collision operator, we used only the fact that electrons are much lighter than the ions. A corresponding treatment therefore holds for collisions between any species with a small mass ratio, for instance for hydrogenic ions (i) colliding with much heavier impurities (Z). Thus, for $m_Z \gg m_i$, we have the collision operator

$$C_{iZ} = \nu_{iZ}(v)\left(\mathcal{L}(f_e) + \frac{m_i \mathbf{v}\cdot\mathbf{V}_Z}{T_i}f_{Mi}\right), \tag{3.32}$$

with the ion–impurity collision frequency

$$\nu_{iZ}(v) \equiv \frac{n_Z e_i^2 e_Z^2}{4\pi m_i^2 \epsilon_0^2 v^3}\ln\Lambda.$$

3.3 Collisions with a Maxwellian background

We now consider the case of collisions between an arbitrary species (a) and a Maxwellian species (b), which we take to be at rest,

$$f_b(\mathbf{v}) = f_{b0}(v) \equiv \frac{n_b}{\pi^{3/2} v_{Tb}^3} e^{-(v/v_{Tb})^2}. \tag{3.33}$$

Here $v_{Tb} = (2T_b/m_b)^{1/2}$ denotes the thermal speed. Since f_b is an isotropic distribution, its Rosenbluth potentials (3.15) and (3.16) can only depend on the magnitude of $\mathbf{v}$ and not on its direction,

$$\varphi_b(\mathbf{v}) = \varphi_b(v), \qquad \psi_b(\mathbf{v}) = \psi_b(v).$$

This circumstance simplifies the collision operator (3.21),

$$C_{ab}(f_a, f_b) = L^{ab} \frac{\partial}{\partial v_k} \left(\frac{m_a}{m_b} \frac{\partial \varphi_b}{\partial v_k} f_a - \frac{\partial^2 \psi_b}{\partial v_k \partial v_l} \frac{\partial f_a}{\partial v_l} \right),$$

since

$$\frac{\partial \varphi_b}{\partial v_k} = \frac{v_k}{v} \varphi_b', \tag{3.34}$$

$$\frac{\partial^2 \psi_b}{\partial v_k \partial v_l} = \frac{\partial^2 v}{\partial v_k \partial v_l} \psi_b' + \frac{v_k v_l}{v^2} \psi_b'' = W_{kl} \psi_b' + \frac{v_k v_l}{v^2} \psi_b'', \tag{3.35}$$

where again $W_{kl} = (v^2 \delta_{kl} - v_k v_l)/v^3$, and a prime denotes a derivative with respect to v. The operator itself can thus be written as

$$C_{ab}(f_a, f_{b0}) = L^{ab} \frac{\partial}{\partial v_k} \left[\frac{m_a}{m_b} \frac{v_k}{v} \varphi_b' f_a - \left(W_{kl} \psi_b' + \frac{v_k v_l}{v^2} \psi_b'' \right) \frac{\partial f_a}{\partial v_l} \right].$$

By analogy with (3.27), we can write the middle term in terms of the Lorentz operator (3.29),

$$\frac{\partial}{\partial v_k} \left(W_{kl} \frac{\partial f_a}{\partial v_l} \right) = \frac{2}{v^3} \mathcal{L}(f_a), \tag{3.36}$$

and by using the vector relation

$$\nabla_v \cdot [A(v)\mathbf{v}] = \frac{1}{v^2} \frac{\partial (v^3 A)}{\partial v}$$

(which follows from the formula for the divergence in spherical coordinates) for the other two terms, we obtain

$$C_{ab}(f_a, f_{b0}) = -\frac{2L^{ab}}{v^3} \psi_b' \, \mathcal{L}(f_a) + \frac{L^{ab}}{v^2} \frac{\partial}{\partial v} \left[v^3 \left(\frac{m_a}{m_b} \frac{\varphi_b'}{v} f_a - \frac{\psi_b''}{v} \frac{\partial f_a}{\partial v} \right) \right], \tag{3.37}$$

where the velocity derivative $\partial/\partial v$ is taken at fixed direction of $\mathbf{v}$. The last term has been simplified by noting that $v_l \partial f_a/\partial v_l = v \partial f_a/\partial v$.

The three terms in (3.37) reflect different effects of collisions on the particle species a. In order to understand their physical significance it is useful to recall the basic results from our discussion of collision dynamics in Section 3.1. In an orthogonal coordinate system such that the first coordinate is aligned with the velocity vector, we found that the effects of Fokker–Planck collisions are fully described by the expectation values

$$A_k^{ab} = -\frac{1}{\Delta t} \left\langle \begin{array}{c} \Delta v_\parallel \\ 0 \\ 0 \end{array} \right\rangle^{ab},$$

$$D_{kl}^{ab} = \frac{1}{2\Delta t} \left\langle \begin{array}{ccc} \Delta v_\parallel^2 & 0 & 0 \\ 0 & \Delta v_\perp^2/2 & 0 \\ 0 & 0 & \Delta v_\perp^2/2 \end{array} \right\rangle^{ab},$$

where we have denoted the direction parallel to the first coordinate by $\parallel$ and the other two directions by $\perp$. Thus, in the notation of Section 3.1, $\Delta v_\parallel = \Delta v_x$ and $\Delta v_\perp^2 = \Delta v_y^2 + \Delta v_z^2$. Evaluating the expectation values (3.17) and (3.18) for an isotropic distribution function using Eqs. (3.34) and (3.35) gives

$$A_k^{ab} = v \left(\begin{array}{c} v_s^{ab} \\ 0 \\ 0 \end{array} \right),$$

$$D_{kl}^{ab} = \frac{v^2}{2} \left(\begin{array}{ccc} v_\parallel^{ab} & 0 & 0 \\ 0 & v_D^{ab} & 0 \\ 0 & 0 & v_D^{ab} \end{array} \right),$$

where we have introduced the three basic collision frequencies

$$v_s^{ab}(v) \equiv -\frac{\left\langle \Delta v_\parallel/v \right\rangle^{ab}}{\Delta t} = L^{ab} \left(1 + \frac{m_a}{m_b} \right) \frac{\varphi_b'(v)}{v}.$$

$$v_D^{ab}(v) \equiv \frac{\left\langle (\Delta v_\perp/v)^2 \right\rangle^{ab}}{2\Delta t} = -\frac{2L^{ab}}{v^3} \psi_b'(v). \tag{3.38}$$

$$v_\parallel^{ab}(v) \equiv \frac{\left\langle (\Delta v_\parallel/v)^2 \right\rangle^{ab}}{\Delta t} = -2L^{ab} \frac{\psi_b''(v)}{v^2}. \tag{3.39}$$

The *slowing-down frequency* v_s^{ab} describes the rate at which a particle of species a is decelerated by collisions with particles of species b. The

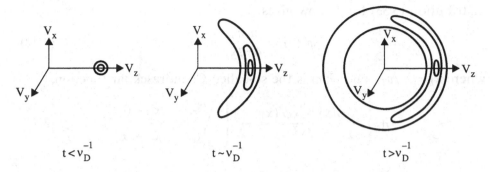

$$t < v_D^{-1} \qquad t \sim v_D^{-1} \qquad t > v_D^{-1}$$

Fig. 3.5. The evolution of a cloud of test particles under the action of the scattering part, $v_D \mathscr{L}$, of the collision operator. The particles diffuse on a sphere of constant energy, and become evenly distributed on this sphere after many scattering times v_D^{-1}. From Trubnikov (1965).

deflection frequency v_D^{ab} determines how quickly the direction of the velocity vector changes, see Fig. 3.5, and $v_{\parallel}^{ab}$ is the parallel velocity diffusion frequency.

With these definitions, we can write the collision operator (3.37) as

$$C_{ab}(f_a, f_{b0}) = v_D^{ab} \mathscr{L}(f_a) + \frac{1}{v^2} \frac{\partial}{\partial v} \left[v^3 \left(\frac{m_a}{m_a + m_b} v_s^{ab} f_a + \frac{1}{2} v_{\parallel}^{ab} v \frac{\partial f_a}{\partial v} \right) \right].$$

(3.40)

Note that the deflection frequency multiplies the Lorentz operator $\mathscr{L}$, which describes diffusion on the surface of a sphere $v = $ constant, thus changing the direction, but not the magnitude, of the velocity vector $\mathbf{v}$.

Our next task is to calculate the collision frequencies v_D^{ab}, v_s^{ab} and $v_{\parallel}^{ab}$ for a Maxwellian distribution function $f_b = f_{b0}$. To this end, it is useful to note the formal similarity between the Rosenbluth potentials and the usual electrostatic potential Φ satisfying

$$\nabla^2 \Phi = -\rho/\epsilon_0 \quad \Rightarrow \quad \Phi(\mathbf{r}) = \int \frac{\rho(\mathbf{r}')}{4\pi\epsilon_0 |\mathbf{r} - \mathbf{r}'|} d^3 r',$$

where ρ is the charge density. A comparison with the definition (3.15) of φ_b shows that

$$\nabla_v^2 \varphi_b = \frac{1}{v^2} \frac{\partial}{\partial v} \left(v^2 \frac{\partial \varphi_b}{\partial v} \right) = f_b(v),$$

(3.41)

where we have used the expression for the Laplacian ∇_v^2 in spherical velocity-space coordinates. Thus, φ_b can be thought of as the 'potential' associated with the 'charge distribution' f_b in velocity space – hence the name Rosenbluth potential. Integrating this equation with a Maxwellian distribution (3.33) for f_b now gives

$$\varphi_b'(v) = \frac{m_b n_b}{4\pi T_b} G(x_b),$$ (3.42)

where $x_b = v/v_{Tb}$ and $G(x)$ is the so-called Chandrasekhar function,

$$G(x) \equiv \frac{\phi(x) - x\phi'(x)}{2x^2} \rightarrow \begin{cases} \frac{2x}{3\sqrt{\pi}}, & x \to 0 \\ \frac{1}{2x^2}, & x \to \infty \end{cases}$$ (3.43)

$$\phi(x) \equiv \frac{2}{\sqrt{\pi}} \int_0^x e^{-y^2} dy.$$

The function $\phi(x)$ is called the error function erf(x).

The second Rosenbluth potential, ψ_b, can be obtained in a similar way, by using its relationship (3.20) to φ_b,

$$\frac{1}{v^2} \frac{d}{dv} \left(v^2 \frac{d\psi_b}{dv} \right) = \varphi_b(v),$$ (3.44)

with the result

$$\frac{d\psi_b}{dv} = -\frac{n_b}{8\pi} \left[\phi(x_b) - G(x_b) \right],$$

as shown in an exercise. Using these results, we can now write down explicit expressions for the three collision frequencies appearing in the collision operator (3.40):

$$v_D^{ab}(v) = \hat{v}_{ab} \frac{\phi(x_b) - G(x_b)}{x_a^3},$$ (3.45)

$$v_s^{ab}(v) = \hat{v}_{ab} \frac{2T_a}{T_b} \left(1 + \frac{m_b}{m_a} \right) \frac{G(x_b)}{x_a},$$ (3.46)

$$v_\parallel^{ab}(v) = 2\hat{v}_{ab} \frac{G(x_b)}{x_a^3},$$ (3.47)

$$\hat{v}_{ab} = \frac{n_b e_a^2 e_b^2 \ln \Lambda}{4\pi \epsilon_0^2 m_a^2 v_{Ta}^3},$$ (3.48)

where the last equation defines a basic collision frequency $\hat{v}_{ab}$, and where $x_a = v/v_{Ta}$.

Runaway electrons

The Chandrasekhar function (3.43) is non-monotonic as can be seen from the asymptotic forms in Eq. (3.43) and from the plot in Fig. 3.6. The fact

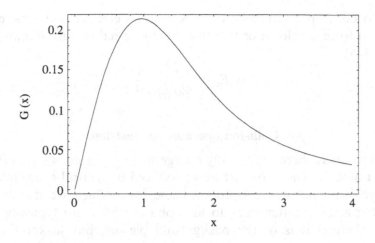

Fig. 3.6. The Chandrasekhar function $G(x)$, which describes the drag force on a particle by collisions with a Maxwellian background.

that $G(x)$ decreases for large x has the remarkable consequence that the average friction force on a particle,

$$\frac{m_a \left\langle \Delta v_\parallel \right\rangle^{ab}}{\Delta t} = -m_a v_\parallel v_s^{ab} \propto G(x_b),$$

decreases with increasing velocity if the latter is sufficiently high. Moreover, this friction force vanishes in the limit of infinite velocity (unless relativistic effects are considered, see below). The physical reason for this can be traced back to the collision dynamics discussed in Section 1.1, where it was shown that the momentum exchanged in a collision decreases with the incident particle's speed if the impact parameter is kept constant.

Now consider what happens if a constant electric field is applied to a plasma. However weak this field may be, it is still larger than the friction force on sufficiently fast electrons. The latter will therefore be accelerated by the electric field to arbitrarily high energy and form a population of so-called runaway electrons. If the electric field is sufficiently strong, ordinary thermal electrons will run away, and the bulk electron distribution will depart significantly from a Maxwellian. This occurs, roughly, when $eE > 2\hat{v}_{ee} m_e v_{Te}$, or, equivalently, when the electric field E exceeds the so-called Dreicer field

$$E_D = \frac{n_e e^3 \ln \Lambda}{4\pi \epsilon_0^2 T_e}.$$

As we shall see in Section 3.6, if relativistic effects are taken into account one finds that the friction does not fall all the way to zero at large energy; some saturation occurs when the velocity approaches the speed of light.

Thus, in order to produce runaway electrons an electric field must exceed the friction force on electrons travelling at the speed of light (Connor and Hastie, 1975),

$$E > E_c \equiv \frac{n_e e^3 \ln \Lambda}{4\pi \epsilon_0^2 m_e c^2}. \tag{3.49}$$

3.4 Collision operator for fast ions

In fusion plasmas there are usually energetic ions present, such as fusion-generated alpha particles or fast ions produced by neutral-beam injection or ion-cyclotron resonance heating. These ions, which we denote by a subscript α and hereafter refer to as alpha particles, are typically faster than the thermal ions of the background plasma, but slower than the electrons,

$$v_{Ti} \ll v_\alpha \ll v_{Te}.$$

In this limit, the largest of the collision frequencies from the preceding section is that of alpha–electron slowing down, which becomes

$$v_s^{\alpha e} \simeq \frac{1}{\tau_s}, \tag{3.50}$$

where the slowing-down time

$$\boxed{\tau_s \equiv \frac{3(2\pi)^{3/2} \epsilon_0^2 m_\alpha T_e^{3/2}}{Z_\alpha^2 e^4 m_e^{1/2} n_e \ln \Lambda}}$$

is calculated by using the small-argument approximation to the Chandrasekhar function (3.43). For drag against ions, the opposite asymptotic, $x \to \infty$, applies for $G(x)$, and we obtain for the slowing-down term in the collision operator (3.40)

$$\frac{m_\alpha}{m_\alpha + m_i} v_s^{\alpha i} = \left(\frac{v_c}{v}\right)^3 \frac{1}{\tau_s}, \tag{3.51}$$

where

$$v_c \equiv \left(\frac{n_i Z_i^2}{n_e} \frac{3\pi^{1/2} m_e}{4 m_i}\right)^{1/3} v_{Te} \tag{3.52}$$

is the critical speed above which the electron drag dominates over ion drag, corresponding to an alpha particle energy of the order of

$$\frac{m_\alpha v_c^2}{2} \sim \frac{m_\alpha}{m_i^{2/3} m_e^{1/3}} T_e \sim 50 \, T_e.$$

Alpha particles are born with a speed, corresponding to the energy 3.5 MeV, that is usually much larger than v_c. At first, an alpha particle is gradually decelerated by friction mainly against the electrons. Later, when it has reached a speed of order v_c, it also experiences friction against ions. Most of the fusion power thus goes to electron heating.

It is not difficult to verify that for collisions between alpha particles and electrons, the deflection frequency $v_D^{\alpha e}$ and velocity diffusion frequency $v_\parallel^{\alpha e}$ are insignificant in comparison with the slowing-down frequency $v_s^{\alpha e}$. The reason for this is of course the large mass difference between alpha particles and electrons. A heavy body moving in a gas of lighter particles can be decelerated by the latter, but is not scattered. This is particularly clear in the limit where the heavy particle is macroscopic. A bicyclist certainly feels the drag force, but is not knocked to the ground, by collisions with molecules in the air.

However, for collisions between alpha particles and ions the deflection frequency is important (since the masses are comparable) and becomes

$$v_D^{\alpha i} = \left(\frac{v_b}{v}\right)^3 \frac{1}{\tau_s}, \tag{3.53}$$

where the critical speed for pitch-angle scattering,

$$v_b \equiv \left(\frac{m_i}{m_\alpha}\right)^{1/3} v_c,$$

is of the same order of magnitude as v_c. Thus, pitch-angle scattering of alpha particles is comparable to ion drag, both being determined by the frequency of α–i collisions.

Collecting the terms describing electron drag (3.50), ion drag (3.51), and scattering against ions (3.53) in the collision operator (3.40), we arrive at the following alpha particle collision operator

$$C_\alpha(f_\alpha) = \frac{1}{v^2 \tau_s} \frac{\partial}{\partial v}[(v^3 + v_c^3)f_\alpha(v)] + \frac{v_b^3}{v^3 \tau_s}\mathscr{L}(f_\alpha).$$

If the alpha particles are born isotropically with a velocity of v_* at a rate of S per unit volume, the kinetic equation is thus

$$\frac{\partial f_\alpha}{\partial t} = \frac{1}{v^2 \tau_s} \frac{\partial}{\partial v}[(v^3 + v_c^3)f_\alpha(v)] + \frac{S\delta(v - v_*)}{4\pi v_*^2},$$

where the last term represents the source term, and the scattering operator vanishes because of isotropy. At velocities above the birth speed v_* there are very few alpha particles. This gives a boundary condition,

$f_\alpha(v > v_*) = 0$, under which it is straightforward to solve for the alpha particle distribution, which becomes

$$f_\alpha(v) = \frac{S\tau_s}{4\pi(v^3 + v_c^3)}$$

for $v < v_*$ in steady state.

3.5 Ion–electron collisions

We now turn our attention to the question of how collisions with electrons affect the ion distribution function. The electrons are much lighter and therefore much faster than the ions unless $T_e \ll T_i$. The ions are thus subject to a bombardment of light, fast particles, and the situation is mathematically similar to Brownian motion, where a relatively large particle undergoes diffusion because of many collisions with small fluid molecules.

The fact that the typical velocity of electrons greatly exceeds that of ions enables us to greatly simplify the ion–electron collision operator. In order to derive this operator, let us write the electron distribution function as a sum,

$$f_e = f_{Me}(\mathbf{v} - \mathbf{V}_i) + f_{e1}(\mathbf{v}),$$

of a Maxwellian with a mean velocity equal to that of the ions, and a remainder f_{e1}, and split the collision operator into similar parts

$$C_{ie}(f_i, f_e) = C_{ie}[f_i, f_{Me}(\mathbf{v} - \mathbf{V}_i)] + C_{ie}(f_i, f_{e1}). \tag{3.54}$$

It is easier to calculate the second term. In the collision operator (3.21), we only need to keep the first term, since this is multiplied by the large factor m_i/m_e. According to (3.17)

$$L^{ie}\frac{m_i}{m_e}\frac{\partial \varphi_e}{\partial v_k} \simeq -\frac{\langle \Delta v_k \rangle^{ie}}{\Delta t} = -\frac{F_k^{ie}(v)}{m_i},$$

where $\mathbf{F}^{ie}(\mathbf{v})$ is the average force acting on an ion that moves with velocity $\mathbf{v}$ as a result of collisions with electrons in the distribution f_{e1}. But this force must be independent of the ion velocity since f_{e1} only varies over velocities of order $v_{Te} \gg v_{Ti}$. $\mathbf{F}^{ie}$ is, in other words, practically the same for all ions, and must therefore be equal to

$$\mathbf{F}^{ie} = \frac{\mathbf{R}_{ie}}{n_i} = -\frac{\mathbf{R}_{ei}}{n_i}.$$

The collision operator (3.21) for the second term in (3.54) thus becomes

$$C_i(f_i, f_{e1}) = -\frac{\partial}{\partial v_k}\left(\frac{F_k^{ie}}{m_i}f_i\right) = \frac{\mathbf{R}_{ei}}{m_i n_i} \cdot \frac{\partial f_i}{\partial \mathbf{v}}. \tag{3.55}$$

Note that the part of the collision operator represented by Eq. (3.55) is largely independent of the electron distribution function. The only information carried by f_{e1} is the average friction force on the ions.

To find the Rosenbluth potentials (3.15) and (3.16) for $f_{Me}(\mathbf{v} - \mathbf{V}_i)$ entering in the first term in (3.54), it is again useful to exploit the fact that ions move much more slowly than electrons. Hence, in the expression (3.42) we may take the limit $G(x \to 0)$, yielding

$$\frac{\partial \varphi_e}{\partial v_k} = \frac{n_e}{3} \left(\frac{m_e}{2\pi T_e} \right)^{3/2} (v_k - V_{ik}).$$

The easiest way to calculate ψ_e is to use its property as a velocity-space potential for φ_e, see (3.20),

$$\nabla_{\mathbf{v}}^2 \psi_e = \varphi_e(\mathbf{v}) \simeq \varphi_e(\mathbf{V}_i) = -\frac{1}{4\pi} \int \frac{f_{Me}(v')}{v'} d^3 v' = -\frac{n_e}{(2\pi)^{3/2}} \sqrt{\frac{m_e}{T_e}},$$

which gives

$$\frac{\partial \psi_e}{\partial v_k \partial v_l} \simeq -\frac{n_e}{3(2\pi)^{3/2}} \sqrt{\frac{m_e}{T_e}} \delta_{kl}.$$

Substituting these two Rosenbluth potentials in (3.21), and adding (3.55) gives the ion–electron collision operator

$$C_{ie}(f_i) = \frac{\mathbf{R}_{ei}}{m_i n_i} \cdot \frac{\partial f_1}{\partial \mathbf{v}} + \frac{m_e n_e}{m_i n_i \tau_{ei}} \frac{\partial}{\partial \mathbf{v}} \cdot \left[(\mathbf{v} - \mathbf{V}_i) f_i + \frac{T_e}{m_i} \frac{\partial f_i}{\partial \mathbf{v}} \right], \tag{3.56}$$

where the electron–ion collision time τ_{ei} was defined in (1.4). The first term on the right represents ion–electron friction, and is inversely proportional to the electron–ion collision time τ_{ei}. The second term describes ion–electron energy exchange, which occurs much more slowly on the ion–electron collision time scale $\tau_{ie} \sim m_i \tau_{ei}/m_e$.

3.6 Collision operator for relativistic particles

So far we have neglected relativistic effects in our discussion of the collision operator. This is normally justified if the plasma temperature is much lower than the electron rest mass, 511 keV. However, sometimes a population of very high energy particles may be present in such a relatively 'cool' plasma, and it is then necessary to consider the relativistic collision operator for these particles. For example, in laboratory plasmas electron cyclotron heating produces mildly relativistic electrons, and runaway electrons in tokamaks frequently reach energies of tens of MeV. It has recently emerged that relativistic runaway electrons may also be produced in thunderstorms, giving rise to discharges propagating upward from thunderstorm clouds toward the ionosphere (Roussel-Dupré and Gurevich, 1996). In space, streams of fast electrons are thought to be responsible for so-called type III radio bursts in solar flares (Tandberg-Hansen and Emslie, 1988), and extremely energetic cosmic rays pervade all corners of our universe.

We are thus led to consider Coulomb collisions between very fast, relativistic particles (a) and much slower, background plasma particles (b). We confine our attention to this case, which is typical of the examples mentioned, rather than the more complicated situation where the velocities of the two particles are comparable. Particle b may thus be taken as practically stationary, so that the collision dynamics is essentially as described in Section 1.1, except that relativistic effects must be taken into account. Furthermore, the energy lost by emission of radiation in the collision (bremsstrahlung) can be ignored as long as the deflection angle of the velocity vector is small (Landau and Lifshitz, 1975).

The initial momentum and energy of the incident particle (a) are equal to

$$p = p_x = \gamma m_a v,$$

$$E_a = \gamma m_a c^2 = \sqrt{m_a^2 c^4 + p^2 c^2},$$

where m_a is the rest mass and $\gamma = (1 - v^2/c^2)^{-1/2}$ the relativistic mass factor, while the initial energy of particle b is just $E_b = m_b c^2$. The post-collision energies are

$$E_a' = \sqrt{m_a^2 c^4 + (p_x + \Delta p_x)^2 c^2 + (\Delta p_\perp)^2 c^2},$$

$$E_b' = \sqrt{m_b^2 c^4 + (\Delta p_x)^2 c^2 + (\Delta p_\perp)^2 c^2},$$

if Δp_x and $\Delta p_\perp$ are the momenta exchanged in the parallel and perpendicular directions (relative to the initial velcoity), respectively, see Fig. 3.7.

As usual, we need only consider mild collisions where these momenta are small, in which case we may use $(1 + \epsilon)^{1/2} \simeq 1 + \epsilon/2$ to obtain

$$E_a' \simeq E_a \left(1 + \frac{2 p_x \Delta p_x + (\Delta p_\perp)^2}{2 E_a^2} c^2 \right),$$

$$E_b' \simeq m_b c^2 \left(1 + \frac{(\Delta p_x)^2 + (\Delta p_\perp)^2}{2 m_b^2 c^2} \right),$$

where the term in E_b' containing $(\Delta p_x)^2$ can be neglected since $\Delta p_x \ll \Delta p_\perp$.

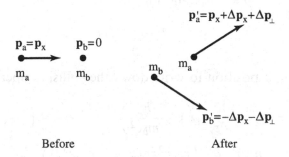

Fig. 3.7. Collision between a relativistic (a) and a stationary (b) particle.

Indeed, energy conservation, $E'_a - E_a + E'_b - E_b = 0$, then implies

$$\Delta p_x = -\left(1 + \frac{E_a}{m_b c^2}\right)\frac{(\Delta p_\perp)^2}{2 p_x},$$

which is much smaller than $\Delta p_\perp$ for mild collisions where $\Delta p_\perp \ll p_x$.

As demonstrated in Section 1.1, the transverse momentum exchanged between the two particles is equal to

$$\Delta p_\perp = \frac{e_a e_b}{2\pi\epsilon_0 b v}.$$

For a test particle of species a flying through a plasma, the cumulative effect of collisions with particles of species b in time Δt is thus described by

$$\frac{\left\langle (\Delta p_\perp)^2 \right\rangle^{ab}}{\Delta t} = \left(\frac{e_a e_b}{2\pi\epsilon_0 v}\right)^2 \int_{b_{\min}}^{\lambda_D} \frac{n_b 2\pi v\, db}{b} = \frac{2 m_a^2 c^3}{v\hat{\tau}_{ab}},$$

$$\frac{\left\langle \Delta p_x \right\rangle^{ab}}{\Delta t} = -\left(1 + \frac{E_a}{m_b c^2}\right)\frac{m_a^2 c^3}{p_x v\hat{\tau}_{ab}}, \tag{3.57}$$

where $n_b v 2\pi b\, db$ is the number of encounters in unit time with impact parameters in the interval $[b, b + db]$, and

$$\frac{1}{\hat{\tau}_{ab}} = \frac{n_b e_a^2 e_b^2 \ln\Lambda}{4\pi\epsilon_0^2 m_a^2 c^3}$$

is the collision frequency at the speed of light. As in Eqs. (3.13) and (3.14), we need to transform this result to arbitrary coordinates,

$$\frac{\left\langle \Delta p_k \right\rangle^{ab}}{\Delta t} = -\left(1 + \frac{E_a}{m_b c^2}\right)\frac{\gamma(m_a c)^3}{\hat{\tau}_{ab}}\frac{p_k}{p^3},$$

$$\frac{\left\langle \Delta p_k \Delta p_l \right\rangle^{ab}}{\Delta t} = \frac{\gamma(m_a c)^3}{\hat{\tau}_{ab}} P_{kl},$$

where we have used $p = \gamma m_a v$ and introduced the tensor

$$P_{kl} = \frac{p^2 \delta_{kl} - p_k p_l}{p^3}.$$

We are now in a position to write down the collision operator (3.3),

$$C_{ab}(f_a) = \frac{(m_a c)^3}{\hat{\tau}_{ab}} \frac{\partial}{\partial p_k} \left[\left(1 + \frac{E_a}{m_b c^2} \right) \frac{\gamma p_k f_a}{p^3} + \frac{\partial}{\partial p_l} \left(\frac{\gamma P_{kl} f_a}{2} \right) \right]$$

$$= \frac{(m_a c)^3}{\hat{\tau}_{ab}} \frac{\partial}{\partial p_k} \left(\frac{m_a}{m_b} \frac{\gamma^2 p_k f_a}{p^3} + \frac{P_{kl}}{2} \frac{\partial(\gamma f_a)}{\partial p_l} \right),$$

where we have used $\partial P_{kl}/\partial p_l = -2p_k/p^3$ in the second line. Finally, employing the formula for divergence in spherical coordinates and noting that

$$\frac{\partial}{\partial p_k} \left(\frac{P_{kl}}{2} \frac{\partial(\gamma f_a)}{\partial p_l} \right) = \frac{\gamma}{p^3} \mathscr{L}(f_a),$$

(cf. Eq. (3.36)) we arrive at the following expression for the relativistic collision operator,

$$\boxed{C_{ab}(f_a) = \frac{(m_a c)^3}{\hat{\tau}_{ab}} \left[\frac{m_a}{m_b p^2} \frac{\partial}{\partial p} \left(\gamma^2 f_a \right) + \frac{\gamma}{p^3} \mathscr{L}(f_a) \right],} \qquad (3.58)$$

where $\gamma = (1 + p^2/m_a^2 c^2)^{1/2}$ and $\hat{\tau}_{ab}$ was defined above.

The first term in this expression describes friction, which causes the fast particle to slow down. This term is small if the fast particle is much lighter than the target particles. For instance, fast electrons suffer very little friction against ions, just as we saw in the non-relativistic case. Indeed, if $m_a \ll m_b$ the first term is comparable to the second one only if $\gamma m_a \sim m_b$, i.e., if particle a is so fast that its relativistic mass γm_a approaches the rest mass of particle b. At lower energies, the friction on fast electrons is mainly caused by collisions with other electrons. This is similar to the non-relativistic case, where friction is described by the term in Eq. (3.40) containing the slowing-down frequency v_s^{ab}. As should be expected, the two results agree in the region of overlap, $v_{Ta} \ll v \ll c$, where Eq. (3.40) can be simplified by using the asymptotic form of the Chandrasekhar function (3.43), and where the non-relativistic limit, $\gamma = 1$, $p = m_a v$, may be used in Eq. (3.58). However, it should be noted that, unlike in the non-relativistic case, the relativistic friction force remains finite at large energies. The friction force (3.57) on a relativistic electron is equal to

$$F_{ee}(p) = -\frac{\langle \Delta p_x \rangle^{ee}}{\Delta t} = (1 + \gamma) \frac{m_e c^3}{\gamma v^2 \hat{\tau}_{ee}},$$

and tends to

$$\lim_{\gamma \to \infty} F_{ee}(p) = -\frac{m_e c}{\hat{\tau}_{ee}},$$

at large energy, where $v \to c$ and $\gamma \to \infty$. It follows that an electric field E must exceed the critical field $E_c = m_e c / e\hat{\tau}_{ee}$, earlier quoted in Eq. (3.49), in order to generate runaway electrons.

The second term in Eq. (3.58) describes pitch-angle scattering. Again, at non-relativistic speeds, it reduces to the analogous (first) term in Eq. (3.40) when this is evaluated for suprathermal particles. In the opposite, ultra-relativistic limit, $m_a c \ll p$, this term becomes small, being proportional to $\gamma(m_a c/p)^3 \simeq \gamma^{-2}$. The reason for this is that the particle a becomes heavier because of the relativistic mass effect, so that a larger force is required to produce a 90-degree scattering of its velocity vector.

3.7 The linearized collision operator

Electron–electron and ion–ion collisions

We have seen repeatedly that it is possible to simplify the collision operator for particles moving with very different characteristic speeds. In the case of collisions between particles of the same species (or thermal species with comparable masses) no such simplification is possible, and we must face up to the full Coulomb collision operator (3.21) or (3.22).

This operator is bilinear, i.e., the relations

$$C_{ab}(f_a + g_a, f_b) = C_{ab}(f_a, f_b) + C_{ab}(g_a, f_b),$$

$$C_{ab}(f_a, f_b + g_b) = C_{ab}(f_a, f_b) + C_{ab}(f_a, g_b),$$

$$C_{ab}(c_a f_a, c_b f_b) = c_a c_b C_{ab}(f_a, f_b),$$

hold for any distribution functions f_a, f_b, g_a, g_b, and constants c_a, c_b. The collision operator $C_{aa}(f_a) \equiv C_{aa}(f_a, f_a)$ for self-collisions is therefore *nonlinear* since, for instance,

$$C_{aa}(2f_a) = C_{aa}(2f_a, 2f_a) = 4C_{aa}(f_a).$$

However, for a system close to local thermodynamic equilibrium the distribution function is nearly Maxwellian,

$$f_a = f_{Ma} + f_{a1}, \qquad f_{a1} \ll f_{Ma},$$

and in this case the collision operator C_{aa} is approximately linear since

$$C_{aa}(f_a) = C_{aa}(f_{Ma} + f_{a1}, f_{Ma} + f_{a1})$$

$$\simeq C_{aa}(f_{a1}, f_{Ma}) + C_{aa}(f_{Ma}, f_{a1}) \equiv C_{aa}^l(f_a),$$

where we have neglected the quadratic term $C_{aa}(f_{a1}, f_{a1})$. Thus, in transport theory for near-equilibrium plasmas, it is only necessary to consider the linearized operator C_{aa}^l, defined in the last equality.

Besides being linear, this operator has a number of other important properties. Perhaps its most fundamental characteristic is that it is *self-adjoint*, i.e., the functional

$$S_{aa}[\hat{f}, \hat{g}] \equiv \int \hat{g}(\mathbf{v}) C_{aa}^l(f_{Ma}\hat{f}) d^3v, \qquad (3.59)$$

acting on any *two* functions $\hat{f}$ and $\hat{g}$, is symmetric,

$$S_{aa}[\hat{f}, \hat{g}] = S_{aa}[\hat{g}, \hat{f}].$$

This circumstance is not only computationally convenient, it also has a profound physical meaning as it implies positive entropy production, which will be demonstrated in the exercises at the end of the chapter. To prove that the linearized Coulomb operator is indeed self-adjoint, we use (3.12) and (3.22) to write it in the form

$$C_{aa}^l(f_a) = C_{aa}(f_{a1}, f_{Ma}) + C_{aa}(f_{Ma}, f_{a1})$$

$$= -\frac{L^{aa}}{8\pi} \frac{\partial}{\partial v_k} \int U_{kl} \left(f_{a1} \frac{\partial f'_{Ma}}{\partial v'_l} - f'_{Ma} \frac{\partial f_{a1}}{\partial v_l} + f_{Ma} \frac{\partial f'_{a1}}{\partial v'_l} - f'_{a1} \frac{\partial f_{Ma}}{\partial v_l} \right) d^3v'$$

$$= -\frac{L^{aa}}{8\pi} \frac{\partial}{\partial v_k} \int U_{kl} f_{Ma} f'_{Ma} \left(\frac{\partial \hat{f}'}{\partial v'_l} - \frac{\partial \hat{f}}{\partial v_l} \right) d^3v', \qquad (3.60)$$

where

$$f_{a1} = f_{Ma}\hat{f},$$

and primes indicate that a function be regarded as dependent on $\mathbf{v}'$ (and not $\mathbf{v}$), i.e., $f' = f(\mathbf{v}')$ and $f = f(\mathbf{v})$. In the above, we have used

$$U_{kl} \left(f_{Ma} \frac{\partial f'_{Ma}}{\partial v'_l} - f'_{Ma} \frac{\partial f_{Ma}}{\partial v_l} \right) = \frac{m_a}{T_a} f_{Ma} f'_{Ma} U_{kl}(v_l - v'_l) = 0,$$

which follows from the relation $U_{kl}u_l = 0$. With the help of the relation (3.60), we can now write the functional S_{aa} as

$$S_{aa}[\hat{f}, \hat{g}] = -\frac{L^{aa}}{8\pi} \int \hat{g} \, d^3v \frac{\partial}{\partial v_k} \int U_{kl} f_{Ma} f'_{Ma} \left(\frac{\partial \hat{f}'}{\partial v'_l} - \frac{\partial \hat{f}}{\partial v_l} \right) d^3v'$$

$$= \frac{L^{aa}}{8\pi} \int d^3v \int d^3v' U_{kl} f_{Ma} f'_{Ma} \frac{\partial \hat{g}}{\partial v_k} \left(\frac{\partial \hat{f}'}{\partial v'_l} - \frac{\partial \hat{f}}{\partial v_l} \right)$$

$$= -\frac{L^{aa}}{8\pi} \int d^3v \int d^3v' U_{kl} f_{Ma} f'_{Ma} \frac{\partial \hat{g}'}{\partial v'_k} \left(\frac{\partial \hat{f}'}{\partial v'_l} - \frac{\partial \hat{f}}{\partial v_l} \right),$$

where the second equality follows from partial integration, and the third one is obtained by interchanging the integration variables v and v'. Note that the tensor U_{kl} is not affected by this operation. Adding the last two expressions for S_{aa} now gives

$$S_{aa}[\hat{f},\hat{g}] = -\frac{L^{aa}}{16\pi} \int d^3v \int d^3v' U_{kl} f_{Ma} f'_{Ma} \left(\frac{\partial \hat{f}'}{\partial v'_l} - \frac{\partial \hat{f}}{\partial v_l}\right)\left(\frac{\partial \hat{g}'}{\partial v'_k} - \frac{\partial \hat{g}}{\partial v_k}\right)$$

$$= S_{aa}[\hat{g},\hat{f}], \tag{3.61}$$

which is obviously symmetric in $\hat{f}$ and $\hat{g}$. We have thus demonstrated the self-adjointness of S_{aa}.

The exact collision operator (3.21), (3.22) conserves particles, momentum and energy. We now demonstrate that the linearized operator C_{aa}^l also has these properties, i.e., that they are not lost in the linearization process. First, we note that the equation

$$C_{aa}^l(f_{Ma}\hat{f}) = 0 \tag{3.62}$$

has the three basic solutions

$$\hat{f} = (1, \mathbf{v}, v^2), \tag{3.63}$$

which may be multiplied by any numerical constants. Thus, the null space of the operator C_{aa}^l consists of the linear combinations of these functions. These solutions are easily verified by direct substitution in the expression (3.60), and can be understood physically in the following way. The full, nonlinear collision operator vanishes when operating on a Maxwellian (2.8), regardless of its density n, mean velocity $\mathbf{V}$, and temperature T. If we let f_{Ma}^0 denote a Maxwellian with $n = n_0$, $\mathbf{V} = 0$, and $T = T_0$, and $f_{Ma}^0 + \delta f$ a Maxwellian with slightly different density, velocity, and temperature, i.e., $n = n_0 + \delta n$, $\mathbf{V} = \delta\mathbf{V}$, and $T = T_0 + \delta T$, we have to the lowest order in the differences

$$\delta f = \left[\frac{\delta n}{n_0} + \frac{m_a}{T_0}\mathbf{v}\cdot\delta\mathbf{V} + \left(\frac{m_a v^2}{2T_0} - \frac{3}{2}\right)\frac{\delta T}{T_0}\right]f_{Ma}^0. \tag{3.64}$$

Since the linearized collision operator must vanish when operating on $f_{Ma}^0 + \delta f$, we must have $C_{aa}^l(\delta f) = 0$ for *any* values of δn, $\delta\mathbf{V}$, and δT. This implies that the equation (3.62) has the solutions (3.63). The conservation of particles, momentum and energy now follows from the self-adjointness of S_{aa}:

$$\int 1\cdot C_{aa}^l(f_{Ma}\hat{f})d^3v = \int \hat{f} C_{aa}^l(f_{Ma}\cdot 1)d^3v = 0,$$

$$\int \mathbf{v} C_{aa}^{l}(f_{Ma}\hat{f})d^{3}v = \int \hat{f} C_{aa}^{l}(f_{Ma}\mathbf{v})d^{3}v = 0,$$

$$\int \frac{mv^{2}}{2} C_{aa}^{l}(f_{Ma}\hat{f})d^{3}v = \int \hat{f} C_{aa}^{l}(f_{Ma}mv^{2}/2)d^{3}v = 0.$$

It is useful to note that the Lorentz operator (3.29) is also self-adjoint, since the functional

$$S_{L}[\hat{f}, \hat{g}] \equiv \int v_{ei}(v)\hat{g}\mathscr{L}(f_{Me}\hat{f})d^{3}v$$

is symmetric

$$S_{L}[\hat{f}, \hat{g}] = -\frac{1}{2} \int_{0}^{\infty} v_{ei}(v)f_{Me}v^{2}dv \int_{0}^{\pi} \sin\theta\, d\theta$$

$$\times \int_{0}^{2\pi} \left(\frac{\partial\hat{f}}{\partial\theta}\frac{\partial\hat{g}}{\partial\theta} + \frac{1}{\sin^{2}\theta}\frac{\partial\hat{f}}{\partial\varphi}\frac{\partial\hat{g}}{\partial\varphi} \right) d\varphi = S_{L}[\hat{g}, \hat{f}]. \quad (3.65)$$

Therefore the electron–ion collision operator (3.28) is self-adjoint. Since thus both the linearized electron–electron collision operator and the electron–ion operator are self-adjoint, it follows that the total, linearized electron collision operator $C_{e} = C_{ee}^{l} + C_{ei}$ also has this property:

$$S_{e}[\hat{f}, \hat{g}] \equiv \int \hat{g}[C_{ee}^{l}(f_{Me}\hat{f}) + C_{ei}(f_{Me}\hat{f})]d^{3}v$$

$$= \int \hat{f}[C_{ee}^{l}(f_{Me}\hat{g}) + C_{ei}(f_{Me}\hat{g})]d^{3}v = S_{e}[\hat{g}, \hat{f}], \quad (3.66)$$

a fact we shall use in the next chapter to prove Onsager symmetry and to derive a variational principle for electron transport.

General linearized operator

For collisions between two *different* species which are both close to local thermodynamic equilibrium,

$$f_{a} = f_{a0} + f_{a1}, \qquad\qquad f_{b} = f_{b0} + f_{b1},$$

it is appropriate to define the linearized collision operator

$$C_{ab}^{l}(f_{a1}, f_{b1}) \equiv C_{ab}(f_{a1}, f_{b0}) + C_{ab}(f_{a0}, f_{b1}), \quad (3.67)$$

which results from neglecting the nonlinear term $C_{ab}(f_{a1}, f_{b1})$ and observing that $C_{ab}(f_{a0}, f_{b0}) = 0$ if the Maxwellian functions f_{a0} and f_{b0} have the same flow velocity and temperature. It is clear that the linearized operator for self-collisions is a special case of this more general operator C_{ab}^l, and it is not difficult to show that the latter has the same conservation properties for particles, momentum, and energy (2.3)–(2.5) as the exact collision operator.

The first term in (3.67) describes collisions of 'test particles' with a fixed background of Maxwellian particles, and has already been treated in some detail in Section 3.3. The second, 'field particle' term is considerably more complicated as it involves the Rosenbluth potentials (3.15) and (3.16) of the non-Maxwellian distribution f_{b1}. The calculation of $C_{ab}(f_{a0}, f_{b1})$ thus involves solving the two partial differential equations defining these potentials, which is of course a difficult task in general. Nevertheless, it is still possible to express the exchange of momentum and energy between two colliding species in a compact way. In the friction force

$$\mathbf{R}_{ab} = \int m_a \mathbf{v} \left[C_{ab}(f_{a1}, f_{b0}) + C_{ab}(f_{a0}, f_{b1}) \right] d^3v,$$

the first term can be written in terms of the slowing-down frequency (3.46) as

$$\int m_a \mathbf{v} C_{ab}(f_{a1}, f_{b0}) \, d^3v = \int m_a \frac{\langle \Delta \mathbf{v} \rangle^{ab}}{\Delta t} f_{a1} \, d^3v = - \int m_a \mathbf{v} v_s^{ab} f_{a1} \, d^3v$$

by recalling the discussion from Section 3.3. The second term can be treated in a similar way after first using momentum conservation to observe that

$$\int m_a \mathbf{v} C_{ab}(f_{a0}, f_{b1}) \, d^3v = - \int m_b \mathbf{v} C_{ba}(f_{b1}, f_{a0}) \, d^3v.$$

Hence the total linearized friction force becomes

$$\mathbf{R}_{ab} = - \int \mathbf{v} \left(m_a v_s^{ab} f_{a1} + m_b v_s^{ba} f_{b1} \right) d^3v.$$

The energy exchanged between the two species is calculated in a similar way by noting that

$$\int \frac{m_a v^2}{2} C_{ab}(f_{a1}, f_{b0}) \, d^3v = \int m_a \frac{\langle (\mathbf{v} + \Delta \mathbf{v})^2 - v^2 \rangle^{ab}}{2\Delta t} f_{a1} \, d^3v$$

$$= \int m_a \frac{\langle 2\mathbf{v} \cdot \Delta \mathbf{v} + \Delta v_\perp^2 + \Delta v_\parallel^2 \rangle^{ab}}{2\Delta t} f_{a1} \, d^3v$$

$$= - \int \frac{m_a v^2}{2} v_E^{ab}(v) f_{a1} \, d^3v,$$

where the energy exchange frequency is defined as

$$v_E^{ab} \equiv 2v_s^{ab} - 2v_D^{ab} - v_\parallel^{ab}.$$

The linearized energy exchange then becomes

$$\int \frac{m_a v^2}{2} C_{ab}^l(f_{a1}, f_{b1})\, d^3v = -\int \frac{v^2}{2}\left(m_a v_E^{ab} f_{a1} + m_b v_E^{ba} f_{b1}\right)\, d^3v.$$

Finally, we comment on the angular eigenfunctions of the operator (3.67). We recall that if (v, θ, φ) are spherical coordinates in velocity space, then Legendre polynomials $P_l(\cos\theta)$, defined by

$$P_l(x) = \frac{1}{2^n n!}\frac{d^n}{dx^n}\left(x^2 - 1\right)^n,$$

are eigenfunctions of the Laplace operator

$$\nabla_v^2 = \frac{1}{v^2}\frac{\partial}{\partial v}v^2\frac{\partial}{\partial v} + \frac{1}{\sin\theta}\frac{\partial}{\partial\theta}\sin\theta\frac{\partial}{\partial\theta} + \frac{1}{\sin^2\theta}\frac{\partial^2}{\partial\varphi^2}$$

$$= \frac{1}{v^2}\frac{\partial}{\partial v}v^2\frac{\partial}{\partial v} + 2\mathscr{L}$$

in the sense

$$\nabla_v^2 P_l(\cos\theta) = -l(l+1)P_l(\cos\theta).$$

(The v- and φ-derivatives in ∇_v^2 do not matter since $P_l(\cos\theta)$ does not depend on these variables.) The most general eigenfunctions of the Laplace operator involve spherical harmonics $Y_{lm}(\theta, \varphi)$, but in magnetized plasmas the azimuthal dependence (φ-variation, typically gyroangle) of the distribution function is often ignorable, in which case spherical harmonics reduce to Legendre polynomials. We therefore restrict our attention to these simpler eigenfunctions.

It follows from Eq. (3.40) that if we let the first, test particle, term of the linearized operator (3.67) act on a function which is proportional to a Legendre polynomial, $f_{a1}(\mathbf{v}) = F_a(v)P_l(\xi)$ where $\xi = \cos\theta$, then the result will be proportional to the same polynomial,

$$C_{ab}\left[F_a(v)P_l(\xi), f_{b0}\right] \propto P_l(\xi).$$

In other words, this piece of the operator does not change the angular structure of the distribution function. This is also true for the second, field particle, term of the operator (3.67). To see this, we recall Eqs. (3.20) and (3.41),

$$\nabla_v^2 \varphi_{b1} = f_{b1}, \qquad \nabla_v^2 \psi_{b1} = \varphi_{b1}. \tag{3.68}$$

These relations show that if the distribution function f_{b1} is of the form $f_{b1}(\mathbf{v}) = F_b(v)P_l(\xi)$ then its Rosenbluth potentials have the same angular structure and can thus be written as

$$\varphi_{b1}(\mathbf{v}) = \Phi_b(v)P_l(\xi),$$

$$\psi_{b1}(\mathbf{v}) = \Psi_b(v)P_l(\xi).$$

Evaluating the expression for the collision operator in terms of Rosenbluth potentials (3.21) for the case where $f_a = f_{a0}$ is Maxwellian gives the field particle operator

$$C_{ab}(f_{a0}, f_{b1}) = (3.21) = L^{ab}\left[\frac{m_a}{m_b}\left(\nabla^2\varphi_{b1} - \frac{m_a v_k}{T_a}\frac{\partial\varphi_{b1}}{\partial v_k}\right)f_{a0}\right.$$

$$\left. + \frac{m_a v_l}{T_a}\frac{\partial\nabla^2\psi_{b1}}{\partial v_l}f_{a0} + \frac{\partial^2\psi_{b1}}{\partial v_k\partial v_l}\frac{\partial}{\partial v_k}\left(\frac{m_a v_l}{T_a}f_{a0}\right)\right]$$

$$= L^{ab}f_{a0}\left[\frac{m_a}{m_b}f_{b1} + \frac{m_a v}{T_a}\left(1 - \frac{m_a}{m_b}\right)\frac{\partial\varphi_{b1}}{\partial v}\right.$$

$$\left. + \frac{m_a}{T_a}\varphi_{b1} - \frac{m_a^2 v^2}{T_a^2}\frac{\partial^2\psi_{b1}}{\partial v^2}\right],$$

where we have used (3.68). Each term in this expression is proportional to $P_l(\xi)$, and this is therefore also true for the full, linearized collision operator,

$$C_{ab}\left[f_{a0}, F_b(v)P_l(\xi)\right] \propto P_l(\xi).$$

3.8 Model operator for self-collisions

The linearized operator (3.60) describing collisions among particles of the same species is much simpler than the full Coulomb collision operator (3.21), (3.22), but is nevertheless quite complicated. It is often useful to have an operator which is less complicated but still describes the effects of collisions reasonably accurately. There are ways to construct such operators in a systematic way from the linearized operator to a high degree of accuracy (Hirshman and Sigmar, 1976). However, for practical purposes, it is frequently adequate to use the yet simpler model operator

$$\boxed{C_{aa}^m(f_a) = v_D^{aa}(v)\left(\mathscr{L}(f_a) + \frac{m_a \mathbf{v} \cdot \mathbf{u}}{T_a}f_{Ma}\right),} \qquad (3.69)$$

where f_{Ma} denotes a stationary Maxwellian, $v_D^{aa}(v)$ is the deflection frequency (3.45), and the constant vector $\mathbf{u}$ is to be chosen appropriately. This operator is based on the observation that the Lorentz scattering

term tends to be the most important ingredient in the full collision operator in transport-theory applications, in particular in neoclassical theory.
However, since Lorentz scattering does not conserve momentum, a term
needs to be added to restore the momentum lost in scattering, which is
essentially the role of the second term on the right in (3.69). Note that
this model operator strongly resembles the electron–ion collision operator
(3.28).

In order to be useful, the model operator should conserve particles,
momentum, and energy. Particles and energy are automatically conserved,

$$\int C_{aa}^m(f_a)d^3v = 0,$$

$$\int \frac{m_a v^2}{2} C_{aa}^m(f_a)d^3v = 0,$$

since the second term in the operator (3.69) is odd in $\mathbf{v}$, and

$$\int F(v)\mathscr{L}f_a(\mathbf{v})d^3v = \int f_a(\mathbf{v})\mathscr{L}F(v)d^3v = 0$$

for any function $F(v)$, including $F(v) = 1$ and $F(v) = m_a v^2/2$, because of
the self-adjointness of the Lorentz operator (3.65). Momentum is conserved
if

$$0 = \int \mathbf{v}C_{aa}^m(f_a)d^3v = \int v_D^{aa}(v)\left(-\mathbf{v}f_a + \frac{m_a(\mathbf{v}\cdot\mathbf{u})\mathbf{v}}{T_a}f_{Ma}\right)d^3v,$$

where the self-adjointness of $\mathscr{L}$ has been used to write

$$\int \mathbf{v}\mathscr{L}(f_a)d^3v = \int f_a\mathscr{L}(\mathbf{v})d^3v = -\int f_a\mathbf{v}d^3v,$$

noting that $\mathscr{L}(\mathbf{v}) = -\mathbf{v}$. Thus, since

$$\int F(v)(\mathbf{u}\cdot\mathbf{v})\mathbf{v}d^3v = \frac{\mathbf{u}}{3}\int F(v)v^2d^3v$$

for any function $F(v)$, the model operator (3.69) conserves momentum if
the vector $\mathbf{u}$ is understood to be

$$\boxed{\mathbf{u} = \int \mathbf{v}v_D^{aa}(v)f_a d^3v \Big/ \int v_D^{aa}(v)\frac{m_a v^2}{3T_a}f_{Ma}d^3v.}$$ (3.70)

This choice of $\mathbf{u}$ also makes the operator self-adjoint since the functional

$$S_m[\hat{f},\hat{g}] \equiv \int \hat{g}C_{aa}^m(f_{Ma}\hat{f})d^3v$$

is symmetric in $\hat{f}$ and $\hat{g}$,

$$S_m[\hat{f}, \hat{g}] = \int v_D^{aa}(v)\hat{g}\left[\mathscr{L}(f_{Ma}\hat{f}) + \frac{m_a \mathbf{v} \cdot \mathbf{u}}{T_a}f_{Ma}\right]d^3v$$

$$= \int v_D^{aa}(v)f_{Ma}\hat{g}\mathscr{L}(\hat{f})d^3v + \frac{\int v_D^{aa}(v)f_{Ma}\hat{g}\frac{m_a\mathbf{v}}{T_a}d^3v \cdot \int v_D^{aa}(v)f_{Ma}\hat{f}\mathbf{v}d^3v}{\int v_D^{aa}(v)\frac{m_a v^2}{3T_a}f_{Ma}d^3v}$$

$$= S_m[\hat{g}, \hat{f}].$$

Thus the model operator (3.69) satisfies particle, momentum, and energy conservation, is self-adjoint, describes pitch-angle scattering correctly, and is Galilean invariant as it must be. Because of these properties it can be expected to be a reasonably accurate approximation of the true collision operator. It is indeed widely useful. We shall use it in Section 4.4 to calculate the electrical conductivity in collisional plasma, and in Chapter 11 to develop neoclassical transport theory for weakly collisional plasmas.

Further reading

Our presentation is influenced by the extensive discussions of the Fokker–Planck collision operator given by Trubnikov (1965) and Sivukhin (1966). Landau's original paper (1936) is also exceptionally clear. Since there are many particles in the Debye sphere when the Coulomb logarithm is large, collisions are strictly speaking not binary events, but involve many-body physics. A more sophisticated collision operator, named after Balescu and Lenard, emphasizes this point. Concise and clear derivations of the Balescu–Lenard operator are presented in Lifshitz and Pitaevskii (1981), and Hazeltine and Waelbroeck (1998). The collision operator for fast alpha particles (including the sometimes neglected pitch-angle scattering) is treated in depth by Hsu, Catto and Sigmar (1990). Hirshman and Sigmar (1976) is a good source for information about general properties of the linearized collision operator and model operators of various degrees of sophistication. An important representation of the exact collision operator for keeping the pitch-angle operator in its exact form but expanding the momentum restoring piece of the operator in Legendre and Laguerre polynomials is given by Taguchi (1988). Many useful details of Coulomb collision operator theory are contained in the pioneering book by Shkarofsky *et al.* (1966).

Exercises

1. Calculate the deflection frequency (3.45) and the parallel velocity diffusion
frequency (3.47) from their definitions (3.38) and (3.39) by using the relation
(3.44).

Solution: From (3.42) follows

$$\frac{d\varphi_b}{dv} = \frac{n_b}{2\pi v_{Tb}^2} G(x_b).$$

To obtain φ_b from this relation, use

$$\int G(x)dx = -\frac{\phi - x\phi'}{2x} - \frac{1}{2}\int \phi'' dx = -\frac{\phi}{2x},$$

which follows from integrating the definition of $G(x)$ by parts. Next, to
solve the differential equation (3.44) for ψ_b, use $\int x\phi' dx = -\phi'/2$ and again
integrate by parts to find

$$2\int x^2 dx \int^x G(x')dx' = -\int x\phi\, dx = -\frac{x^2}{2}\phi + \int \frac{x^2}{2}\phi' dx$$

$$= -\frac{x^2}{2}\phi - \frac{x\phi'}{4} + \int \frac{\phi'}{4}dx$$

$$= -\frac{x^2(\phi - G)}{2}.$$

Hence it follows that

$$\frac{d\psi_b}{dv} = -\frac{n_b}{8\pi}\left[\phi(x_b) - G(x_b)\right],$$

and the deflection frequency (3.38) becomes equal to (3.45). In order to
calculate (3.39), note that $\phi'' = -2x\phi'$ and

$$G'(x) = \frac{d}{dx}\left(\frac{\phi(x) - x\phi'(x)}{2x^2}\right) = \left(1 + \frac{1}{x^2}\right)\phi' - \frac{\phi}{x^3} = \phi' - \frac{2G}{x},$$

so that $\phi' - G' = 2G/x$, and Eq. (3.47) follows.

2. For viscosity calculations, it is useful to know at what rate the pressure
anisotropy

$$p_{a\parallel} - p_{a\perp} \equiv \int m_a\left(v_\parallel^2 - v_\perp^2/2\right) f_a\, d^3v$$

decays by collisions with a Maxwellian background species f_{b0}. Calculate
this rate along the same lines as used for the friction force and momentum
exchange in Section 3.7.

Solution:

$$\frac{d}{dt}\left(p_{a\parallel} - p_{a\perp}\right)\Big|_{\text{collisions}} = \int m_a \left(v_\parallel^2 - v_\perp^2/2\right) C_{ab}(f_a, f_{b0})\, d^3v$$

$$= \int \frac{m_a}{\Delta t} \left\langle (v_\parallel + \Delta v_\parallel)^2 - v_\parallel^2 - \Delta v_\perp^2/2 \right\rangle^{ab} f_a\, d^3v$$

$$= \int \frac{m_a}{\Delta t} \left\langle 2v\Delta v_\parallel + \Delta v_\parallel^2 - \Delta v_\perp^2/2 \right\rangle^{ab} f_a\, d^3v$$

$$= -\int m_a v^2 v_T^{ab} f_a\, d^3v,$$

where the anisotropy relaxation frequency is $v_T^{ab} = 2v_s^{ab} - v_\parallel^{ab} + v_D^{ab} = v_E^{ab} + 3v_D^{ab}$.

3. Entropy. The entropy s per particle of a species with the distribution function f and density n is defined by

$$s \equiv -\frac{1}{n} \int f \ln f\, d^3v.$$

(a) Show that $s/n = \frac{3}{2}\ln(p/n^{5/3}) + \text{const.}$ for a Maxwellian species, as claimed at the end of Section 2.2.

Solution: Straightforward substitution of (2.8) into the definition of s gives this result.

(b) Calculate the rate of change in the total entropy $S = \int ns\, d^3r$ from the kinetic equation (2.1).

Solution: We have

$$\frac{dS}{dt} = -\int \frac{\partial f}{\partial t}(1 + \ln f)d^3v\, d^3r,$$

where the first term vanishes because of particle conservation. From the kinetic equation in conservation form

$$\frac{\partial f}{\partial t} = -\nabla \cdot (\mathbf{v}f) - \nabla_v \cdot (\mathbf{F}f/m) + C(f),$$

where $\mathbf{F}$ is the Lorentz force (note that $\nabla_v \cdot \mathbf{F} = 0$), it is apparent that the first two terms do not contribute to entropy production since

$$\int \ln f [\nabla \cdot (\mathbf{v}f) + \nabla_v \cdot (\mathbf{F}f/m)]d^3r\, d^3v$$

$$= \int [\nabla \cdot (\mathbf{v}f \ln f) + \nabla_v \cdot (\mathbf{F}f \ln f/m)]d^3r\, d^3v = 0.$$

This integral vanishes because, by Gauss's theorem, it can be converted to a surface integral at infinity, where the distribution function can be assumed to vanish. Entropy production is thus entirely caused by collisions,

$$\frac{dS}{dt} = -\int C(f)\ln f\, d^3r\, d^3v.$$

(c) Show that self-collisions increase the entropy for a distribution function close to equilibrium.

Solution: If $f = f_0(1 + \hat{f})$, where f_0 is Maxwellian and $\hat{f}$ is small, then $\ln f \simeq \ln f_0 + \hat{f}$, and the local entropy production rate due to collisions is

$$\frac{\partial n s}{\partial t} = -\int C(f)\ln f\, d^3v \simeq -\int \hat{f} C^l(\hat{f}f_0)d^3v = -S_{aa}[\hat{f},\hat{f}],$$

where S_{aa} is the functional defined in (3.59). But from (3.61) it is apparent that $S_{aa}[\hat{f},\hat{f}] \le 0$ for any function $\hat{f}$ since for any vector $\mathbf{a}$

$$U_{kl}a_k a_l = u^{-3}[u^2 a^2 - (\mathbf{u}\cdot\mathbf{a})^2] \ge 0,$$

by the Schwartz inequality. Thus, linearized self-collisions always increase entropy.

(d) Show that the full, nonlinearized, self-collision operator increases entropy (Boltzmann's H-theorem) by a procedure similar to that leading to (3.61).

Solution: Multiplying the Landau operator (3.22) for self-collisions by $\ln f_a$, integrating over velocity space and dropping the species index gives

$$\frac{\partial n s}{\partial t} = -\int \ln f\, C(f)d^3v$$

$$= -\frac{L}{8\pi}\int \frac{\partial \ln f}{\partial v_k}d^3v \int U_{kl}\left(f\frac{\partial f'}{\partial v'_l} - f'\frac{\partial f}{\partial v_l}\right)d^3v',$$

where we have integrated by parts in $\mathbf{v}$. Going through steps identical to those in the text now gives

$$\frac{\partial n s}{\partial t} = \frac{L}{16\pi}\int d^3v \int d^3v' f f' U_{kl}\left(\frac{\partial \ln f'}{\partial v'_k} - \frac{\partial \ln f}{\partial v_k}\right)\left(\frac{\partial \ln f'}{\partial v'_l} - \frac{\partial \ln f}{\partial v_l}\right),$$

which is positive for exactly the same reason as presented in the previous exercise.

(e) Show that the entropy production from self-collisions vanishes only when the distribution function is locally Maxwellian.

Solution: The Schwartz inequality invoked under (c) becomes an equality if and only if the vector $\mathbf{a}$ is parallel to $\mathbf{u}$. Therefore $\partial n s/\partial t = 0$ only if

$$\frac{\partial \ln f'}{\partial v'_k} - \frac{\partial \ln f}{\partial v_k} = c_0(v_k - v'_k),$$

where c_0 is a constant. Hence $\partial \ln f/\partial v_k = -c_0(v_k - V_k)$, where V_k is another constant, and f must be of the form

$$f = c_1 e^{-c_0(\mathbf{v}-\mathbf{V})^2}.$$

4

Plasma fluid equations

4.1 Outline of closure in the case of short mean-free path

If the collisionality is high it is possible to close the system of fluid equations (2.15)–(2.17) by employing the method devised by Chapman (1916) and Enskog (1917, 1922) in the kinetic theory of gases with minor modifications. This was done in full detail in a famous review article by Braginskii (1965), whose work has since become a paradigm of transport theory, being a practically complete theory that is rigorous and derived entirely from first principles.

The basic assumption underlying Chapman and Enskog's method of solving the kinetic equation is the assumption of high collisionality. The collision frequency v is assumed to be high enough that local thermodynamic equilibrium is attained on a time scale faster than the relaxation of macroscopic gradients due to transport. Thus the following orderings are assumed to be satisfied:

$$\frac{\partial}{\partial t} \ll v, \tag{4.1}$$

$$\lambda \ll L, \tag{4.2}$$

where $\lambda = v_T/v$ is the mean-free path, and $L = \nabla^{-1}$ is the length scale of the variation of macroscopic plasma parameters (density, temperature, magnetic field, etc.). In a magnetized plasma, gradients are often very different along and across the field lines, so it is useful to introduce separate parallel ($L_\parallel$) and perpendicular ($L_\perp$) length scales. In addition, the gyrofrequency is usually much larger than the collision frequency, i.e.,

$$\Delta \equiv v/\Omega \ll 1, \tag{4.3}$$

implying that the Larmor radius is smaller than the mean-free path, $\rho \ll \lambda$. When this is the case, the ordering (4.2) perpendicular to the field may be relaxed,

$$\lambda \ll L_\|, \qquad \delta \equiv \rho/L_\perp \ll 1. \tag{4.4}$$

In other words, if the plasma is *magnetized*, the mean-free path need only be short in comparison with the parallel length scale. Throughout this book, we deal with magnetized plasmas, and it is the latter spatial ordering (4.3), (4.4), rather than Eq. (4.2), we have in mind when considering plasma fluid equations with collisional closure, the so-called Braginskii equations. The temporal (4.1) and spatial (4.4) orderings are not entirely independent since weak gradients ensure that collisional relaxation is slow, of order $\nu(\lambda/L_\|)^2$ or $\nu(\rho/L_\perp)^2$ for diffusion processes. However, the ordering (4.1) also implies that no other phenomena on the collisional time scale can be described by Braginskii's equations.

The nature of the ordering (4.4) can be understood in the following way. The particles in a plasma are free to travel a mean-free path in the direction parallel to the magnetic field and a Larmor radius across the field. These are the bounds within which the orbit is confined without experiencing collisions. Only if the plasma is nearly homogeneous over this region, i.e., if Eqs. (4.4) are satisfied, is it possible to construct a transport theory independent of the magnetic field geometry. This condition is typically *not* satisfied for hydrogenic species in the hot core of a tokamak but often holds for heavy impurities in the cooler edge plasma.

When the above orderings are satisfied, the distribution function f_a can be expanded in the corresponding small parameters, $f_a = f_{a0} + f_{a1} + \cdots$, characterizing a small deviation from local thermodynamic equilibrium. As we shall see, it then follows from the kinetic equation (2.1) that to lowest order each species is distributed according to a local Maxwellian,

$$f_{a0}(\mathbf{r}, \mathbf{v}, t) = n_a(\mathbf{r}, t) \left(\frac{m_a}{2\pi T_a(\mathbf{r}, t)} \right)^{3/2} \exp\left\{ \frac{m_a[\mathbf{v} - \mathbf{V}_a(\mathbf{r}, t)]^2}{2T_a(\mathbf{r}, t)} \right\}, \tag{4.5}$$

basically because the collision operator is the largest term in the kinetic equation (2.1). The density n_a, flow velocity $\mathbf{V}_a$, and temperature T_a are allowed to vary slowly (but otherwise arbitrarily) in space in the sense dictated by (4.4). To obtain the quantities π, $\mathbf{R}$, $\mathbf{q}$, and Q needed to close the fluid equations (2.15)–(2.17), one must go to the next order in the expansion for f_a and solve the kinetic equation for f_{a1}.

The present chapter is devoted to this problem. In the sections that follow, we discuss various methods of solving kinetic equations and calculating transport fluxes in a collisional plasma. These techniques are useful beyond the present context, for instance in stability calculations, and are widely used in the literature. We begin by showing how to derive fluid equations for a plasma with highly charged ions (the so-called

Lorentz limit), and then proceed to describe increasingly accurate calculations, including that of Braginskii. His treatment is limited to two species: electrons and one species of ions. The extension to several ion species has been systematically developed and reviewed by, e.g., Hirshman and Sigmar (1981), and is of importance for transport in deuterium–tritium fusion plasmas, and in plasmas consisting of a mixture of light and heavy 'impurity' ions, which is the standard situation in many laboratory plasma experiments.

Before proceeding with the analysis, we comment on the relation between Braginskii's equations and other possible short-mean-free-path closure schemes of the fluid equations. In the present chapter as well as in Braginskii's article, the flow velocity is assumed to be of the same order as the ion thermal velocity

$$V_a \sim v_{Ti}. \tag{4.6}$$

This allows for a faster particle flux than what occurs naturally in many plasmas of practical importance. In particular, by this ordering the ion particle flux is regarded as larger than the conductive heat flux, $V_i \gg q_i/p_i$, since the lowest order distribution function does not allow for a heat flux, which must therefore be carried by the next term, f_{i1}, in the expansion. Thus, while the flow velocity can in principle always be comparable to the thermal speed, the conductive heat flux is small in a collisional plasma. Transport theory derived under the assumption of small flows,

$$V_a \sim \delta v_{Ti},$$

does not necessarily lead to the same closure relations as Braginskii's, and we shall see in Chapter 12 that the viscosity is indeed different in the two cases.

4.2 Lorentz plasma

Consider a plasma where the Larmor radius is small and the mean-free path is short, i.e., a plasma satisfying the orderings (4.4). Let us estimate the relative magnitude of the terms in the electron kinetic equation (2.1),

$$C_e(f_e) + \Omega_e(\mathbf{b} \times \mathbf{v}) \cdot \frac{\partial f_e}{\partial \mathbf{v}} = \mathbf{v} \cdot \nabla f_e - \frac{e\mathbf{E}}{m_e} \cdot \frac{\partial f_e}{\partial \mathbf{v}} + \frac{\partial f_e}{\partial t}, \tag{4.7}$$

where we have written $\mathbf{b} \equiv \mathbf{B}/B$. Note that $\Omega_e = e_e B/m_e = -eB/m_e$ is negative. The terms on the left are of order $\nu_e \sim \nu_{ei}$ and Ω_e, respectively, and the first two on the right are both of order $v_{Te}/L_\parallel$, if the parallel electric field is assumed to be of the order $T/eL_\parallel$ (since $E_\parallel = -\nabla_\parallel \Phi$ and we expect the electrostatic potential to be comparable to the electron temperature). They are, in other words, small compared with the collision term,

$$\frac{v_\| \nabla_\| f_e}{C_e(f_e)} \sim \frac{(eE_\|/m_e)\partial f_e/\partial v_\|}{v_e f_e} \sim \frac{\lambda}{L_\|} \ll 1,$$

where $\lambda \equiv v_T/v_e$ denotes the mean-free path and $L_\| = \nabla_\|^{-1}$ the parallel length scale. When the gradient and the electric field are taken in the perpendicular direction, these terms are small in comparison with the magnetic term,

$$\frac{\mathbf{v}_\perp \cdot \nabla f_e}{\Omega_e(\mathbf{b} \times \mathbf{v}) \cdot (\partial f_e/\partial \mathbf{v})} \sim \frac{(e\mathbf{E}_\perp/m_e)\partial f_e/\partial \mathbf{v}}{\Omega_e(\mathbf{b} \times \mathbf{v}) \cdot (\partial f_e/\partial \mathbf{v})} \sim \frac{\rho}{L_\perp} \ll 1,$$

if the perpendicular electric field is assumed to be of the order $E_\perp \sim (T_e/eL_\perp)$. Under these assumptions, the first two terms on the right-hand side of (4.7) are thus small in comparison with the terms on the left-hand side. The third term, the time derivative $\partial f_e/\partial t$, is smaller still if it arises because of collisional relaxation; it is then of order $v_e(\lambda/L_\|)^2$ or $v_e(\rho/L_\perp)^2$. Thus, in Eq. (4.7), the large terms are collected on the left side and the small ones on the right. We expand the distribution function accordingly,

$$f_e = f_{e0} + f_{e1} + \cdots.$$

To lowest order in $\lambda/L_\|$ and $\rho/L_\perp$, the left-hand side of (4.7) must vanish, which it does for a Maxwellian,

$$f_{e0}(\mathbf{v}) = n_e \left(\frac{m_e}{2\pi T_e}\right)^{3/2} e^{-m_e v^2/2T_e}, \tag{4.8}$$

since with this choice for f_{e0}, both terms in the collision operator $C_e(f_{e0}) = C_{ee}(f_{e0}) + C_{ei}(f_{e0})$ and the magnetic term vanish separately. For simplicity, we assume the ions to be stationary, $\mathbf{V}_i = 0$, since the ion velocity is usually small in comparison with the electron thermal speed. We may therefore neglect the mean velocity in the Maxwellian for f_{e0}. It is not difficult to see that the Maxwellian (4.5) is the only possible lowest-order solution. Introducing spherical coordinates (v, θ, φ) in velocity space aligned with the magnetic field, we can rewrite the magnetic term in (4.7) so that

$$C_e(f_{e0}) + \Omega_e \frac{\partial f_{e0}}{\partial \varphi} = 0.$$

Multiplying this equation by $\ln f_{e0}$ and integrating over velocity space gives

$$\int C_e(f_{e0}) \ln f_{e0} d^3 v = 0,$$

which implies that f_{e0} is Maxwellian according to Exercise 3 in the previous chapter.

In next order we have

$$C_e(f_{e1}) + \Omega_e(\mathbf{b} \times \mathbf{v}) \cdot \frac{\partial f_{e1}}{\partial \mathbf{v}} = \mathbf{v} \cdot \mathbf{Q}(v), \qquad (4.9)$$

where $C_e(f_{e1}) = C_{ee}^l(f_{e1}) + C_{ei}(f_{e1})$ is the sum of the linearized e–e collision operator,

$$\mathbf{Q}(v) \equiv \left[\mathbf{A}_1 + \left(\frac{m_e v^2}{2T_e} - \frac{5}{2} \right) \mathbf{A}_2 \right] f_{e0},$$

and we have introduced the 'thermodynamic forces'

$$\mathbf{A}_1 \equiv \nabla \ln p_e + e\mathbf{E}/T_e, \qquad \mathbf{A}_2 \equiv \nabla \ln T_e, \qquad (4.10)$$

which act as drives for the transport.

It is not difficult to solve (4.9) if the collision operator is approximated by a Lorentz operator,

$$C_e(f_{e1}) = v_{ei}(v)\mathscr{L}(f_{e1}).$$

This 'Lorentz approximation' is exact in the limit of highly charged ions, $Z_i \to \infty$, since e–e collisions may then be ignored, and the e–i collision operator is given by (3.28) with $\mathbf{V}_i = 0$. Note that C_{ei} is proportional to Z_i^2 and will therefore dominate over C_{ee} if $Z_i^2 \gg 1$. Later in this chapter, we investigate a more exact treatment by including e–e collisions by means of the model operator C_{ee}^m introduced in Section 3.8. Noting that $\mathscr{L}(\mathbf{v}) = -\mathbf{v}$ and making the Ansatz $f_{e1}(\mathbf{v}) \equiv \mathbf{v} \cdot F(v)$, reduces Eq. (4.9) to an algebraic equation,

$$-v_{ei}(v)\mathbf{v} \cdot F(v) + \Omega_e(\mathbf{b} \times \mathbf{v}) \cdot F(v) = \mathbf{v} \cdot \mathbf{Q}(v).$$

Since this equation must be satisfied for all $\mathbf{v}$, we have

$$v_{ei}F(v) + \Omega_e\mathbf{b} \times F(v) = -\mathbf{Q}(v).$$

The parallel component of F can be read off directly, $F_\parallel = -Q_\parallel/v_{ei}$, and the perpendicular components are easily obtained by taking the vectorial product with $\mathbf{b}$ and solving the ensuing system of equations. The distribution function becomes

$$f_{e1}(\mathbf{v}) = -\left(\frac{\mathbf{v}_\parallel}{v_{ei}} + \frac{v_{ei}\mathbf{v}_\perp - \Omega_e\mathbf{b} \times \mathbf{v}}{v_{ei}^2 + \Omega_e^2} \right) \cdot \mathbf{Q}(v). \qquad (4.11)$$

To this particular solution, any homogenous solutions to (4.9) could be added. However, the homogeneous solutions are just Maxwellians and could then just as well have been incorporated in f_{e0}, which is better than adding them to f_{e1}. Otherwise there is a density n_1 and a temperature T_1 associated with f_{e1}. It is more logical to have n_e and T_e in (4.8) denote the *total* electron density and temperature, so that $n_1 = 0$ and $T_1 = 0$.

This completes the solution of the kinetic equation. Note that we have not yet made use of the smallness of the parameter $\Delta \equiv v_{ei}/\Omega_e = \rho_e/\lambda$, introduced in Eq. (4.3). Our solution (4.11) is valid for arbitrary Δ, but is simplified somewhat if $\Delta \ll 1$,

$$f_{e1}(\mathbf{v}) = -\left(\frac{\mathbf{v}_{\parallel}}{v_{ei}} - \frac{\mathbf{b} \times \mathbf{v}}{\Omega_e} + \frac{v_{ei}\mathbf{v}_{\perp}}{\Omega_e^2} \right) \cdot \mathbf{Q}(v) \equiv f_{\parallel} + f_{\wedge} + f_{\perp}, \qquad (4.12)$$

where it is noted that $f_{\parallel} : f_{\wedge} : f_{\perp} \sim 1 : \Delta : \Delta^2$. The different magnitudes of these terms reflect the circumstance that transport takes place on widely disparate time scales in different directions. The first term $(f_{\parallel})$ is proportional to the parallel gradients in $Q_{\parallel}$, and also describes transport in this direction, which is very fast. The second term $(f_{\wedge})$ contains the so-called diamagnetic flows, which we shall return to later in this chapter. Note that this term is independent of collision frequency (in the limit $v_e \ll \Omega_e$). It has actually nothing to do with collisional transport and is entirely caused by gyromotion. Finally, the third term $(f_{\perp})$ describes transport across the magnetic field, which is proportional to B^{-2} as anticipated in Eqs. (1.8) and (1.9).

The so-called 'classical' particle and heat fluxes that arise in response to the thermodynamic forces contained in $\mathbf{Q}(v)$ can now be calculated by taking appropriate moments of f_{e1}. They are

$$n_e\mathbf{u} \equiv \int f_{e1}\mathbf{v}d^3v = n_e(\mathbf{u}_{\parallel} + \mathbf{u}_{\wedge} + \mathbf{u}_{\perp}) \qquad (4.13)$$

and

$$\mathbf{q} \equiv \int f_{e1}\left(\frac{m_e v^2}{2} - \frac{5T_e}{2} \right) \mathbf{v}d^3v = \mathbf{q}_{\parallel} + \mathbf{q}_{\wedge} + \mathbf{q}_{\perp}, \qquad (4.14)$$

where

$$n_e u_{\parallel} = -r_1 \frac{\tau_{ei}}{m_e} \left(\nabla_{\parallel} p_e + n_e e E_{\parallel} + r_2 n_e \nabla_{\parallel} T_e \right), \qquad (4.15)$$

$$n_e\mathbf{u}_{\wedge} = \frac{\mathbf{b} \times \nabla p_e}{m_e \Omega_e} + n_e \frac{\mathbf{E} \times \mathbf{B}}{B^2}, \qquad (4.16)$$

$$n_e\mathbf{u}_{\perp} = -\frac{1}{m_e \Omega_e^2 \tau_{ei}} \left(\nabla_{\perp} p_e + n_e e \mathbf{E}_{\perp} - \frac{3}{2} n_e \nabla_{\perp} T_e \right), \qquad (4.17)$$

and

$$q_\parallel = r_2 n_e T_e u_\parallel - r_3 \frac{n_e T_e \tau_{ei}}{m_e} \nabla_\parallel T_e, \tag{4.18}$$

$$\mathbf{q}_\wedge = \frac{5 n_e T_e}{2 m_e \Omega_e} \mathbf{b} \times \nabla T_e, \tag{4.19}$$

$$\mathbf{q}_\perp = -\frac{3 n_e T_e}{2 \Omega_e \tau_{ei}} \mathbf{b} \times \mathbf{u}_\wedge - r_4 \frac{n_e T_e}{m_e \Omega_e^2 \tau_{ei}} \nabla_\perp T_e. \tag{4.20}$$

We have used (3.31) for the electron–ion collision frequency, and introduced the numbers

$$r_1 = \frac{32}{3\pi}, \qquad r_2 = \frac{3}{2}, \qquad r_3 = \frac{128}{3\pi}, \qquad r_4 = \frac{13}{4},$$

which are related to the matrix elements of C_{ei} defined later in this chapter.

These transport laws will later be seen to have exactly the same structure as those of Braginskii. Only the coefficients r_1–r_4 differ, which is a consequence of our neglect of electron–electron collisions. Braginskii's values are correct if $Z = 1$; the ones derived here are correct in the so-called Lorentz limit $Z \to \infty$.

Since the ions are assumed to be stationary, the plasma current is equal to $\mathbf{j} = -n_e e \mathbf{u}$, and (4.13), (4.15)–(4.17) can be thought of as Ohm's law. The middle terms in (4.15) and (4.17) are of a familiar form and give a current proportional to the electric field. The electrical conductivity is different along and across the magnetic field. The other terms produce currents by other mechanisms. It is not surprising that an electron pressure gradient produces a current when the ions are held fixed since, intuitively, one would expect the electron population to expand in the direction of $-\nabla p_e$, like an ordinary gas. Its ability to do so is very different in the directions along and across the magnetic field. We also note that an electron temperature gradient can drive a current in Eqs. (4.15) and (4.17) – apparently a kind of thermoelectric effect. The reason for this phenomenon will be explained later in this section.

The term in the flow velocity denoted by $\mathbf{u}_\wedge$ is called the diamagnetic velocity and will be encountered again in Eq. (4.64); similarly $\mathbf{q}_\wedge$ is the diamagnetic heat flux (4.76). These fluxes are carried by the term $f_\wedge$ in the distribution function, and are perpendicular to the gradients that drive them and to the magnetic field. To understand the origin of these fluxes, it is instructive to note that $f_\wedge$ can be written as

$$f_\wedge = -\boldsymbol{\rho} \cdot \nabla f_{e0}, \tag{4.21}$$

where $\boldsymbol{\rho} = \mathbf{b} \times \mathbf{v} / \Omega_e$ is the Larmor radius vector. In other words, $f_{e0} + f_\wedge$ is approximately equal to the lowest-order distribution f_{e0} evaluated at

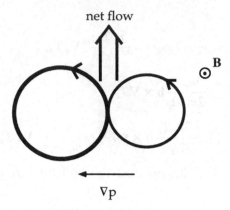

Fig. 4.1.　Diamagnetic particle flux.

the position of the guiding centre $\mathbf{R} = \mathbf{r} - \rho$,

$$f_{e0}(\mathbf{r}) + f_{\wedge}(\mathbf{r}) = (1 - \rho \cdot \nabla)f_{e0}(\mathbf{r}) \simeq f_{e0}(\mathbf{R}).$$

The reason for this can be found in the ordering $\Delta \ll 1$. According to this assumption, the collision frequency is smaller than the gyrofrequency, so the distribution function must be constant along the Larmor orbit and can therefore be a function only of the guiding-centre location. This will be discussed in more detail in Section 6.5. Consider again Fig. 1.2, and recall that there are more particles in the left Larmor orbit than in the right one. As a result there must be a vertical particle flux, although the guiding centres do not move, see Fig. 4.1. This is the diamagnetic flux $n_e \mathbf{u}_{\wedge}$, and the heat flux $\mathbf{q}_{\wedge}$ can be understood in similar terms from Fig. 1.3.

The force $\mathbf{R}_e$ between electrons and ions can be related to the particle flux (4.15)–(4.17) by considering the electron momentum equation (2.16),

$$0 = m_e n_e \Omega_e \mathbf{u} \times \mathbf{b} - n_e e \mathbf{E} - \nabla p_e + \mathbf{R}_e,$$

where inertia and viscosity are neglected. Inserting the expressions for the flow velocity $\mathbf{u} = \mathbf{u}_{\|} + \mathbf{u}_{\wedge} + \mathbf{u}_{\perp}$ from above gives the following expression for the 'classical friction force' $\mathbf{R}_e$ on the electrons,

$$\mathbf{R}_e = -\frac{m_e n_e(\mathbf{u}_{\|}/r_1 + \mathbf{u}_{\wedge})}{\tau_{ei}} - r_2 n_e \nabla_{\|} T_e + \frac{3n_e}{2\Omega_e \tau_e} \mathbf{b} \times \nabla T_e,$$

in agreement (to lowest order in Δ) with Braginskii's results presented later in this chapter in Eqs. (4.36)–(4.38), except for the numerical values of the coefficients r_1 and r_2.

The first term in $\mathbf{R}_e$ represents the drag force between electrons and ions. As might be expected, it is proportional to the collision frequency and the relative velocity $\mathbf{u}$. It is perhaps surprising that the friction is smaller in the direction of $\mathbf{u}_\parallel$ than in the direction of $\mathbf{u}_\wedge$. This can be understood from the form (4.12) of the electron distribution function. Since $f_\parallel$ is inversely proportional to $\nu_{ei}(v) \sim v^{-3}$, it has an elevated tail

$$f_\parallel \sim v^4 e^{-m_e v^2/2T_e}$$

at high energies. Because of their high speed, the tail electrons contribute substantially to the total flow velocity $u_\parallel$ but are less collisional than thermal electrons and therefore do not experience much friction. In the direction of the perpendicular flow $\mathbf{u}_\wedge$, the distribution function does not have a tail. Thus, the electrons carrying the flux in the parallel direction are less collisional, and experience less friction, than the ones carrying the perpendicular particle flux.

The second term in $\mathbf{R}_e$ is called the thermal force and also arises because of the velocity dependence of the electron–ion collision frequency. If the macroscopic parallel flow of electrons vanishes, $\mathbf{u} = 0$, but there exists a parallel temperature gradient, electrons coming from the hotter region (from the direction of $\nabla_\parallel T_e$) are typically slightly more energetic than the ones travelling in the opposite direction. The mean energy difference is of order $\lambda \nabla_\parallel T_e$. The hotter electrons collide less frequently with the ions than do the colder electrons travelling in the opposite direction. The ions therefore experience a force in the direction of $\nabla_\parallel T_e$, and the electrons are pushed in the opposite direction, producing a thermoelectric current.

The third term in $\mathbf{R}_e$ is the thermal force perpendicular to the field, and is smaller than the parallel force by a factor $1/\Omega_e \tau_e = \rho/\lambda$, since the electrons are only free to move the distance ρ_e across the field.

4.3 Onsager symmetry and a variational principle

Onsager symmetry

The transport laws we have just derived exhibit important symmetries (named after Lars Onsager, Nobel laureate 1969), to which we now turn our attention. In the parallel direction, we can write the fluxes of particles (4.15) and heat (4.18) in terms of the thermodynamic forces

$$\mathbf{A}_1 \equiv \nabla \ln p_e + e\mathbf{E}/T_e, \qquad \mathbf{A}_2 \equiv \nabla \ln T_e,$$

as

$$n_e u_\parallel = -\frac{n_e T_e \tau_{ei}}{m_e}(L_{11}\mathbf{A}_1 + L_{12}\mathbf{A}_2)_\parallel, \qquad (4.22)$$

$$\frac{q_\parallel}{T_e} = -\frac{n_e T_e \tau_{ei}}{m_e}(L_{21}\mathbf{A}_1 + L_{22}\mathbf{A}_2)_\parallel, \qquad (4.23)$$

where we have introduced the transport coefficients

$$L_{11} = r_1, \qquad L_{12} = L_{21} = r_1 r_2, \qquad L_{22} = r_1 r_2^2 + r_3.$$

Similarly, in the perpendicular directions we have

$$n_e \mathbf{u}_\wedge = \frac{n_e T_e}{m_e \Omega_e} \mathbf{b} \times \mathbf{A}_1,$$

$$\frac{\mathbf{q}_\wedge}{T_e} = \frac{n_e T_e}{m_e \Omega_e} \mathbf{b} \times \frac{5 \mathbf{A}_2}{2},$$

and

$$n_e \mathbf{u}_\perp = -\frac{n_e T_e}{m_e \Omega_e^2 \tau_{ei}} \left(\mathbf{A}_1 - \frac{3}{2} \mathbf{A}_2 \right)_\perp,$$

$$\frac{\mathbf{q}_\perp}{T_e} = -\frac{n_e T_e}{m_e \Omega_e^2 \tau_{ei}} \left(-\frac{3}{2} \mathbf{A}_1 + r_4 \mathbf{A}_2 \right)_\perp.$$

All these transport laws are symmetric in the off-diagonal terms. In the parallel direction, we have $L_{12} = L_{21}$; in the diamagnetic direction ($\wedge$) there are no off-diagonal terms in either the expression for $n_e \mathbf{u}_\wedge$ or $\mathbf{q}_\wedge$; and in the other perpendicular direction ($\perp$) the coefficients 3/2 are symmetrically distributed. These symmetries are *exact*, i.e., not an artefact of the simplified collision operator we have employed. To understand these symmetries, and indeed to prove them rigorously, let us again consider the equation for the perturbed electron distribution (4.9). For simplicity, we focus on the parallel transport laws, and thus only retain the parallel gradients in the source term on the right, but now include both electron–ion and electron–electron collisions. The kinetic equation (4.9) then becomes

$$C_e^l(f_{e1}) = v_\parallel \left[A_{1\parallel} + \left(\frac{mv^2}{2T} - \frac{5}{2} \right) A_{2\parallel} \right] f_{e0}. \qquad (4.24)$$

In this equation, whose solution is sometimes called the 'Spitzer problem', the left-hand side,

$$C_e^l(f_{e1}) = C_{ee}(f_{e1}, f_{e0}) + C_{ee}(f_{e0}, f_{e1}) + v_{ei} \mathscr{L}(f_{e1}),$$

is the full linearized collision operator. As we have seen in (3.66), this operator is self-adjoint, i.e., the functional (defined only slightly differently from the functional S_e in (3.66))

$$S[\varphi, \psi] \equiv \int \frac{\varphi}{f_{e0}} C_e^l(\psi) d^3 v$$

is symmetric: for any two functions φ and ψ

$$S[\varphi, \psi] = S[\psi, \varphi].$$

This has two remarkable consequences.

First, it makes it possible to prove the symmetry of the transport coefficients L_{jk} in (4.22) and (4.23). The linearity of Eq. (4.24) shows that

$$f_{e1} = A_{1\parallel}h_1 + A_{2\parallel}h_2,$$

where h_1 and h_2 are the solutions to the prototypical 'Spitzer equations'

$$C_e^l(h_1) = v_\parallel f_{e0},$$

$$C_e^l(h_2) = \left(\frac{mv^2}{2T} - \frac{5}{2}\right) v_\parallel f_{e0}.$$

Relating the particle and heat fluxes to these functions gives

$$nu_\parallel = \int (A_{1\parallel}h_1 + A_{2\parallel}h_2)v_\parallel d^3v,$$

$$\frac{q_\parallel}{T_e} = \int (A_{1\parallel}h_1 + A_{2\parallel}h_2)\left(\frac{mv^2}{2T} - \frac{5}{2}\right) v_\parallel d^3v.$$

Thus the transport fluxes are linear combinations of the thermodynamic forces. A comparison with Eqs. (3.66) and (4.23) shows that

$$L_{12} = -\frac{m_e}{n_e T_e \tau_{ei}} \int h_2 v_\parallel d^3v = -\frac{m_e}{n_e T_e \tau_{ei}} \int \frac{h_2}{f_{e0}} C_e^l(h_1) d^3v,$$

$$L_{21} = -\frac{m_e}{n_e T_e \tau_{ei}} \int h_1 \left(\frac{mv^2}{2T} - \frac{5}{2}\right) v_\parallel d^3v = -\frac{m_e}{n_e T_e \tau_{ei}} \int \frac{h_1}{f_{e0}} C_e^l(h_2) d^3v.$$

By the definition and self-adjointness of S, we can write these relations as

$$L_{12} = -\frac{m_e}{n_e T_e \tau_{ei}} S[h_2, h_1] = -\frac{m_e}{n_e T_e \tau_{ei}} S[h_1, h_2] = L_{21},$$

proving Onsager symmetry $L_{12} = L_{21}$.

A variational principle

The second useful consequence of the self-adjointness of S is the possibility of calculating the transport coefficients L_{ij} by means of a variational principle. This can be done with the help of the functionals

$$P[\varphi] \equiv \int \varphi v_\parallel \left[A_{1\parallel} + \left(\frac{mv^2}{2T_e} - \frac{5}{2}\right) A_{2\parallel}\right] d^3v, \qquad (4.25)$$

$$R[\varphi] \equiv 2P[\varphi] - S[\varphi, \varphi].$$

The variation of any functional $F[\varphi]$ with respect to a perturbation $\delta\varphi$ is defined as

$$\delta F \equiv \frac{d}{dx}F[\varphi + x\delta\varphi]_{x=0},$$

so that if $\delta\varphi$ is small

$$\delta F \simeq F[\varphi + \delta\varphi] - F[\varphi].$$

For small deviations from the solution f_{e1} to the Spitzer problem (4.24), the self-adjointness of S implies

$$\delta S \equiv S[f_{e1} + \delta f, f_{e1}] + S[f_{e1}, f_{e1} + \delta f] - S[f_{e1}, f_{e1}] = 2S[\delta f, f_{e1}].$$

Thus

$$\delta R = 2P[\delta f] - 2S[\delta f, f_{e1}]$$

$$= 2\int \left\{ v_\| \left[A_{1\|} + \left(\frac{mv^2}{2T} - \frac{5}{2} \right) A_{2\|} \right] - \frac{C_e^l(f_{e1})}{f_{e0}} \right\} \delta f d^3v = 0.$$

In other words, arbitrary variations, $\delta f \neq 0$, in R around the solution f_{e1} to (4.24) vanish, so $R[\varphi]$ has an extremum for $\varphi = f_{e1}$. This means that solving the Spitzer problem (4.24) is mathematically equivalent to finding the extremum of R. The function that extremizes R is the solution to Eq. (4.24).

This property is useful because it enables an approximate calculation of the transport in the following manner. Suppose we could guess the solution f_{e1} with reasonable accuracy, e.g., by solving the kinetic equation (4.24) with an approximate collision operator, as was done in the previous section. The guess, or 'trial function', which we denote by F, will then differ slightly from the correct solution, i.e.,

$$F = f_{e1} + \delta f,$$

where δf is small. Since the first variation of $R[\varphi]$ vanishes for $\varphi = f_{e1}$, $R[F]$ differs from $R[f_{e1}]$ only by a quadratic term,

$$R[F] = R[f_{e1}] + O(\delta f^2).$$

In other words, using a good guess F for f_{e1} gives a better guess $R[F]$ for the functional $R[f_{e1}]$! If F is good to first order, $R[F]$ is good to second order. But $R[f_{e1}] = P[f_{e1}]$ is a linear combination of parallel particle and heat fluxes. According to Eqs. (4.22) and (4.26),

$$R[f_{e1}] = P[f_{e1}] = A_{1\|}n_e u_\| + A_{2\|}q_\|/T_e$$

$$= -\frac{n_e T_e \tau_{ei}}{m_e} (L_{11}A_{1\|}^2 + 2L_{12}A_{1\|}A_{2\|} + L_{22}A_{2\|}^2),$$

where we have also used the symmetry $L_{12} = L_{21}$ and the definitions (4.13) and (4.14). On the other hand, since any reasonable trial function F is linear in A_1 and A_2, evaluating $R[F]$ also gives a quadratic form,

$$R[F] = -\frac{n_e T_e \tau_{ei}}{m_e} \left(l_{11} A_{1\parallel}^2 + 2 l_{12} A_{1\parallel} A_{2\parallel} + l_{22} A_{2\parallel}^2 \right).$$

which closely approximates $R[f_{e1}]$. The numbers l_{jk} thus obtained are therefore good approximations of the transport coefficients L_{jk}.

This procedure can be streamlined slightly by using the functional

$$Q[\varphi] \equiv \frac{P^2[\varphi]}{S[\varphi, \varphi]}$$

instead of $R[\varphi]$. If f_{e1} satisfies (4.24) then

$$P[f_{e1}] = S[f_{e1}, f_{e1}],$$

and thus also

$$P[f_{e1}] = Q[f_{e1}].$$

Differentiating the definition of Q (keeping only terms that are linear in the perturbation) shows that

$$\delta Q \equiv Q[f_{e1} + \delta f] - Q[f_{e1}] = 2\delta P - \delta S,$$

so that $\delta Q = \delta R$. The functionals Q and R thus have the same variational property, but Q has the additional advantage of being independent of the overall magnitude of the distribution function. For instance, a typical trial function might look like

$$F = a f_{e0}(v)(1 + bv^n) v_\parallel,$$

so that the term bv^n creates a high-energy tail. Transport coefficients are found by varying the numbers a and b so as to extremize the functional R or Q. The latter functional has the advantage of being independent of a, which reduces the computational burden somewhat.

4.4 Spitzer conductivity

Unlike the analysis presented in Section 4.2, the variational principle we have just described is not restricted to Lorentz plasmas (i.e., plasmas with highly charged ions), where electron–electron collisions can be ignored. The variational principle is thus useful for calculating transport coefficients in a plasma with arbitrary ion charge. This method yields very accurate results if good trial functions are used, but can result in rather lengthy calculations. Alternatively, the model operator (3.69) introduced in Section 3.8 may be used to account for the effects of electron–electron collisions in an approximate way (Connor *et al.* 1973). We demonstrate this procedure by solving the kinetic equation (4.24) to obtain the parallel electrical conductivity.

The transport coefficient L_{11}, defined in (4.22), determines the electrical conductivity of the plasma along the magnetic field since the electron current that arises in response to a parallel electric field is

$$j_\| = -n_e e u_\| = \frac{n_e e^2 \tau_{ei} L_{11}}{m_e} E_\| \equiv \sigma_\| E_\|.$$

Hence the parallel 'Spitzer' conductivity becomes

$$\sigma_\| = \frac{n_e e^2 \tau_{ei}}{m_e} L_{11} = \frac{12\pi^{3/2}}{2^{1/2}} \frac{T_e^{3/2} \epsilon_0^2}{Z e^2 m_e^{1/2} \ln \Lambda} L_{11},$$

where we have used Eq. (1.4) for the collision time, and (4.22) for the flow velocity. Note that $\sigma_\|$ is independent of density apart from a weak logarithmic dependence in the Coulomb logarithm $\ln \Lambda$, but increases strongly ($\sim T_e^{3/2}$) with electron temperature.

The conductivity is determined up to the constant L_{11} by the expression above. The calculation of this constant is the task of kinetic theory. In Section 4.2, we saw that in the Lorentz limit $Z \to \infty$, where electron–electron collisions may be ignored, it becomes $L_{11} \to 32/3\pi \simeq 3.2$. In order to calculate the coefficient when $Z = O(1)$, we solve the Spitzer problem

$$C_{ee}(f_{e1}) + C_{ei}(f_{e1}) = \frac{e E_\|}{T_e} v_\| f_{e0},$$

with the model operator (3.69) for C_{ee}. Hence we have

$$\left(v_D^{ee} + v_D^{ei} \right) \mathcal{L} f_{e1} + \frac{m_e v_D^{ee} u}{T_e} v_\| f_{e0} = \frac{e E_\|}{T_e} v_\| f_{e0},$$

where we have recognized that the vector **u** is entirely in the parallel direction. This equation has the simple solution

$$f_{e1} = \frac{v_D^{ee} u + u_* }{v_D^{ee} + v_D^{ei}} \frac{m_e v_\|}{T_e} f_{e0},$$

where $u_* \equiv -e E_\|/m_e$. The quantity u is determined by momentum conservation (3.70) as in Section 3.8,

$$\int v_\| C_{ee}(f_{e1}) d^3 v = 0.$$

In the present situation, this requires

$$\int v_D^{ee}\left(-\frac{v_D^{ee}u + u_*}{v_D^{ee} + v_D^{ei}} + u\right)\frac{m_e v_\parallel^2}{T_e}f_{e0}d^3v = 0,$$

where $v_\parallel^2 d^3v = 4\pi v^4 dv \xi^2 d\xi$ with $\xi = v_\parallel/v$. Thus u becomes

$$u = \frac{\left\{\dfrac{v_D^{ee}}{v_D^{ee} + v_D^{ei}}\right\}}{\left\{\dfrac{v_D^{ee}v_D^{ei}}{v_D^{ee} + v_D^{ei}}\right\}}u_*,$$

where we have introduced the velocity-space average

$$\{F(v)\} \equiv \int F\frac{mv_\parallel^2}{nT}f_M d^3v = \frac{8}{3\sqrt{\pi}}\int_0^\infty F(x)e^{-x^2}x^4 dx, \qquad (4.26)$$

where $x^2 = mv^2/2T$. This notation is generally useful in transport theory and frequently appears in the literature. The electric current is

$$j_\parallel = -e\int f_{e1}v_\parallel d^3v = -n_e e\left\{\frac{v_D^{ee}u + u_*}{v_D^{ee} + v_D^{ei}}\right\}$$

$$= \left(\frac{\left\{\dfrac{v_D^{ee}}{v_D^{ee} + v_D^{ei}}\right\}^2}{\left\{\dfrac{v_D^{ee}v_D^{ei}}{v_D^{ee} + v_D^{ei}}\right\}} + \left\{\frac{1}{v_D^{ee} + v_D^{ei}}\right\}\right)\frac{n_e e^2 E_\parallel}{m_e},$$

so the conductivity becomes

$$\sigma_\parallel = \left(\frac{\left\{\dfrac{v_D^{ee}}{v_D^{ee} + v_D^{ei}}\right\}^2}{\left\{\dfrac{v_D^{ee}v_D^{ei}\tau_{ei}}{v_D^{ee} + v_D^{ei}}\right\}} + \left\{\frac{1}{\tau_{ei}(v_D^{ee} + v_D^{ei})}\right\}\right)\frac{n_e e^2 \tau_{ei}}{m_e} = L_{11}\frac{n_e e^2 \tau_{ei}}{m_e}. \qquad (4.27)$$

Here the quantities appearing in the velocity-space integrals are dimensionless, and these integrals can be evaluated numerically, recalling the definition (3.45) of the deflection frequency. The table gives the result for various values of the ion charge Z, along with the exact result. The agreement is fairly good (error of 20%) for $Z = 1$ and gets progressively better as Z increases and the collision operator approaches a pure Lorentz operator.

Z	1	2	4	16	∞
$\dfrac{L_{11}}{32/3\pi}$ (exact)	0.58	0.68	0.78	0.92	1
$\dfrac{L_{11}}{32/3\pi}$ (Eq. (4.27))	0.70	0.81	0.89	0.97	1

4.5 Expansion in orthogonal polynomials

In this section we describe yet another way to solve kinetic problems, namely, to expand the distribution function in a series of orthogonal polynomials and then use this expansion to solve the kinetic equation. This approach, which was the one used by Braginskii, has the advantage of being systematic and being able to deliver results to any degree of accuracy.

In principle, the distribution function can be expanded in any complete set of functions. A particularly useful set of such functions are the so-called *Sonine polynomials* or generalized Laguerre polynomials,

$$L_j^{(m)}(x) \equiv \frac{1}{j!} \frac{e^x}{x^m} \frac{d^j}{dx^j} \left(x^{j+m} e^{-x} \right),$$

which are widely used in kinetic theory of gases and plasmas. The lower index is an integer, $j = 0, 1, \ldots$, and here we shall only need polynomials with $m = 3/2$ for the upper index,

$$L_0^{(3/2)}(x) = 1,$$

$$L_1^{(3/2)}(x) = \frac{5}{2} - x,$$

$$L_2^{(3/2)}(x) = \frac{35}{8} - \frac{7x}{2} + \frac{x^2}{2}.$$

Sonine polynomials are orthogonal in the sense

$$\int_0^\infty L_j^{(m)}(x) L_k^{(m)}(x) x^m e^{-x} dx = \frac{\Gamma(j + m + 1)}{j!} \delta_{jk},$$

where $\Gamma(n)$ is the Gamma function, which satisfies $\Gamma(1/2) = \pi^{1/2}$, $\Gamma(n + 1) = n\Gamma(n) = n!$. In terms of the velocity-space average (4.26), the orthogonality relation for Sonine polynomials can be expressed as

$$\left\{ L_j^{(3/2)}(x^2) L_k^{(3/2)}(x^2) \right\} = \frac{\Gamma(j + 5/2)}{j! \Gamma(5/2)} \delta_{jk}. \tag{4.28}$$

While Sonine polynomials are not eigenfunctions of the Coulomb collision operator, they are specially tailored for kinetic transport problems. For instance, the integral

$$\int f_a(\mathbf{v})v_{\parallel}L_j^{(3/2)}\left(x_a^2\right)d^3v,$$

with $x_a^2 = m_a v^2/2T_a$, is equal to the particle flux of species a if $j = 0$, and is equal to the negative heat flux divided by T if $j = 1$. Accordingly, if f_{a1} is expanded as

$$f_{a1} = f_a - f_{a0} = f_{a0}\frac{m_a v_{\parallel}}{T_a}\sum_j u_{aj}L_j^{(3/2)}(x_a^2), \qquad (4.29)$$

where f_{a0} is a stationary Maxwellian, then the first two coefficients u_{aj} are very simply related to the parallel mean velocity $V_{\parallel}$ and the conductive heat flux $q_{a\parallel}$ by

$$u_{a0} = V_{a\parallel}, \qquad u_{a1} = -\frac{2q_{a\parallel}}{5p_a}.$$

Next, let us define the Braginskii matrix elements of the collision operator

$$M_{ab}^{jk} \equiv \frac{\tau_{ab}}{n_a}\int v_{\parallel}L_j^{(3/2)}(x_a^2)C_{ab}\left[\frac{m_a v_{\parallel}}{T_a}L_k^{(3/2)}(x_a^2)f_{a0}, f_{b0}\right]d^3v,$$

$$N_{ab}^{jk} \equiv \frac{\tau_{ab}}{n_a}\int v_{\parallel}L_j^{(3/2)}(x_a^2)C_{ab}\left[f_{a0}, \frac{m_b v_{\parallel}}{T_b}L_k^{(3/2)}(x_b^2)f_{b0}\right]d^3v, \qquad (4.30)$$

where $\tau_{ab} = 3\pi^{1/2}/4\hat{v}_{ab}$, and $\hat{v}_{ab}$ was defined in Eq. (3.48), so that τ_{ab} agrees with our earlier definitions (1.4) and (1.5). The calculation of the matrix elements from the full Coulomb collision operator is tedious but straightforward if one uses the generating function for Sonine polynomials,

$$(1-y)^{-m-1}\exp\left(-\frac{xy}{1-y}\right) = \sum_{j=0}^{\infty}y^j L_j^{(m)}(x).$$

We give the results in the Appendix at the end of the book.

Knowing these quantities, the Spitzer problem (4.24) for parallel transport in a multi-species plasma,

$$\sum_b [C_{ab}(f_{a1}, f_{b0}) + C_{ab}(f_{a0}, f_{b1})] = v_{\parallel}\left[A_{a1}L_0^{(3/2)}(x_a^2) - A_{a2}L_1^{(3/2)}(x_a^2)\right]f_{a0},$$

can be transformed into an algebraic system of equations for the coefficients u_{aj} by multiplying by $(m_a v_{\parallel}/T_a)L_j^{(3/2)}(x_b^2)$ and integrating over velocity-space

$$\sum_{b,k}\frac{m_a}{\tau_{ab}T_a}\left(M_{ab}^{jk}u_{ak} + N_{ab}^{jk}u_{bk}\right) = A_{a1}\delta_{j0} - \frac{5}{2}A_{a2}\delta_{j1}, \qquad (4.31)$$

where we have used the orthogonality of the Sonine polynomials. Note that the 'driving terms' on the right only appear in the first two equations. As a result of this property and because of the variational nature of the transport coefficients, the Sonine polynomial expansion tends to converge very quickly. Although the system of equations is infinite, it can usually be truncated after only a few (typically three) equations (thus keeping only three terms in the expansion for f_{a1}) with very little loss of accuracy, typically a few per cent compared with a full numerical solution of the kinetic equation.

4.6 Braginskii's equations

In Section 4.2 we simplified the transport problem by taking the ions to be stationary and highly charged, $Z \gg 1$. As a result, we could calculate electron transport coefficients relatively easily. In this section we consider the more difficult, and interesting, problem of transport in a hydrogen plasma, $Z = 1$, where the ion dynamics is fully taken into account. This means that we must solve a kinetic equation for ions as well as for electrons, and carefully treat collisions between similar particles (ion–ion and electron–electron collisions). In this section, we first show how this transport problem is set up, following Braginskii (1965), and then present his results for transport in a singly charged plasma.

Kinetic problem

Unlike the situation in Section 4.2, the electrons no longer collide on a stationary background when the ion dynamics is taken into account. It is therefore appropriate to transform the kinetic equation of each species a to a coordinate frame moving with the flow velocity $\mathbf{V}_a(\mathbf{r}, t)$ of that species. While the kinetic equation has the form (2.1) when the velocity is measured in the laboratory frame, we need to re-express this equation in a form where the velocity is measured relative to the rest frame $\mathbf{V}_a$ of the moving fluid. In a coordinate transformation $(\mathbf{r}, \mathbf{v}, t) \rightarrow (\mathbf{r}, \mathbf{v}'_a, t)$ where $\mathbf{v}'_a = \mathbf{v} - \mathbf{V}_a$, the derivatives transform as

$$\frac{\partial}{\partial t} \rightarrow \frac{\partial}{\partial t} - \frac{\partial \mathbf{V}_a}{\partial t} \cdot \frac{\partial}{\partial \mathbf{v}'_a},$$

$$\frac{\partial}{\partial \mathbf{r}} \rightarrow \frac{\partial}{\partial \mathbf{r}} - \frac{\partial V_{aj}}{\partial \mathbf{r}} \frac{\partial}{\partial v'_{aj}},$$

$$\frac{\partial}{\partial \mathbf{v}} \rightarrow \frac{\partial}{\partial \mathbf{v}'_a},$$

and the kinetic equation (2.1) thus becomes

$$\frac{df_a}{dt} + \mathbf{v}'_a \cdot \nabla f_a + \left[\frac{e_a}{m_a} \left(\mathbf{E}' + \mathbf{v}'_a \times \mathbf{B} \right) - \frac{d\mathbf{V}_a}{dt} \right] \cdot \frac{\partial f_a}{\partial \mathbf{v}'_a} - v'_{aj} \frac{\partial V_{ak}}{\partial r_j} \frac{\partial f_a}{\partial v'_{ak}} = C_a(f_a),$$

where $\mathbf{E}' = \mathbf{E} + \mathbf{V}_a \times \mathbf{B}$ is the electric field measured in the moving frame, and

$$\frac{d}{dt} = \frac{\partial}{\partial t} + \mathbf{V}_a \cdot \nabla$$

is the convective derivative. As in Section 4.2, the largest terms are those involving the collision operator and the magnetic field, which reflects the ordering (4.4) of high collision frequency and small gyroradius. Collecting these large terms on the left-hand side gives the following equation for electrons ($e_e = -e$),

$$C_{ee}(f_e) + C^0_{ei}(f_e) + \frac{e}{m_e} \mathbf{v}'_e \times \mathbf{B} \cdot \frac{\partial f_e}{\partial \mathbf{v}'_e}$$

$$= \frac{df_e}{dt} + \mathbf{v}'_e \cdot \nabla f_e - \left(\frac{e}{m_e} \mathbf{E}' + \frac{d\mathbf{V}_e}{dt} \right) \cdot \frac{\partial f_e}{\partial \mathbf{v}'_e} - v'_{ej} \frac{\partial V_{ek}}{\partial r_j} \frac{\partial f_e}{\partial v'_{ek}} - C^1_{ei}(f_e),$$

where we have considered the term C^1_{ei} in the collision operator, i.e., the second term in Eq. (3.28), as small and put it on the right-hand side.

The next step is to solve this equation perturbatively by expanding the distribution function, $f_e = f_{e0} + f_{e1} + \cdots$. In lowest order, the left-hand side must vanish, which forces the distribution function to be a Maxwellian at rest in the moving frame,

$$f_{e0} - n_e \left(\frac{m_e}{2\pi T_e} \right)^{3/2} e^{-x^2},$$

where $x^2 = m_e v'^2_e / 2T_e$. In the next order, we insert this distribution on the right-hand side and obtain an equation for the correction f_{e1} to the Maxwellian

$$C_{ee}(f_{e1}) + C^0_{ei}(f_{e1}) + \frac{e}{m_e} \mathbf{v} \times \mathbf{B} \cdot \frac{\partial f_{e1}}{\partial \mathbf{v}}$$

$$= \left[\frac{d \ln n_e}{dt} + \left(x^2 - \frac{3}{2} \right) \frac{d \ln T_e}{dt} + \mathbf{v} \cdot \nabla \ln n_e + \left(x^2 - \frac{3}{2} \right) \mathbf{v} \cdot \nabla \ln T_e \right.$$

$$\left. + \frac{m_e \mathbf{v}}{T_e} \cdot \left(\frac{e}{m_e} \mathbf{E}' + \frac{d\mathbf{V}_e}{dt} \right) + \frac{m_e v_j v_k}{T_e} \frac{\partial V_{ek}}{\partial r_j} + v_{ei} \frac{m_e \mathbf{v} \cdot \mathbf{u}}{T_e} \right] f_{e0}, \qquad (4.32)$$

where we have now written $\mathbf{v}$ instead of $\mathbf{v}'_e$ and where $\mathbf{u} = \mathbf{V}_e - \mathbf{V}_i$ is the mean velocity of the electrons relative to the ions. This equation has a feature we have not encountered earlier: the driving terms on the right contain time derivatives of the density, velocity, and temperature. These derivatives can, however, be eliminated by taking moments of the equation. Indeed, integrating Eq. (4.32) over velocity space gives the continuity equation (2.15), which implies that $d \ln n_e/dt$ can be replaced by $-\nabla \cdot \mathbf{V}_e$. The $m_e\mathbf{v}$-moment gives the momentum equation (2.16), but without the viscosity term, and can be used to eliminate $d\mathbf{V}_e/dt$ and $\mathbf{E}'$ by

$$\frac{d\mathbf{V}_e}{dt} + \frac{e\mathbf{E}'}{m_e} = \frac{\mathbf{R}_e - \nabla(n_e T_e)}{m_e n_e}.$$

The energy moment, finally, gives the energy equation (2.17), but without heat conduction, viscous heating and energy exchange. The energy equation obtained is thus Eq. (2.17) with the right side set to zero,

$$\frac{3}{2}\frac{d \ln T_e}{dt} + \nabla \cdot \mathbf{V}_e = 0,$$

and this can be used to eliminate dT_e/dt in Eq. (4.32). The reason why certain terms in the full fluid equations (2.15)–(2.17) do not appear in the moments of Eq. (4.32) is that they are small in the ordering assumed. For instance, the viscosity is formally smaller than the friction force and the pressure gradient. This circumstance, which can be verified *a posteriori*, hints, in fact, at a weakness of Braginskii's theory. When the viscosity is eventually calculated (from the appropriate moment of f_1, not f_0), it is more accurate than the friction force. Ideally, one would like to know the friction as accurately as the viscosity, but Braginskii's analysis does not allow this.

In any case, eliminating time derivatives from Eq. (4.32) in this manner gives the following kinetic equation for f_{e1} in the electron rest frame,

$$C_{ee}(f_{e1}) + C_{ei}^0(f_{e1}) + \frac{e}{m_e}\mathbf{v} \times \mathbf{B} \cdot \frac{\partial f_{e1}}{\partial \mathbf{v}}$$

$$= \left[\left(x^2 - \frac{5}{2}\right)\mathbf{v} \cdot \nabla \ln T_e + \mathbf{v} \cdot \left(\frac{\mathbf{R}_e}{p_e} + \frac{m_e v_{ei}\mathbf{u}}{T_e}\right)\right.$$

$$\left. + \frac{m_e}{2T_e}\left(v_j v_k - \frac{v^2}{3}\delta_{jk}\right)W_{jk}^e\right]f_{e0}, \tag{4.33}$$

where

$$W_{jk}^a \equiv \frac{\partial V_{aj}}{\partial r_k} + \frac{\partial V_{ak}}{\partial r_j} - \frac{2}{3}(\nabla \cdot \mathbf{V}_a)\delta_{jk}, \tag{4.34}$$

is the so-called rate-of-strain tensor. Note that there are three driving terms on the right: the temperature gradient, which we have encountered before as a 'thermodynamic force', see Eq. (4.10); the term involving the friction force, which replaces the earlier combination (4.10) of pressure gradient and electric field in Braginskii's point of view; and the tensor W_{jk} which measures how rapidly the flow velocity varies in space and, as we shall see, gives rise to plasma viscosity.

The ion analysis is slightly simpler since the entire ion–electron collision operator can be regarded as small, which implies that the friction $\mathbf{R}_{ie}$ is negligible in the ion kinetic problem. The analogue of Eq. (4.33) for ions is thus

$$
C_{ii}(f_{i1}) - \frac{e}{m_i} \mathbf{v} \times \mathbf{B} \cdot \frac{\partial f_{i1}}{\partial \mathbf{v}}
$$

$$
= \left[\left(x^2 - \frac{5}{2} \right) \mathbf{v} \cdot \nabla \ln T_i + \frac{m_i}{2T_i} \left(v_j v_k - \frac{v_i^2}{3} \delta_{jk} \right) W_{jk}^i \right] f_{i0}, \qquad (4.35)
$$

Results: transport in a magnetized hydrogen plasma

Equations (4.33) and (4.35) constitute the kinetic problem that needs to be solved to calculate transport in a two-component plasma. This can be done by using the variational principle derived in Section 4.3 or the Sonine polynomial expansion from Section 4.5. Since we have already outlined the methodology of such calculations, we do not go into further details here. The interested reader is referred to Robinson and Bernstein (1962) for the variational treatment and to Braginskii (1965) for details of the expansion in polynomials. Instead, we merely present the results. Because of their importance, we treat these results in greater detail than we did for the case of Lorentz plasma. Here we present transport coefficients for both ions and electrons, giving expressions not only for friction and heat fluxes but also for viscosity and electron–ion energy exchange. This is all the information needed for closure of the fluid equations (2.15)–(2.17), which then form a complete model of plasma behaviour. It is assumed that $\Delta \ll 1$ for all species, see Eq. (4.3), so that the gyrofrequency exceeds the collision frequency. Following Braginskii, we use the collision times

$$
\tau_e = \tau_{ei},
$$
$$
\tau_i = \sqrt{2}\tau_{ii},
$$

and only consider the strongly magnetized limit $\Omega_a \tau_a \to \infty$, which is the case of greatest practical interest. Braginskii also derived transport coefficients for finite $\Omega_a \tau_a$, but some of these results should be treated with caution. Epperlein and Haines (1986) have shown that some of his transport coefficients are inaccurate at intermediate values of $\Omega_a \tau_a$, and others have the wrong asymptotic behaviour as $\Omega_a \tau_a \to \infty$. However, none of these conclusions affect the results quoted here.

The force $\mathbf{R}_e$ acting on electrons consists of a drag force and a thermal force

$$\mathbf{R}_e = \mathbf{R}_u + \mathbf{R}_T, \tag{4.36}$$

$$\mathbf{R}_u = -\frac{m_e n_e}{\tau_e}(0.51\mathbf{u}_\parallel + \mathbf{u}_\perp), \tag{4.37}$$

$$\mathbf{R}_T = -0.71 n_e \nabla_\parallel T_e + \frac{3n_e}{2\Omega_e \tau_e}\mathbf{b} \times \nabla T_e, \tag{4.38}$$

where again $\mathbf{u} \equiv \mathbf{V}_e - \mathbf{V}_i$, and, because of momentum conservation in Coulomb collisions, $\mathbf{R}_i = -\mathbf{R}_e$. The parallel friction coefficient in Eq. (4.37) is seen to be smaller than the perpendicular one by a factor of 0.51. As already mentioned, this has to do with the fact that the collision frequency decreases with velocity, $(\tau_e \sim v^3)$, causing electrons with large parallel velocities to be more distorted from the Maxwellian distribution than slower ones. The fast electrons then contribute more to the relative velocity, and less to the friction since their collisionality is relatively low. The parallel thermal force (4.38) is also a consequence of the circumstance that the collision frequency falls off with increasing energy, and the second term in (4.38) is smaller than the first one by a factor of $1/\Omega_e \tau_e \ll 1$.

The electron heat flux also has two pieces

$$\mathbf{q}_e = \mathbf{q}_u^e + \mathbf{q}_T^e, \tag{4.39}$$

$$\mathbf{q}_u^e = 0.71 n_e T_e \mathbf{u}_\parallel - \frac{3n_e T_e}{2\Omega_e \tau_e}\mathbf{b} \times \mathbf{u}_\perp, \tag{4.40}$$

$$\mathbf{q}_T^e = -\kappa_\parallel^e \nabla_\parallel T_e - \kappa_\wedge^e \mathbf{b} \times \nabla T_e - \kappa_\perp^e \nabla_\perp T_e, \tag{4.41}$$

where the heat conductivities are

$$\kappa_\parallel^e = 3.16\frac{n_e T_e \tau_e}{m_e}, \tag{4.42}$$

$$\kappa_\wedge^e = -\frac{5n_e T_e}{2m_e \Omega_e}, \tag{4.43}$$

$$\kappa_\perp^e = 4.66\frac{n_e T_e}{m_e \Omega_e^2 \tau_e}. \tag{4.44}$$

It should be remembered that $\Omega_e = -eB/m_e$ is negative. Note the Onsager symmetry between $\mathbf{q}_u$ and $\mathbf{R}_T$. Physically, the first term in Eq. (4.41) has to do with the distortion of the distribution of fast electrons from a Maxwellian. In a coordinate system where $\mathbf{V}_e = 0$, more fast electrons travel in the direction of $\mathbf{u}$ and more slow electrons in the direction of $-\mathbf{u}$, which gives rise to a heat flux. This effect does not arise for ions since the ion–electron frequency is independent of ion energy. The ion heat flux therefore only has terms related to ∇T_i,

$$\mathbf{q}_i = -\kappa^i_\parallel \nabla_\parallel T_i + \kappa^i_\wedge \mathbf{b} \times \nabla T_i - \kappa^i_\perp \nabla_\perp T_i, \tag{4.45}$$

where

$$\kappa^i_\parallel = 3.9 \frac{n_i T_i \tau_i}{m_i}, \tag{4.46}$$

$$\kappa^i_\wedge = \frac{5 n_i T_i}{2 m_i \Omega_i}, \tag{4.47}$$

$$\kappa^i_\perp = 2 \frac{n_i T_i}{m_i \Omega_i^2 \tau_i}. \tag{4.48}$$

Note that the conductivities $\kappa_\parallel$, $\kappa_\wedge$ and $\kappa_\perp$ are consecutively separated by the large factor $\Omega \tau$ for both species. The first and third terms in the expressions for $\mathbf{q}^e_T$ and $\mathbf{q}_i$ are parallel to the gradients that drive them, and will therefore tend to relax these gradients. This relaxation occurs on widely separate time scales in the two directions. Perpendicular to the magnetic field, the ion contribution is dominant, and scales according to the estimate (1.8), indicating a random walk with the step size ρ_i. Parallel to the field, on the other hand, the electron heat flux dominates, and scales as Eq. (1.9) with a random-walk step size of λ. Of intermediate magnitude is the diamagnetic heat flux term $\mathbf{q}_\wedge = \pm \kappa_\wedge \mathbf{b} \times \nabla T$, carrying heat across the field perpendicular to the gradient. As already mentioned, this flux is not caused by collisions, as is clear from the fact that the collision frequency does not appear in these terms. It will be discussed in greater detail in the next section.

The heat exchanged between the species,

$$Q_i = -Q_e - \mathbf{R}_e \cdot \mathbf{u} = \frac{3 n_e m_e}{m_i \tau_e} (T_e - T_i), \tag{4.49}$$

is a result of temperature equilibration on the slow time scale τ_{ie}, and frictional generation of heat.

We now turn to the viscosity, π, which is the flux of momentum. When the flow velocity is large in the sense of the ordering (4.6), the viscosity is driven by gradients of this flow, just as in the Navier–Stokes equation for an ordinary fluid the viscosity tensor is

$$\pi_{jk} = \eta W_{jk} + \zeta \delta_{jk} \nabla \cdot \mathbf{V}.$$

Here η and ζ are viscosity coefficients, and W_{jk} is the rate-of-strain tensor (4.34). The last term is non-zero only for flow velocity fields with non-vanishing divergence, i.e., for compressible fluid flows.

In a magnetized plasma, the viscosity tensor is more complicated as the transport of momentum occurs at very different rates in different directions. Each species has its own rate-of-strain tensor, and as we shall show in the next section, the viscosity tensors are of the form

$$\pi_{xx} = -\frac{\eta_0}{2}(W_{xx} + W_{yy}) - \frac{\eta_1}{2}(W_{xx} - W_{yy}) - \eta_3 W_{xy}, \tag{4.50}$$

$$\pi_{yy} = -\frac{\eta_0}{2}(W_{xx} + W_{yy}) - \frac{\eta_1}{2}(W_{yy} - W_{xx}) + \eta_3 W_{xy}, \tag{4.51}$$

$$\pi_{xy} = \pi_{yx} = -\eta_1 W_{xy} + \frac{\eta_3}{2}(W_{xx} - W_{yy}), \tag{4.52}$$

$$\pi_{xz} = \pi_{zx} = -\eta_2 W_{xz} - \eta_4 W_{yz}, \tag{4.53}$$

$$\pi_{yz} = \pi_{zy} = -\eta_2 W_{yz} + \eta_4 W_{xz}, \tag{4.54}$$

$$\pi_{zz} = -\eta_0 W_{zz}, \tag{4.55}$$

where the coordinate system (x, y, z) is taken to be aligned with the magnetic field, $\mathbf{b} = (0, 0, 1)$.

All terms in the viscosity tensor are small (by factors of $\rho/L_\perp$ or $\lambda/L_\parallel$) in comparison with the other terms in the momentum equation,

$$\nabla \cdot \boldsymbol{\pi} \ll \nabla p \sim \mathbf{R}, \tag{4.56}$$

since the lowest-order Maxwellian distribution (4.5) does not contribute to the viscosity. Nevertheless, this does not mean that viscosity is unimportant. For instance, in a tokamak the parallel pressure gradient vanishes in leading order and parallel viscosity plays a crucial role in determining the parallel flow velocity. The viscosity coefficients for ions are

$$\eta_0^i = 0.96 n_i T_i \tau_i, \tag{4.57}$$

$$\eta_1^i = \frac{3 n_i T_i}{10 \Omega_i^2 \tau_i} = \eta_2^i/4, \tag{4.58}$$

$$\eta_3^i = \frac{n_i T_i}{2 \Omega_i} = \eta_4^i/2, \tag{4.59}$$

and for electrons

$$\eta_0^e = 0.73 n_e T_e \tau_e, \tag{4.60}$$

$$\eta_1^e = 0.51 \frac{n_e T_e}{\Omega_e^2 \tau_e} = \eta_2^e/4, \tag{4.61}$$

$$\eta_3^e = \frac{n_e T_e}{2 \Omega_e} = \eta_4^e/2. \tag{4.62}$$

As in the case of the heat flux, we see that the viscosity contains terms with different dependencies of the collision frequency. First, there is a piece proportional to $\eta_0 \sim nT\tau$, similar to that for an ordinary gas. This scaling is identical to that of parallel heat conduction (1.9) and reflects transport of momentum due to a random walk with a step size equal to the mean-free path. Second, there are entries in the viscosity tensor proportional to η_1 and η_2, which scale as the perpendicular heat conduction coefficient (1.8). They come from random-walk transport of momentum with a step size equal to the Larmor radius, and are therefore smaller than the parallel viscosity η_0 by a factor of $(\Omega\tau)^2$. Finally, there is the contribution from 'gyroviscosity', proportional to η_3 and η_4. This contribution is inversely proportional to the gyrofrequency but is completely independent of the collision frequency. Gyroviscosity represents a diamagnetic flow of momentum across the field, and is analogous to the term $\mathbf{q}_\wedge$ in the heat flux.

As mentioned in the introduction to this chapter, the viscosity comes out different from Braginskii's if a small flow ordering, $V = O(\delta v_T)$ is adopted. A recent reference containing these results is Claassen *et al.* (2000).

4.7 Diamagnetic flows

In Section 4.2 we derived electron transport laws by solving the kinetic equation for a plasma with highly charged ions. It was found that the perpendicular fluxes of heat and particles are predominantly diamagnetic, and that these fluxes are independent of the collision frequency. The same was seen to be true in Braginskii's equations. This suggests that diamagnetic fluxes are not dependent on collisions at all. Indeed, as we shall see in the present section, these fluxes can be obtained directly from moment equations, without using the kinetic equation. The presence of a small parameter $\delta = \rho/L_\perp$ makes it possible partly to solve the (partial differential) moment equations for a magnetized plasma by purely algebraic means. In this manner it is possible to calculate the diamagnetic flow of particles, momentum, and energy, and to gain additional insight into the nature of the diamagnetic heat flow (4.44) and the gyroviscosity (4.59), (4.62).

Particle flux

First consider the momentum equation (2.16). Taking the vectorial product with $\mathbf{B}$ immediately gives the perpendicular velocity as a sum of the $\mathbf{E} \times \mathbf{B}$ drift and the diamagnetic drift velocity to leading order in δ:

$$\mathbf{V}_\perp = \frac{\mathbf{E} \times \mathbf{B}}{B^2} + \frac{\mathbf{b} \times (\nabla p + \nabla \cdot \pi - \mathbf{R} + mn\, d\mathbf{V}/dt)}{mn\Omega} \qquad (4.63)$$

$$\simeq \boxed{\frac{\mathbf{E} \times \mathbf{B}}{B^2} + \frac{\mathbf{b} \times \nabla p}{mn\Omega}}. \qquad (4.64)$$

Note that this perpendicular velocity is smaller than thermal, $\mathbf{v}_\perp \sim \delta v_T$. The terms neglected in the last expression are usually small. In particular, if Braginskii's orderings are assumed, the perpendicular force and the viscosity are smaller (by a factor Δ or δ) than the pressure gradient according to Eqs. (4.37), (4.38) and (4.56). More generally, the viscous force is smaller than the pressure gradient whenever the distribution function is close to a Maxwellian. As we shall see in neoclassical theory, this occurs whenever the magnetic field forms closed flux surfaces, even if the mean-free path is long. The perpendicular force in Eq. (4.63) $\mathbf{R}_\perp \sim v m n \mathbf{V}_\perp$ is small if $\Delta = v/\Omega \ll 1$. If, in addition, the flow velocity is much smaller than the thermal speed and time variation is weak, it is legitimate to neglect the inertial term $d\mathbf{V}/dt$. If the flow velocity is as large as the thermal speed, the term $mn(\mathbf{V} \cdot \nabla)\mathbf{V}$ must be included. This is, for instance, the case in a toroidally rotating tokamak plasma with $V \sim v_{Ti}$, where this term represents the centrifugal force.

The flow velocity (4.63) is in general compressible, i.e., $\nabla \cdot \mathbf{V}_\perp \neq 0$. (But in the special case of constant magnetic field and $\mathbf{E} = -\nabla\Phi$ the flows of (4.64) are incompressible.) Since in steady state the continuity equation (2.15) requires an incompressible flow, a parallel flow arises with a divergence balancing that of the perpendicular flow, $\nabla \cdot (\mathbf{V}_\parallel + \mathbf{V}_\perp) = 0$. As we shall see in a later chapter, this so-called return flow plays a fundamental role in Pfirsch–Schlüter diffusion in a torus.

Momentum flux – viscosity

The viscosity tensor can be derived in a way entirely analogous to that used to find Eq. (4.64). Indeed, it is simple to derive Eqs. (4.50)–(4.62) from the kinetic equation (2.1) if the collision operator is approximated by a so-called BGK operator (Bhatnagar, Gross and Krook, 1954; Welander, 1954),

$$C(f) = v(f_0 - f), \tag{4.65}$$

which is frequently used in kinetic gas theory. This operator models relaxation to local thermodynamic equilibrium f_0 at a rate determined by the 'collision frequency' v, which we take to be equal to the inverse collision time, $v = \tau^{-1}$. Since the collision operator is simplified, this yields only approximate collisional viscosity coefficients η_0, η_1, and η_2, but gives the gyroviscosity, η_3 and η_4, exactly, as well as the full structure of the viscosity tensor. Following Kaufman (1960), we show the procedure here as a simple and lucid example of transport theory.

We begin by taking the $v'_l v'_m$-moment of the kinetic equation (2.1), which we write as

$$\frac{\partial f}{\partial t} + \frac{\partial}{\partial x_i}\left[(V_i + v_i')f\right] + \frac{\partial}{\partial v_i}\left[\left(\frac{eE_i}{m} + \epsilon_{ijk}(V_j + v_k')\Omega_k\right)f\right] = C(f),$$

(4.66)

where $\mathbf{v}' = \mathbf{v} - \mathbf{V}$ is the velocity relative to the mean velocity $\mathbf{V}$ for the species in question, $\Omega_k = eB_k/m$, and the Levi–Civita symbol ϵ_{jkl} is the completely antisymmetric unit tensor in three dimensions, i.e., $\epsilon_{ijk} = 1$ if (ijk) is an even permutation of the numbers (123), $\epsilon_{ijk} = -1$ for odd permutations, and $\epsilon_{ijk} = 0$ if any two indices are the same. The $v_l'v_m'$-moment of this equation is

$$\frac{\partial P_{lm}}{\partial t} + \frac{\partial}{\partial x_i}(V_i P_{lm} + q_{ilm}) + \frac{\partial V_l}{\partial x_i}P_{im} + \frac{\partial V_m}{\partial x_i}P_{il}$$
$$- \Omega_k(\epsilon_{ljk}P_{jm} + \epsilon_{mjk}P_{jl}) = \nu(p\delta_{lm} - P_{lm}),$$

(4.67)

with

$$P_{lm} \equiv \int mv_l'v_m'f\,d^3v,$$

$$q_{ilm} \equiv \int mv_i'v_l'v_m'f\,d^3v.$$

Note that the non-diagonal elements of the pressure tensor P_{lm} coincide with those of the viscosity tensor (2.12),

$$P_{lm} = p\delta_{lm} + \pi_{lm}.$$

We now apply Braginskii's short-mean-free-path orderings to the evolution equation (4.67) for P_{lm}. This singles out the terms proportional to the frequencies Ω and ν as being larger than all other terms. (We do not yet invoke $\Omega \gg \nu$.) To lowest order the distribution function then becomes Maxwellian, $f = f_0$, so that

$$P_{lm} = p\delta_{lm}, \qquad q_{ilm} = 0.$$

Going to the next order, we substitute these expressions in the small terms of Eq. (4.67) and so obtain an equation for the next-order correction to P_{lm} in the terms proportional to Ω and ν,

$$\mathsf{K}(\mathsf{P}) \equiv \mathsf{S}/\Omega,$$

(4.68)

where S is the tensor defined by

$$S_{lm} = \left[\frac{\partial p}{\partial t} + \frac{\partial}{\partial x_i}(V_i p)\right]\delta_{lm} + p\left(\frac{\partial V_l}{\partial x_m} + \frac{\partial V_m}{\partial x_l}\right) + \nu(P_{lm} - p\delta_{lm}),$$

(4.69)

and the operator K is defined by

$$K(P) = P \times \mathbf{b} + \text{transpose} = \frac{\Omega_k}{\Omega} \left(\epsilon_{ljk} P_{jm} + \epsilon_{mjk} P_{lj} \right)$$

$$= \begin{pmatrix} 2P_{xy} & P_{yy} - P_{xx} & P_{yz} \\ P_{yy} - P_{xx} & -2P_{xy} & -P_{xz} \\ P_{yz} & -P_{xz} & 0 \end{pmatrix}. \tag{4.70}$$

In this matrix notation, we have taken the z-axis to coincide with the magnetic field, so that $\Omega_k/\Omega = \delta_{k3}$. It is now useful to note two things. First, since $\pi_{lm} = P_{lm} - p\delta_{lm}$ by definition, we have $K(P) = K(\pi)$. Second, since the viscosity, the heat flux and the collisional terms in the entropy equation (2.18) (which is equivalent to the trace of Eq. (4.67)) vanish to the lowest order, entropy is conserved, $d \ln p/dt = (5/3)d \ln n/dt$, so that the first term in (4.69) becomes

$$\frac{\partial p}{\partial t} + \nabla \cdot (\mathbf{V}p) = p \left(\frac{d \ln p}{dt} + \nabla \cdot \mathbf{V} \right) = -\frac{2p}{3} \nabla \cdot \mathbf{V},$$

and the expression for S can be simplified to

$$S = pW + v\pi, \tag{4.71}$$

where W was defined in (4.34). Equation (4.68) with (4.70) and (4.71) now gives the following system of algebraic equations

$$\begin{pmatrix} 2\pi_{xy} & \pi_{yy} - \pi_{xx} & \pi_{yz} \\ & -2\pi_{xy} & -\pi_{xz} \\ & & 0 \end{pmatrix} = \frac{1}{\Omega} (pW + v\pi),$$

which can be solved for π. The entries in the lower left corner of the matrix are omitted as these simply reproduce those in the upper right one. Solving this system of six equations gives viscosity elements π_{lm} with exactly the same structure as those of Braginskii presented in Eqs. (4.50)–(4.55), if the viscosity coefficients are

$$\eta_0 = p\tau,$$
$$\eta_1 = \frac{\eta_0}{1 + 4\Omega^2\tau^2},$$
$$\eta_2 = \frac{\eta_0}{1 + \Omega^2\tau^2},$$
$$\eta_3 = 2\eta_4 = \frac{2\Omega\tau}{1 + 4\Omega^2\tau^2}\eta_0.$$

where $\tau = v^{-1}$. These coefficients are very similar to those of Braginskii in Eqs. (4.57)–(4.62). Thus, not only is the form of the viscosity tensor reproduced exactly, but the viscosity coefficients come out to depend on the gyrofrequency and the collision frequency in a qualititatively correct way. The numerical coefficients differ from Braginskii, however, because of the simplified collision operator (4.65) used. Nevertheless, in the collisionless limit $\Omega\tau \gg 1$, the gyroviscosity, which dominates the non-diagonal elements in the viscosity tensor, becomes

$$\pi_{xy} = \pi_{yx} = \frac{p}{4\Omega}(W_{xx} - W_{yy}), \tag{4.72}$$

$$\pi_{xz} = \pi_{zx} = -\frac{p}{\Omega}W_{yz}, \tag{4.73}$$

$$\pi_{yz} = \pi_{zy} = \frac{p}{\Omega}W_{xz}. \tag{4.74}$$

This result agrees exactly with Braginskii since $p/2\Omega$ coincides with the gyroviscosity coefficient η_3 when $\Omega\tau \gg 1$. Gyroviscosity is caused by the Larmor gyration of the plasma particles, and not by collisions, and is therefore insensitive to the collision frequency (or even the collision operator) in the limit $\Omega\tau \gg 1$, where gyromotion is not interrupted by collisions.

Heat flux

The diamagnetic flow of heat, Eqs. (4.44), (4.48) can be calculated in an entirely analogous manner, by considering the $(mv^2/2)\mathbf{v}$-moment of the kinetic equation (4.66). For the important class of transport problems where the flow velocity $\mathbf{V}$ arises in response to gentle gradients ($\delta \ll 1$), we can assume that the flow velocity is small, $V \sim \delta v_T \ll v_T$, so that we do not need to make any distinction between $\mathbf{v}$ and $\mathbf{v}' = \mathbf{v} - \mathbf{V}$ in the lowest order in the pressure tensor. Then

$$\frac{\partial \mathbf{Q}}{\partial t} + \nabla \cdot \mathbf{r} - \frac{e}{m}\left(\mathbf{E} \cdot \mathbf{P} + \frac{3}{2}p\mathbf{E} + \mathbf{Q} \times \mathbf{B}\right) = \mathbf{G}, \tag{4.75}$$

where the total (convective + conductive) heat flux is

$$\mathbf{Q} \simeq \mathbf{q} + \frac{5p\mathbf{V}}{2}$$

from (2.14) since $\Pi_{jk} = p\delta_{jk}$ in a Maxwellian plasma. We have introduced the 'heat stress' tensor

$$r_{jk} \equiv \int \frac{mv^2}{2}v_j v_k f d^3v,$$

and the collisional rate of change of total heat flux,

$$\mathbf{G} \equiv \int \frac{mv^2}{2}\mathbf{v}C(f)d^3v.$$

For a distribution function close to a Maxwellian, $r_{jk} = (5pT/2m)\delta_{jk}$, and $P_{jk} = p\delta_{jk}$. By taking the vectorial product with $\mathbf{B}$ we thus obtain

$$\mathbf{Q}_\perp = \frac{5p}{2}\left(\frac{\mathbf{E}\times\mathbf{B}}{B^2} + \frac{\mathbf{b}\times\nabla p}{mn\Omega}\right) + \frac{5p}{2m\Omega}\mathbf{b}\times\nabla T,$$

where we have neglected $\partial\mathbf{Q}/\partial t$ and $\mathbf{G}\times\mathbf{b}/\Omega \sim \mathbf{q}_\perp v/\Omega$, and recognize the lowest-order velocity (4.64) within the parentheses. For a particle species with small flow velocity, the total (conductive + convective) heat flux is $\mathbf{Q} = \mathbf{q} + (5p/2)\mathbf{V}$, as follows from (2.14) since the two last terms on the right-hand side are smaller than the two first ones. The conductive part of the diamagnetic heat flux thus becomes

$$\boxed{\mathbf{q}_\perp = \frac{5p}{2m\Omega}\mathbf{b}\times\nabla T} \tag{4.76}$$

in lowest order.

Further reading

A generation of plasma physicists has learned the topics covered in this chapter from Braginskii's outstanding review article (1965), which remains the best reference in the area. His earlier paper (1958) gives details of the calculation of matrix elements, as does Balescu (1988). Our discussion of the Lorentz plasma is partly inspired by Lifshitz and Pitaevskii (1981). Hazeltine and Waelbroeck (1998) give a very thoughtful introduction to transport theory, including the variational aspect, which is treated more fully by Robinson and Bernstein (1962). This topic goes right back to Enskog's beautiful thesis (1917), which laid the foundation for modern kinetic gas theory. The classic reference on this is Chapman and Cowling (1970). The Spitzer conductivity is accurately calculated in Hirshman (1978b) using a Laguerre-polynomial expansion. While the parallel Spitzer conductivity links the current parallel to the magnetic field to the parallel electric field in an intuitively obvious way, the perpendicular conductivity deserves special discussion in a magnetized plasma where the $\mathrm{E}\times\mathrm{B}$-drift is the same for electrons and ions thus giving zero current. This is given by Hazeltine and Waelbroeck (1998, pp. 294–296.)

Exercises

1. Calculate the resistivity $\eta = 1/\sigma_\parallel$ of a hydrogen plasma, and compare with copper, $\eta_{\mathrm{Cu}} = 1.7\cdot 10^{-8}\ \Omega\ \mathrm{m}$.

   Solution:

   $$\eta = \left(\frac{1\,\mathrm{eV}}{T_e}\right)^{3/2}\ln\Lambda\ 5.3\cdot 10^{-5}\ \Omega\,\mathrm{m}.$$

Thus, at temperatures above a few keV a hydrogen plasma is a better conductor than copper.

2. Verify Braginskii's result (4.49) by calculating the rate of energy transfer between Maxwellian ions and electrons of different temperatures, using the ion–electron collision operator (3.56).

Solution:

$$Q_i = \int \frac{m_i v^2}{2} C_{ie}(f_{i0}) \, d^3v = \frac{m_e}{m_i \tau_{ei}} \int \frac{m_i v^2}{2} \frac{\partial}{\partial \mathbf{v}} \cdot \left(\mathbf{v} f_{i0} + \frac{T_e}{m_i} \frac{\partial f_{i0}}{\partial \mathbf{v}} \right) d^3v$$

$$= \frac{m_e}{m_i \tau_{ei}} \left(\frac{T_e}{T_i} - 1 \right) \int m_i v^2 f_{i0} d^3v = (4.49)$$

3. Demonstrate that the transport matrix L_{jk} introduced in Section 4.3 must be positive definite in order to comply with the second law of thermodynamics.

Solution: According to Exercise 3c in the previous chapter, the entropy production rate is equal to $-S[f_{e1}, f_{e1}]$. We saw in Section 4.3 that

$$S[h_j, h_k] = -\frac{n_e T_e \tau_{ei}}{m_e} L_{jk},$$

so since $f_{e1} = A_{1\parallel} h_1 + A_{2\parallel} h_2$ the entropy production rate is

$$\frac{n_e T_e \tau_{ei}}{m_e} \sum_{j,k} A_{j\parallel} L_{jk} A_{k\parallel},$$

which is positive if and only if L_{jk} is positive definite, i.e.,

$$L_{11} > 0, \qquad L_{22} > 0, \qquad L_{11} L_{22} - L_{12}^2 > 0.$$

4. Calculate the ion thermal conductivity in a pure hydrogen plasma by the method given in Section 4.5.

Solution: Only ion–ion collisions need to be considered, and the system of equations (4.31) reduces to

$$\frac{m_i}{T_i \tau_{ii}} \sum_k c_{jk} u_{ik} = -\frac{5}{2} \delta_{j1} \nabla_\parallel \ln T_i,$$

where $c_{jk} = M_{ii}^{jk} + N_{ii}^{jk}$. The heat flux is

$$q_{i\parallel} = -\frac{5 p_i}{2} u_{i1} = -\kappa_\parallel^i \nabla_\parallel T_i.$$

Since $c_{0k} = c_{k0} = 0$ for all k, the lowest possible truncation of the system is to include only $j = k = 1$, which gives

$$\kappa_\parallel^i = \frac{25}{8} \frac{p_i \tau_i}{m_i},$$

where we have used $c_{11} = -2^{1/2}$ and $\tau_i = 2^{1/2} \tau_{ii}$. Keeping also the $j \leq 2, k \leq 2$ terms gives Braginskii's result (4.46), which is a 25% correction.

5

Transport in a cylindrical plasma

In this chapter, we show how to use fluid equations to calculate cross-field fluxes of particles, momentum, and heat in a cylindrical plasma embedded in a straight magnetic field. As long as the orderings adopted in the preceding chapter are satisfied ($\lambda \ll L_\parallel$, $\rho \ll L_\perp$), the initial density, flow-velocity, and temperature profiles may be prescribed at will. The fluid equations (2.15)–(2.17) then determine the evolution of these profiles on the slow transport time scale. Intuitively, one might expect that the transport in a cylinder would closely approximate that in a torus with very large aspect ratio (major radius divided by minor radius). However, as we shall see in the chapter on 'Pfirsch–Schlüter' transport, toroidal effects are important even in this limit. Analysis of the cylindrical case is nonetheless valuable since it sheds light on the basic physics and general methodology without additional geometrical effects. Of course, the transport can also be calculated directly from the kinetic equation, and we shall show how to do this in a later chapter in connection with so-called orbit squeezing.

5.1 Particle transport

Consider a cylindrical plasma consisting of three species, electrons (e), hydrogen ions (i), and heavier ions (Z), in an axial magnetic field $\mathbf{B} = B\hat{z}$. We use (r, θ, z) to denote the usual cylindrical coordinates, and take the density $n_a(r)$ and temperature $T_a(r)$ profiles, as well as the electrostatic potential $\Phi(r)$, to depend only on the radius, see Fig. 5.1. The flow velocity is assumed to arise in response to the corresponding gradients. Because of azimuthal (or 'poloidal') symmetry all θ-derivatives vanish, $\partial/\partial\theta = 0$. It follows from Eq. (4.64) that the flow is in the poloidal (θ) direction,

$$
V_{a\theta} = \frac{\Phi'(r)}{B} + \frac{p_a'(r)}{m_a n_a \Omega_a},
\tag{5.1}
$$

90

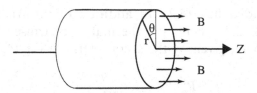

Fig. 5.1. Plasma cylinder embedded in a straight magnetic field.

to lowest order in δ. Primes denote radial derivatives. In order to obtain the radial flux $\Gamma_a = n_a V_{ar}$, we must therefore include contributions from the smaller terms in Eq. (4.63),

$$V_{ar} = \frac{-(\nabla \cdot \pi_a)_\theta + R_{a\theta} - m_a n_a\, d\mathbf{V}_{a\theta}/dt}{m_a n_a \Omega_a} \simeq \frac{R_{a\theta}}{m_a n_a \Omega_a}. \qquad (5.2)$$

Note that the force $R_{a\theta}$ is perpendicular to $\mathbf{B}$. As will be shown in the next section, the terms related to viscosity and inertia are small but interesting. They do not contribute much to the flow, but change the nature of it in a qualitative manner. For the time being we neglect them. This makes the total radial current vanish,

$$j_r = \sum_a e_a \Gamma_a = \sum_a \frac{R_{a\theta}}{B} = 0, \qquad (5.3)$$

since $\Sigma R_a = 0$ because of momentum conservation in Coulomb collisions. In other words, in the steady state radial transport is *automatically ambipolar* (to lowest order in δ), regardless of the magnitude of the radial electric field.

The force $\mathbf{R}_a$ is the sum of ordinary friction and thermal forces, as shown in Eqs. (4.36)–(4.38) for a pure plasma. Ordinary friction,

$$\mathbf{R}_{ab\perp}^{\text{friction}} = -\frac{m_a n_a}{\tau_{ab}} (\mathbf{V}_a - \mathbf{V}_b)_\perp, \qquad (5.4)$$

(with $\tau_{el} = \tau_e$ for the usual electron ion friction (4.37)) arises because different species have different flow velocities (5.1) within the flux surface. Since these diamagnetic flows are perpendicular to the magnetic field, we need only the friction force in this direction. It is not difficult to see why a radial electric field $E_r = -\Phi'(r)$ is unable to drive a radial current (5.3). The resulting $\mathbf{E} \times \mathbf{B}$-velocity (the first term in (5.1)) is the same for all species, and produces no friction force which could drive a radial flux (5.2).

Pure plasma

In a pure hydrogen plasma, the radial ion and electron fluxes are equal because of ambipolarity, and the poloidal electron–ion force $R_{ei\theta}$ completely determines the radial flux (5.2). In addition to the friction force given by Eqs. (5.1) and (5.4), there is a thermal force caused by the electron temperature gradient, given by the second term in Eq. (4.38),

$$R_{ei\theta}^{\text{thermal}} = \frac{3n_e T_e'(r)}{2\Omega_e \tau_e}. \tag{5.5}$$

Adding the contributions of these forces in Eq. (5.2) gives the radial flux

$$\Gamma_i = \Gamma_e = -n_e \frac{T_e}{m_e \Omega_e^2 \tau_e} \left(\frac{p_e' + p_i'}{p_e} - \frac{3}{2} \frac{T_e'}{T_e} \right). \tag{5.6}$$

Note that the (classical) diffusion coefficient here,

$$D_e = \frac{T_e}{m_e \Omega_e^2 \tau_e} = \frac{\rho_e^2}{2\tau_e}, \tag{5.7}$$

scales with the gyroradius $\rho_e = v_{Te}/|\Omega_e|$ and the collision time as predicted by the simple random-walk argument presented in Chapter 1. Also note that the flux does not depend only on the density gradient. The electron and ion temperature gradients also contribute through the pressure gradients. In addition, there is the term solely related to the electron temperature gradient, which acts to reduce the transport if T_e decreases with radius. This is sometimes referred to as the 'temperature screening effect'.

Impure hydrogen plasma

In practice, a fusion plasma is never pure; the main hydrogenic ion species is always diluted with impurity ions of higher atomic numbers Z originating from sputtering from the wall by impact of energetic hydrogen ions. The main ions therefore experience friction against both electrons and impurity ions,

$$\mathbf{R}_i = \mathbf{R}_{ie} + \mathbf{R}_{iZ}.$$

The relative magnitude is estimated by

$$\frac{\mathbf{R}_{ie}}{\mathbf{R}_{iZ}} \sim \frac{m_e n_e v_{ei}}{m_i n_i v_{iZ}} \sim \frac{1}{\alpha} \sqrt{\frac{m_e}{m_i}}, \tag{5.8}$$

where $\alpha \equiv n_Z Z^2 / n_i$. In a fusion plasma α is usually in the range 0.5–2, which is much larger than $\sqrt{m_e/m_i}$, implying that R_{ie} can be neglected. Therefore, collisional ion particle transport is dominated by collisions with impurity ions, and not with electrons.

Because of the smallness of m_i/m_Z for most impurity ions ($Z \gg 1$), $\mathbf{R}_{iZ}$ is the force arising from collisions of a light and a heavy particle species, similar to the electron–ion force. Since the ion–impurity collision operator (3.32) is similar to the electron–ion operator, the ion–impurity thermal force can be shown to be exactly analogous to that between electrons and ions, Eqs. (4.38), (5.5),

$$\mathbf{R}_{iZ}^{\text{thermal}} = \frac{3n_i}{2\Omega_i\tau_{iZ}}\mathbf{b}\times\nabla T_i = \frac{3n_i T_i'(r)}{2\Omega_i\tau_{iZ}}\hat{\theta},$$

where $\hat{\theta} = r\nabla\theta$ is the unit vector in the azimuthal direction. Substituting this relation and (5.4) in the expression (5.2) for the radial ion flux, we find

$$\Gamma_i = -n_i\frac{T_i}{m_i\Omega_i^2\tau_{iZ}}\left(\frac{p_i'}{p_i} - \frac{T_z}{ZT_i}\frac{p_Z'}{p_Z} - \frac{3}{2}\frac{T_i'}{T_i}\right). \qquad (5.9)$$

This corresponds to a diffusion coefficient

$$D_i = \frac{T_i}{m_i\Omega_i^2\tau_{iZ}},$$

which is much larger than that for electrons (5.7). The electron flux is therefore much smaller than the ion flux, in contrast to the pure plasma case where we found $\Gamma_e = \Gamma_i$. The impurity flux can now be deduced from the ambipolarity condition (5.3),

$$\sum_a e_a\Gamma_a = e(\Gamma_i + Z\Gamma_Z - \Gamma_e) = 0 \;\Rightarrow\; \Gamma_Z = -\frac{1}{Z}\Gamma_i + O(\Gamma_e). \qquad (5.10)$$

This result implies *inward* impurity diffusion if the main ions move outward! Main ions and impurities diffuse in opposite directions, while the electrons are practically stationary. The reason for the inward flow of heavy ions is that the friction they experience is opposite to the friction force acting on the ions, $\mathbf{R}_i = -\mathbf{R}_Z$. (Friction against electrons is small.) The resulting radial drift is therefore directed inward if the main ions diffuse outward.

Therefore, impurity ions tend to accumulate in the centre of the plasma column if their transport is governed by classical diffusion. In a steady state, the radial fluxes (5.9) and (5.10) vanish, $\Gamma_i = \Gamma_Z = 0$, so that

$$(\ln p_i)' - (\ln p_Z^{1/Z})' - (\ln T_i^{3/2})' = 0.$$

If the ion temperatures are equal, $T_i = T_Z$. Since $Z > 1$, the impurity density profile thus becomes very peaked,

$$\frac{n_Z(r)}{n_Z(0)} = \left(\frac{n_i(r)}{n_i(0)}\right)^Z\left(\frac{T_i(0)}{T_i(r)}\right)^{Z/2-1}.$$

For instance, if the temperature profiles are flat, $T_i' = T_Z' = 0$, and the main ion density at the edge is one tenth of that in the core, an enormous core accumulation of impurities results, with a core density 10^Z times larger than the edge density (Taylor, 1961).

Because of charge neutrality, $\Sigma n_a e_a = 0$, a few impurity ions n_Z (with $Z \gg 1$) can displace many hydrogen ions n_i at fixed electron density n_e. If alpha particles are also present in the plasma,

$$n_e = n_i + Z n_Z + 2n_\alpha,$$

the fusion reaction rate becomes

$$P_f = \frac{n_i^2}{4} \langle \sigma_f v \rangle E_\alpha = \frac{n_e^2}{4} \left(1 - \frac{Z n_Z}{n_e} - \frac{2n_\alpha}{n_e} \right)^2 \langle \sigma_f v \rangle E_\alpha,$$

where $\langle \sigma_f v \rangle$ is the fusion reaction cross section. This shows a deleterious quadratic dilution factor caused by impurity and alpha particle accumulation in the core. In quiescent (fluctuation-free) plasmas, central accumulation following near (neo-)classical predictions has been observed.

5.2 The influence of viscosity on ambipolarity

In the previous section, the terms related to viscosity and inertia were neglected in the radial flow (5.2). When they are retained, the transport is no longer automatically ambipolar, and the radial current density (5.3) becomes

$$j_r = \sum_a e_a n_a V_{ar} = -\frac{1}{B} \sum_a \left[(\nabla \cdot \boldsymbol{\pi}_a)_\theta + m_a n_a \frac{dV_{a\theta}}{dt} \right]. \tag{5.11}$$

In Chapter 3 we have already remarked that Braginskii's expression for the viscosity tensor $\boldsymbol{\pi}$ is only accurate when the flow velocity is of the order of the thermal speed, $V = O(v_T)$. In the present case, this is not true. The velocity (5.1) is much smaller than thermal, $V_\theta = O(\delta v_T)$. To keep the presentation simple, we shall nevertheless use Braginskii's result, thus sacrificing accuracy for transparency.

In the Braginskii viscosity tensor (4.50)–(4.55), we identify the coordinates as $x = r$, $y = \theta$, $z = z$. Since the only non-vanishing derivatives are in the radial (x-) direction because of azimuthal and axial symmetry, we conclude that the only significant components of the rate-of-strain tensor W are

$$W_{r\theta} = W_{\theta r} = \frac{\partial V_\theta}{\partial r}$$

because $V_r \ll V_\theta$. The viscosity tensor thus becomes

$$\pi = \begin{pmatrix} -\eta_3 W_{r\theta} & -\eta_1 W_{r\theta} & 0 \\ -\eta_1 W_{r\theta} & \eta_3 W_{r\theta} & -\eta_4 W_{r\theta} \\ 0 & -\eta_4 W_{r\theta} & 0 \end{pmatrix},$$

so that the poloidal viscous force is

$$(\nabla \cdot \pi)_\theta = -\frac{1}{r}\frac{\partial}{\partial r} r\eta_1 \frac{\partial V_\theta}{\partial r}, \tag{5.12}$$

where we have recalled the expression for divergence in cylindrical coordinates. This force is larger for ions than for electrons, and it is small compared with friction, which justifies its neglect in the previous section. Indeed, its approximate magnitude is

$$(\nabla \cdot \pi_i)_\theta \sim \frac{n_i T_i}{a^2 \Omega_i^2 \tau_{ii}} V_{i\theta} \sim \left(\frac{m_i}{m_e}\right)^{1/2} \left(\frac{\rho_i}{a}\right)^2 R_{i\theta},$$

which is smaller than the friction $R_{e\theta} = -R_{e\theta} \sim m_e n_e V_{i\theta}/\tau_{ei} \sim (m_e/m_i)^{1/2}$ $m_i n_i V_{i\theta}/\tau_{ii}$ appearing, e.g., in (5.2) if the Larmor radius $\rho_i = v_{Ti}/\Omega_i$ is sufficiently smaller than the radial scale length a. The current (5.11) is, in other words, small in comparison with the lowest-order, ambipolar flux of particles in the radial direction.

The existence of a radial current, however small, is important since it is linked to a radial electric field by Ampère's law

$$\nabla \times \mathbf{B} = \mu_0 \mathbf{j} + \mu_0 \epsilon_0 \frac{\partial \mathbf{E}}{\partial t}.$$

In the radial component of this relation, the left-hand side vanishes since $\partial/\partial\theta = \partial/\partial z = 0$. Thus combining (5.11) and (5.12) gives

$$\frac{\partial E_r}{\partial t} = -\frac{j_r}{\epsilon_0} = \frac{1}{\epsilon_0 B}\left(-\frac{1}{r}\frac{\partial}{\partial r} r\eta_1^i \frac{\partial V_{i\theta}}{\partial r} + m_i n_i \frac{dV_{i\theta}}{dt}\right),$$

where the convective derivative is

$$\frac{dV_\theta}{dt} = \frac{\partial V_\theta}{\partial t} + (\mathbf{V} \cdot \nabla \mathbf{V})_\theta = \frac{\partial V_\theta}{\partial t} + \frac{V_r}{r}\frac{\partial(rV_\theta)}{\partial r}$$

in cylindrical coordinates. Now recall the expression (5.1) for the poloidal flow, which we write as

$$V_{i\theta} = -\frac{E_r - E_{r0}}{B},$$

with

$$E_{r0} \equiv \frac{p_i'(r)}{n_i e_i}.$$

We then obtain

$$\frac{d(E_r - E_{r0})}{dt} + \frac{v_A^2}{c^2}\frac{dE_{r0}}{dt} = \frac{1}{m_i n_i r}\frac{\partial}{\partial r}r\eta_1\frac{\partial(E_r - E_{r0})}{\partial r},$$

where

$$\frac{\epsilon_0 B^2}{m_i n_i} = \frac{\Omega_i^2}{\omega_{pi}^2} = \frac{v_A^2}{c^2} \ll 1,$$

in a typical fusion plasma. Here $\omega_{pi} \equiv (n_i e^2/m_i\epsilon_0)^{1/2}$ is the ion plasma frequency, $c = (\epsilon_0\mu_0)^{-1/2}$ the speed of light, and $v_A \equiv B/\sqrt{\mu_0/m_i n_i}$ is the Alfvén velocity. Thus, finally, we obtain

$$\frac{d(E_r - E_{r0})}{dt} = \frac{1}{m_i n_i r}\frac{\partial}{\partial r}r\eta_1\frac{\partial(E_r - E_{r0})}{\partial r},$$

which is a diffusion equation for the radial electric field, implying that the latter relaxes on the viscous time scale

$$\frac{1}{\tau_{\text{visc}}} \sim \frac{\eta_1^i}{a^2} \sim \frac{\rho_i^2}{a^2}v_{ii},$$

after which the radial current (5.11) vanishes. Not surprisingly, this time scale, which dictates how quickly poloidal momentum diffuses radially, is comparable to the time scale of radial transport of ion energy and axial momentum, as we shall see in the next section. Strictly speaking, this implies that the temperature cannot be regarded as constant during the evolution of the radial electric field, so that E_{r0} also changes in time. In practice, the viscous time scale may be so long that the electric field does not have time to relax to its asymptotic value during a discharge.

In summary, we have found that the radial ion flux can be written

$$\Gamma_i = \Gamma_i^{\text{friction}} + \Gamma_i^{\text{viscous}},$$

where the first term is automatically ambipolar and independent of E_r, and the second term is smaller by a factor of $(m_i/m_e)^{1/2}(\rho_i/a)^2$ and dies out on the slow viscous time scale.

5.3 Transport of momentum and heat

For the sake of completeness, we conclude this chapter with a brief discussion of momentum and heat diffusion in a cylindrical plasma. They are both dominated by the ions because of their large mass and gyroradius.

Momentum diffusion

Suppose there is a substantial axial flow of plasma, i.e., a flow of the order of the sound speed, $V_z \sim v_T$ along the cylinder. Recalling that the velocities in the other directions are relatively small,

$$V_\theta \sim \frac{\rho_i}{a} v_T \ll v_T,$$

$$V_r \sim \frac{V_\theta}{\Omega \tau} \ll V_\theta,$$

from Eqs. (5.1) and (5.2), we conclude that the only important entries in the rate-of-strain tensor are $W_{rz} = W_{zr} = \partial V_z / \partial r$. The viscosity tensor is then

$$\pi = \begin{pmatrix} 0 & 0 & -\eta_2 W_{rz} \\ 0 & 0 & \eta_4 W_{rz} \\ -\eta_2 W_{rz} & \eta_4 W_{rz} & 0 \end{pmatrix},$$

and gives rise to the parallel viscous force

$$(\nabla \cdot \pi_i)_z = -\frac{1}{r} \frac{\partial}{\partial r} r \eta_2^i \frac{\partial V_z}{\partial r},$$

acting on the ions. The viscosity coefficient η_2^i (4.58) is of the order

$$\eta_2^i \sim \rho_i^2 m_i n_i / \tau_{ii}.$$

The parallel component of the ion momentum equation,

$$m_i n_i \frac{\partial V_{iz}}{\partial t} = \frac{1}{r} \frac{\partial}{\partial r} r \eta_2^i \frac{\partial V_{iz}}{\partial r},$$

shows a diffusion of parallel momentum in the radial direction, with a decay rate

$$\frac{1}{\tau_{\text{mom}}^{\text{class}}} \sim \frac{\rho_i^2}{a^2 \tau_{ii}},$$

typical of classical transport. This is quite long for typical tokamak parameters, and the observed momentum decay time is much shorter, $\tau_{\text{mom}}^{\text{exp}} \sim 10^{-2} \tau_{\text{mom}}^{\text{class}}$. In other words, classical transport is much too slow to account for the observed viscosity in tokamaks.

Heat diffusion

The classical transport of heat for ions is simply given by Braginskii's expression (4.45), where the first two terms vanish because of axial and azimuthal symmetry,

$$q_r^i = -\kappa_\perp^i \frac{\partial T_i}{\partial r}.$$

The diffusion coefficient scales in the usual way for classical transport,

$$\kappa_\perp^i \sim \frac{\rho_i^2}{a^2 \tau_{ii}},$$

and is thus larger than the corresponding one for electrons. The electron heat flux (4.40) consists of two pieces, $\mathbf{q}_u = (4.40)$ driven by the poloidal flow (5.1), and $\mathbf{q}_T = (4.41)$ driven by the temperature gradient,

$$q_r^e = -\frac{3 n_e T_e}{2 \Omega_e \tau_e}(V_{e\theta} - V_{i\theta}) - \kappa_\perp^e \frac{\partial T_e}{\partial r} = -\kappa_\perp^e \frac{\partial T_e}{\partial r} + \frac{3 T_e}{2 m_e \Omega_e^2 \tau_e} \frac{\partial(p_e + p_i)}{\partial r}.$$

(5.13)

When comparing this expression with the corresponding particle flux (5.6), one notes Onsager symmetry. The particle flux has a term proportional to the temperature gradient, and the heat flux has a term proportional to the pressure gradient. The corresponding transport coefficients are equal, which becomes apparent when the transport is displayed in matrix form,

$$\begin{pmatrix} \Gamma^e/n_e \\ q^e/n_e T_e \end{pmatrix} = -\frac{T_e}{m_e \Omega_e^2 \tau_e} \begin{pmatrix} 1 & -3/2 \\ -3/2 & 4.66 \end{pmatrix} \begin{pmatrix} (p_e' + p_i')/p_e \\ T_e'/T_e \end{pmatrix}.$$

6

Particle motion

6.1 Equations of motion

We begin this chapter by reviewing the analytical mechanics of charged particle motion. The Lagrange function of a particle with mass m and charge Ze in an electromagnetic field

$$\mathbf{B} = \nabla \times \mathbf{A}, \quad \mathbf{E} = -\nabla\Phi - \partial\mathbf{A}/\partial t,$$

is

$$L(\mathbf{r}, \dot{\mathbf{r}}, t) = \frac{m|\dot{\mathbf{r}}|^2}{2} + Ze\mathbf{A}(\mathbf{r}, t) \cdot \dot{\mathbf{r}} - Ze\Phi(\mathbf{r}, t). \tag{6.1}$$

It is a function of the position $\mathbf{r}(t)$, the velocity $\dot{\mathbf{r}}(t)$, and time t. The particle moves in such a way as to minimize the action integral

$$S[\mathbf{r}(t)] \equiv \int_{t_0}^{t_1} L(\mathbf{r}, \dot{\mathbf{r}}, t) \, dt,$$

under the constraints $\delta\mathbf{r}(t_0) = \delta\mathbf{r}(t_1) = 0$. This means that out of all conceivable paths $\mathbf{r}(t)$, $t_0 < t < t_1$, between the points $\mathbf{r}(t_0)$ and $\mathbf{r}(t_1)$, which are held fixed, the particle selects the trajectory that minimizes $S[\mathbf{r}(t)]$. This implies that the first-order variation of S vanishes, $\delta S = 0$. The variation is, explicitly,

$$\delta S = \int_{t_0}^{t_1} \left(\frac{\partial L}{\partial \dot{\mathbf{r}}} \cdot \delta\dot{\mathbf{r}} + \frac{\partial L}{\partial \mathbf{r}} \cdot \delta\mathbf{r} \right) dt = [\mathbf{p} \cdot \delta\mathbf{r}]_{t_0}^{t_1} - \int_{t_0}^{t_1} \left(\frac{d\mathbf{p}}{dt} - \frac{\partial L}{\partial \mathbf{r}} \right) \cdot \delta\mathbf{r} \, dt,$$

where we have integrated by parts, and introduced the so-called generalized momentum $\mathbf{p} \equiv \partial L/\partial \dot{\mathbf{r}}$. It follows that if the end points are held fixed, $\delta\mathbf{r}(t_0) = \delta\mathbf{r}(t_1) = 0$, the Lagrangian satisfies the Euler–Lagrange equation

$$\frac{d}{dt}\left(\frac{\partial L}{\partial \dot{\mathbf{r}}} \right) = \frac{\partial L}{\partial \mathbf{r}}. \tag{6.2}$$

By calculating the next-order variation, $\delta^2 S$, it is possible to show that the action integral has a minimum (rather than a maximum or a saddle point) for the path $\mathbf{r}(t)$ taken by the particle, provided that the time interval $t_1 - t_0$ is sufficiently short.

Substituting the Lagrangian (6.1) in (6.2) gives

$$\frac{d}{dt}(m\dot{\mathbf{r}} + Ze\mathbf{A}) = Ze\,\nabla(\mathbf{A}\cdot\dot{\mathbf{r}} - \Phi). \tag{6.3}$$

Since the left-hand side of this equation is $m\ddot{\mathbf{r}} + Ze(\partial\mathbf{A}/\partial t + \dot{\mathbf{r}}\cdot\nabla\mathbf{A})$, and on the right-hand side $\nabla(\mathbf{A}\cdot\dot{\mathbf{r}}) = \dot{\mathbf{r}}\times(\nabla\times\mathbf{A}) + (\dot{\mathbf{r}}\cdot\nabla)\mathbf{A}$, the resulting equation of motion becomes

$$m\ddot{\mathbf{r}} = Ze(\mathbf{E} + \dot{\mathbf{r}}\times\mathbf{B}), \tag{6.4}$$

where the right-hand side is the Lorentz force.

The particle motion can also be given a Hamiltonian formulation. In general, given a Lagrangian $L(\mathbf{r},\dot{\mathbf{r}},t)$, the Hamiltonian is defined by

$$H(\mathbf{p},\mathbf{r},t) \equiv \mathbf{p}\cdot\dot{\mathbf{r}} - L(\mathbf{r},\dot{\mathbf{r}},t), \tag{6.5}$$

where $\dot{\mathbf{r}}$ is eliminated in favour of the generalized momentum $\mathbf{p}$. The equations of motion then become

$$\dot{\mathbf{r}} = \frac{\partial H}{\partial\mathbf{p}}, \quad \dot{\mathbf{p}} = -\frac{\partial H}{\partial\mathbf{r}}. \tag{6.6}$$

In the present situation, with the Lagrangian (6.1), the momentum is

$$\mathbf{p} \equiv \partial L/\partial\dot{\mathbf{r}} = m\dot{\mathbf{r}} + Ze\mathbf{A}(\mathbf{r},t), \tag{6.7}$$

the Hamiltonian becomes

$$H(\mathbf{p},\mathbf{r},t) \equiv \frac{m|\dot{\mathbf{r}}|^2}{2} + Ze\Phi = \frac{1}{2m}(\mathbf{p} - Ze\mathbf{A})^2 + Ze\Phi,$$

and the second of Hamilton's equations (6.6) is identical to Eq. (6.3). Note that the canonical momentum $\mathbf{p}$ is not simply equal to $m\mathbf{v}$, but also has a magnetic term $Ze\mathbf{A}$.

6.2 Nearly periodic motion

In a uniform, constant magnetic field, a charged particle gyrates around a magnetic field line while moving freely along it. The 'Larmor' radius of the gyration is $\rho = v_\perp/\Omega$, where $\Omega \equiv ZeB/m$ is the gyrofrequency. The guiding centre (cf. Section 1.2),

$$\mathbf{R} \equiv \mathbf{r} - \frac{1}{\Omega}\mathbf{b}\times\mathbf{v},$$

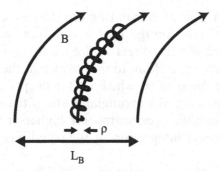

Fig. 6.1. Gyromotion in a slowly varying magnetic field.

where $\mathbf{b} \equiv \mathbf{B}/B$, remains on the same field line, and moves only in the parallel direction

$$\dot{\mathbf{R}} = \mathbf{b} v_{\parallel}.$$

Now suppose that the magnetic field is non-uniform, but varies *slowly* in space and time, i.e., the characteristic length and times scales of the magnetic field,

$$L_B \sim |\nabla \ln \mathbf{B}|^{-1}, \quad \tau_B \sim |\partial \ln \mathbf{B}/\partial t|^{-1},$$

are long compared with the gyroradius and the gyrofrequency, see Fig. 6.1

$$\frac{\rho}{L_B} \ll 1, \quad \Omega \tau_B \gg 1.$$

We can formalize this by introducing a small parameter

$$\delta \equiv \max\left(\frac{\rho}{L_B}, \frac{1}{\Omega \tau_B}\right) \ll 1. \tag{6.8}$$

Then the particle orbit does not quite close on itself after one gyration around the magnetic field, and the guiding centre drifts slowly across the field. Since the field is approximately constant, the deviation from perfect Larmor gyration is small, and its effect is noticeable only after many gyrations. By then the particle may have drifted far from its original position. We shall show that, in this process, the magnetic moment

$$\mu \equiv \frac{m v_{\perp}^2}{2B} \tag{6.9}$$

remains nearly constant. This important result follows from a general property of Hamiltonian systems; μ is an 'adiabatic invariant' of guiding-centre motion.

The most common way of calculating the drift motion of the guiding centre is to take the gyroaverage of the equation of motion (6.3). Although this procedure is very direct and easy to understand, it has the disadvantage of ignoring the Hamiltonian nature of the problem. We shall instead follow another procedure, which strikes deeper at the problem and preserves the Hamiltonian and Lagrangian form of the equations. This has two advantages: it simplifies the search for higher-order invariants, and it can facilitate numerical integration of the equations of motion (White, 1989).

The method we adopt was developed by Whitham (1974) for the study of nonlinear waves, and consists in taking an average of the Lagrangian (6.1) rather than the equation of motion. In order to illustrate the method, we first consider a simple example of a particle in a two-dimensional potential well of the form $V(x, y, t) = \omega^2(y, t)x^2/2$. We assume that $\partial\omega/\partial y$ and $\partial\omega/\partial t$ are small, so that the shape of the potential well changes slowly with time t and with the coordinate y along the well. The Lagrangian is equal to the difference between kinetic and potential energies, which is

$$L = \frac{\dot{x}^2 + \dot{y}^2}{2} - \frac{\omega^2(y, t)x^2}{2} \tag{6.10}$$

for a particle of unit mass. If ω were completely constant the particle would execute sinusoidal oscillations in the x-direction. When ω varies slowly, we may thus expect that $x(t)$ can be written as

$$x(t) = A(t) \sin \theta(t),$$

where the amplitude $A(t)$ varies slowly with time and the phase θ varies more rapidly. Substituting this expression into the Lagrangian (6.10) and neglecting dA/dt gives

$$L = \frac{A^2}{2}(\dot{\theta}^2 \cos^2 \theta - \omega^2 \sin^2 \theta) + \frac{\dot{y}^2}{2}.$$

The variational principle described in the previous section states that the time integral of the Lagrangian assumes a minimum for the path taken by the system. Since the phase θ varies nearly linearly with time, to a first approximation a similar statement must hold for the average of L over θ,

$$\bar{L} \equiv \frac{1}{2\pi} \int_0^{2\pi} L \, d\theta = \frac{A^2}{4}(\dot{\theta}^2 - \omega^2) + \frac{\dot{y}^2}{2}.$$

In other words, we may use $\bar{L}(A, \dot{A}, \theta, \dot{\theta}, y, \dot{y}, t)$ as an approximate Lagrangian for motion on time scales longer than the oscillation period in the x-direction, which is $2\pi/\omega$. The Euler–Lagrange equation with respect to the variable A,

$$\frac{d}{dt}\left(\frac{\partial \bar{L}}{\partial \dot{A}}\right) = \frac{\partial \bar{L}}{\partial A},$$

then implies

$$\frac{\partial \bar{L}}{\partial A} = \frac{A^2}{2}(\dot{\theta}^2 - \omega^2) = 0,$$

which means that $\dot{\theta} = \pm\omega$. More interestingly, the corresponding equation for the angle θ,

$$\frac{d}{dt}\left(\frac{\partial \bar{L}}{\partial \dot{\theta}}\right) = \frac{\partial \bar{L}}{\partial \theta},$$

shows that

$$\mu = \frac{A^2\omega}{2}$$

is a constant of motion. Although A and ω individually vary with time, this product remains approximately constant. This is the adiabatic invariant of our system and corresponds to the magnetic moment associated with guiding centres.

In the case when ω depends only on y but not on time, the adiabatic invariance of μ implies that the particle may be trapped in the y-direction of the potential well. Since $\partial \bar{L}/\partial y = -(A^2\omega/2)\partial\omega/\partial y$, the equation of motion for y is

$$\ddot{y} = -\mu\frac{\partial\omega}{\partial y},$$

which can be integrated to yield

$$\frac{d}{dt}\left(\frac{\dot{y}^2}{2} + \mu\omega(y)\right) = 0,$$

if $\partial\omega(y,t)/\partial t = 0$. This last equation implies that the particle cannot move into a region where $\omega(y)$ is very large. Indeed, the inequality

$$\mu\omega(y) \le \left(\frac{\dot{y}^2}{2} + \mu\omega(y)\right)_{t=0}$$

must be satisfied at all times. This is analogous to trapping of charged particles in regions of weak magnetic field.

6.3 Guiding-centre motion

Lagrangian

Guided by this simple example, we are now ready to apply the same procedure to the more complicated problem of guiding-centre motion. We expect that in a slowly varying magnetic field the motion of a charged particle should consist of rapid gyromotion superimposed on a slower drift, so that we may write

$$\mathbf{r} = \mathbf{R} + \boldsymbol{\rho},$$

where the gyromotion is described by the vector

$$\boldsymbol{\rho} = \rho(\hat{\mathbf{x}} \cos \vartheta + \hat{\mathbf{y}} \sin \vartheta).$$

Here $\rho = v_\perp/\Omega$ is the gyroradius, $\hat{\mathbf{x}}$ and $\hat{\mathbf{y}}$ are unit vectors in the directions perpendicular to the magnetic field, which is in the local z-direction ($\hat{\mathbf{z}} = \hat{\mathbf{x}} \times \hat{\mathbf{y}}$), and ϑ denotes the gyrophase, which is analogous to the phase variable θ in the previous section.

In the Lagrangian (6.1), the gyroaverage of the first and third terms is

$$\frac{m\overline{|\dot{\mathbf{r}}|^2}}{2} - Ze\overline{\Phi(\mathbf{r}, t)} \simeq \frac{m}{2}[(\mathbf{b} \cdot \dot{\mathbf{R}})^2 + (\rho\dot{\vartheta})^2] - Ze\Phi(\mathbf{R}, t), \qquad (6.11)$$

to the lowest order in ρ/L_B. For the second term in (6.1) we need to be more careful. This term is larger than the others by a factor of δ^{-1} since the magnetic field is assumed to be strong, see Eq. (6.8). Thus, when calculating the gyroaverage of this term we need to retain higher-order Larmor radius corrections, in order to achieve the same precision as in the other two terms. We therefore expand around $\mathbf{R}$ in the Larmor radius

$$\mathbf{A} \cdot \dot{\mathbf{r}} \simeq \left(\mathbf{A}(\mathbf{R}) + \frac{\partial \mathbf{A}}{\partial x}\rho \cos \vartheta + \frac{\partial \mathbf{A}}{\partial y}\rho \sin \vartheta\right) \cdot \left[\dot{\mathbf{R}} + \rho\dot{\vartheta}(-\hat{\mathbf{x}} \sin \vartheta + \hat{\mathbf{y}} \cos \vartheta)\right],$$

so that the average becomes

$$\overline{\mathbf{A} \cdot \dot{\mathbf{r}}} = \mathbf{A} \cdot \dot{\mathbf{R}} + \frac{\rho^2\dot{\vartheta}}{2}\left(\frac{\partial A_y}{\partial x} - \frac{\partial A_x}{\partial y}\right) = \mathbf{A} \cdot \dot{\mathbf{R}} + \frac{B\rho^2\dot{\vartheta}}{2}, \qquad (6.12)$$

where all quantities are evaluated at the guiding-centre position $\mathbf{R}$ and we have used $\nabla \times \mathbf{A} = \mathbf{B}$. Collecting the results (6.11) and (6.12) gives the gyroaveraged Lagrangian

$$\bar{L}(\mathbf{R}, \rho, \vartheta, \dot{\mathbf{R}}, \dot{\rho}, \dot{\vartheta}, t) = \frac{m(\mathbf{b}(\mathbf{R}) \cdot \dot{\mathbf{R}})^2}{2} + \frac{m\rho^2\dot{\vartheta}^2}{2} + Ze\mathbf{A}(\mathbf{R}) \cdot \dot{\mathbf{R}}$$

$$+ \frac{Ze\rho^2\dot{\vartheta}}{2}B(\mathbf{R}) - Ze\Phi(\mathbf{R}). \qquad (6.13)$$

We now inspect the Euler–Lagrange equations associated with this Lagrangian. The equation for ρ is $\partial\bar{L}/\partial\rho = 0$, which implies

$$\dot{\vartheta} = -ZeB/m = -\Omega,$$

which is an expected result. The minus sign implies that a positively charged particle gyrates in the direction of a left-handed screw along **B**. The equation for the gyrophase ϑ is

$$\frac{d}{dt}\left(\frac{\partial \bar{L}}{\partial \dot{\vartheta}}\right) = \frac{d}{dt}\left(m\rho^2\dot{\vartheta} + \frac{ZeB\rho^2}{2}\right) = 0,$$

which implies that $\rho^2\Omega$ is constant. Since $\rho = v_\perp/\Omega$ this means that the magnetic moment (6.9) is constant, and we have thus proved the adiabatic invariance of $\mu = mv_\perp^2/2B$ to lowest order. Just as in the simple example considered in the previous section, the constancy of μ implies particle trapping in stationary fields. For instance, if there is no electric field, $\mathbf{E} = 0$, and $\partial \mathbf{B}/\partial t = 0$, then the energy $H = mv^2/2$ is constant, and a particle with magnetic moment μ cannot enter regions where $\mu B > H$.

Finally we have the Euler–Lagrange equations for the guiding-centre coordinates **R**,

$$\frac{d}{dt}\left(\frac{\partial \bar{L}}{\partial \dot{\mathbf{R}}}\right) = \frac{\partial \bar{L}}{\partial \mathbf{R}}, \tag{6.14}$$

which are going to give us the expression for the drift velocity. At this point we note that the second term in the Lagrangian (6.13) does not depend on either **R** or $\dot{\mathbf{R}}$, and can therefore be dropped from the guiding-centre Lagrangian, which then becomes (Taylor, 1964)

$$\boxed{\bar{L}(\mathbf{R}, \dot{\mathbf{R}}, t) = \frac{m(\mathbf{b}\cdot\dot{\mathbf{R}})^2}{2} + Ze\mathbf{A}\cdot\dot{\mathbf{R}} - \mu B - Ze\Phi,} \tag{6.15}$$

where the functions **b**, B, **A**, and Φ are taken at the guiding-centre position **R**.

Each term in the Lagrangian (6.15) has a simple and intuitive physical interpretation. In general, the Lagrangian is equal to the difference between the kinetic and potential energies. The kinetic energy of the guiding centre is $m(\mathbf{b}\cdot\dot{\mathbf{R}})^2/2$ since the drift across the magnetic field is slow, so that $\dot{\mathbf{R}} \simeq \mathbf{v}_\parallel$. The potential energy is seen to consist not only of the usual magnetic term $-Ze\mathbf{A}\cdot\dot{\mathbf{R}}$ and the electrostatic energy $Ze\Phi$, but also of a another magnetic piece, $-\mu B$. This is the energy of a magnetic dipole μ antiparallel to a magnetic field B. Note that a gyrating particle carries a current $i = Ze\Omega/2\pi$, which produces a magnetic moment $\mu = \pi\rho^2 i$ with a field opposite to **B**.

Equations of motion

It is straightforward to derive the guiding-centre equations of motion from Eqs. (6.14) and (6.15). The left-hand side of (6.14) is equal to

$$\frac{d}{dt}\left(\frac{\partial \bar{L}}{\partial \dot{\mathbf{R}}}\right) = \left(\frac{\partial}{\partial t} + \dot{\mathbf{R}} \cdot \nabla\right)\frac{\partial \bar{L}}{\partial \dot{\mathbf{R}}}$$

$$= m\dot{v}_\parallel \mathbf{b} + mv_\parallel \left(\frac{\partial}{\partial t} + \dot{\mathbf{R}} \cdot \nabla\right)\mathbf{b} + Ze\left(\frac{\partial}{\partial t} + \dot{\mathbf{R}} \cdot \nabla\right)\mathbf{A},$$

with $v_\parallel = \mathbf{b} \cdot \dot{\mathbf{R}}$, and by using the vector rule for the divergence of a scalar product,

$$\nabla(\mathbf{C} \cdot \mathbf{X}) = (\mathbf{C} \cdot \nabla)\mathbf{X} + \mathbf{C} \times (\nabla \times \mathbf{X}),$$

where $\mathbf{C}$ is constant and $\mathbf{X}$ varies, we find that the right-hand side of (6.14) is

$$\frac{\partial \bar{L}}{\partial \mathbf{R}} = mv_\parallel[(\dot{\mathbf{R}} \cdot \nabla)\mathbf{b} + \dot{\mathbf{R}} \times (\nabla \times \mathbf{b})]$$

$$+ Ze[(\dot{\mathbf{R}} \cdot \nabla)\mathbf{A} + \dot{\mathbf{R}} \times \mathbf{B}] - \mu\nabla B - Ze\nabla\Phi.$$

The Euler–Lagrange equation (6.14) thus becomes

$$m\dot{v}_\parallel \mathbf{b} = Ze(\mathbf{E} + \dot{\mathbf{R}} \times \mathbf{B}) - \mu\nabla B + mv_\parallel \dot{\mathbf{R}} \times (\nabla \times \mathbf{b}) - mv_\parallel\frac{\partial \mathbf{b}}{\partial t}. \tag{6.16}$$

In this equation, the last term on the right is small compared with the electric field since

$$mv_\parallel\frac{\partial \mathbf{b}}{\partial t} \sim \frac{mv_\parallel}{B}\frac{\partial \mathbf{B}}{\partial t} = -\frac{Zev_\parallel}{\Omega}\nabla \times \mathbf{E} \sim ZeE\frac{\rho}{L_B}.$$

In the term preceding it, we may approximate $\dot{\mathbf{R}}$ by $v_\parallel\mathbf{b}$ and introduce the magnetic field curvature,

$$\boldsymbol{\kappa} \equiv (\mathbf{b} \cdot \nabla)\mathbf{b} = -\mathbf{b} \times (\nabla \times \mathbf{b}), \tag{6.17}$$

(which follows from $\nabla(\mathbf{b} \cdot \mathbf{b}) = 0$) to write

$$m\dot{v}_\parallel \mathbf{b} = Ze(\mathbf{E} + \dot{\mathbf{R}} \times \mathbf{B}) - \mu\nabla B - mv_\parallel^2\boldsymbol{\kappa}. \tag{6.18}$$

This is the equation of motion for the guiding centre. Its parallel component is

$$\boxed{m\dot{v}_\parallel = ZeE_\parallel - \mu\nabla_\parallel B,} \tag{6.19}$$

where the last term is the so-called mirror force. It reflects particles trying to enter regions of strong magnetic field. It is responsible for confinement in magnetic mirror machines, and for the phenomenon of particle trapping in tokamaks and in the Van Allen belts of the Earth's magnetosphere.

Evaluating $\mathbf{b}\times$(6.18) gives the cross-field drift velocity

$$\mathbf{v}_d \equiv \dot{\mathbf{R}}_\perp = \frac{\mathbf{E} \times \mathbf{B}}{B^2} + \frac{v_\perp^2}{2\Omega}\mathbf{b} \times \nabla \ln B + \frac{v_\parallel^2}{\Omega}\mathbf{b} \times \boldsymbol{\kappa}. \tag{6.20}$$

This is a famous result, first discovered by Alfvén (1940, Nobel prize in 1970), according to which the guiding-centre drift across a strong magnetic field consists of three terms, usually called the $\mathbf{E} \times \mathbf{B}$ drift, the grad-B drift, and the curvature drift.

It should be noted that the drift velocity (6.20) is only accurate for weak electric fields since we have assumed that the orbit nearly closes on itself after one Larmor period. This is true only if the total $\mathbf{E} \times \mathbf{B}$ drift over one Larmor period is smaller than the Larmor radius, i.e., $E/B \ll \rho/\Omega$, or

$$E/B \ll v_\perp. \tag{6.21}$$

If this condition is not satisfied, the $\mathbf{E} \times \mathbf{B}$ drift dominates over the magnetic drifts, and there are additional terms in the drift velocity that should be added to the latter.

To understand the curvature $\boldsymbol{\kappa}$, we use the steady-state relation $\nabla \times \mathbf{B} = \mu_0 \mathbf{j}$ between the magnetic field and the current density $\mathbf{j}$, to write

$$\boldsymbol{\kappa} = -\mathbf{b} \times \left(\nabla \times \frac{\mathbf{B}}{B}\right) = \frac{\mu_0 \mathbf{j} \times \mathbf{B}}{B^2} + \frac{\nabla_\perp B}{B}. \tag{6.22}$$

In a plasma equilibrium with small flows, $\mathbf{j} \times \mathbf{B} = \nabla p$, and we have

$$\boldsymbol{\kappa} = \frac{\mu_0}{B^2}\nabla_\perp \left(p + \frac{B^2}{2\mu_0}\right).$$

The curvature is thus proportional to the perpendicular gradient of the total (plasma+magnetic) pressure. In equilibrium, the force that this pressure gradient represents is balanced by the magnetic field tension associated with the curvature $\boldsymbol{\kappa}$. In particular, in a low $\beta = 2\mu_0 p/B^2$ plasma,

$$\boldsymbol{\kappa} \simeq \frac{\nabla_\perp B}{B}. \tag{6.23}$$

Alternative forms of the guiding-centre equations

The guiding-centre equations of motion can be re-written in an elegant way by introducing the modified vector potential

$$\mathbf{A}^* = \mathbf{A} + \frac{mv_\parallel}{Ze}\mathbf{b},$$

and the corresponding modified fields

$$\mathbf{B}^* = \nabla \times \mathbf{A}^* = \mathbf{B} + \frac{mv_\|}{Ze}\nabla \times \mathbf{b},$$

$$\mathbf{E}^* = -\frac{\partial \mathbf{A}^*}{\partial t} - Ze\nabla\Phi = \mathbf{E} - \frac{mv_\|}{Ze}\frac{\partial \mathbf{b}}{\partial t},$$

where we have treated $v_\|$ as a constant in the differentiations. The difference between the actual magnetic and electric fields and these modified fields is of order $\delta = \rho/L_B \ll 1$. Note that the modified fields depend on the parallel velocity of the particle. Equation (6.16) can now be expressed as

$$m\dot{v}_\|\mathbf{b} = Ze(\mathbf{E}^* + \dot{\mathbf{R}} \times \mathbf{B}^*) - \mu\nabla B,$$

which looks like an equation of motion in the modified fields. Taking the vector product of this equation with $\mathbf{b}$ gives the guiding-centre velocity

$$\dot{\mathbf{R}} = \frac{1}{B_\|^*}\left(v_\|\mathbf{B}^* + \mathbf{E}^* \times \mathbf{b} + \frac{\mu\mathbf{b} \times \nabla B}{Ze}\right), \qquad (6.24)$$

where $B_\|^* = \mathbf{b} \cdot \mathbf{B}^*$.

Alfvén's result (6.20) is recovered from (6.24) if $B_\|^*$ is replaced by B and $\mathbf{E}^*$ by $\mathbf{E}$. While this approximation is perfectly admissible in the accuracy considered, it has the disadvantage of destroying the Hamiltonian nature of the drift equations. Equation (6.24) is Hamiltonian since it follows exactly from the Lagrangian (6.15), while (6.20) is not because of the approximation made following Eq. (6.16).

If the electric and magnetic fields are *time-independent*, so that the energy

$$H = \frac{mv_\|^2}{2} + \mu B + Ze\Phi$$

is conserved, it is possible to simplify the drift equations further. The parallel velocity may then be regarded as a function of energy, magnetic moment and position, so that if H and μ are held constant

$$v_\|\nabla v_\| = -(\mu\nabla B + Ze\nabla\Phi)/m.$$

Thus, if we define a new modified magnetic field by

$$\mathbf{B}^\times = \nabla \times \mathbf{A}^* = \mathbf{B}^* - \frac{m\mathbf{b} \times \nabla v_\|}{Ze} = \mathbf{B}^* + \frac{\mathbf{E} - (\mu/Ze)\nabla B}{v_\|} \times \mathbf{b},$$

which differs from $\mathbf{B}^*$ since $v_\|$ is no longer treated as a constant in the differentiation, we can write the guiding-centre velocity (6.24) in the remarkably simple form

$$\dot{\mathbf{R}} = \frac{v_\|\mathbf{B}^\times}{B_\|^\times}.$$

Another way of formulating this result is to express the cross-field drift as

$$\mathbf{v}_d = \frac{v_\parallel}{B_\parallel^\times} \nabla \times \left(\frac{v_\parallel \mathbf{B}}{\Omega} \right) \simeq v_\parallel \nabla \left(\frac{v_\parallel}{\Omega} \right) \times \mathbf{b} + \frac{\mu_0 v_\parallel^2}{\Omega B} \mathbf{j}, \qquad (6.25)$$

where, again, the curl is taken at constant magnetic moment μ and energy H.

6.4 Other adiabatic invariants

Adiabatic invariants in general

The motion of a charged particle in a strong magnetic field (in the sense of Eq. (6.8)) is an example of a Hamiltonian system which executes rapid oscillations superimposed on a slow drift. In such systems, which are quite common in physics, there always exists an adiabatic invariant. In describing such a system mathematically, suppose that we choose coordinates so that the rapid oscillations occur only in one coordinate, q, say, while the others, $\mathbf{x}$, evolve more slowly. The Lagrangian, $L(q, \dot{q}, \mathbf{x}, \dot{\mathbf{x}}, t)$, is thus assumed to depend weakly on $\mathbf{x}$, $\dot{\mathbf{x}}$ and t. If we, for convenience, lump this weak dependence into a single vector $\lambda(t)$, we are left with a Lagrangian $L(q, \dot{q}, \lambda)$ describing a one-dimensional dynamical system with a slow dependence on some parameters $\lambda(t)$. Most advanced textbooks on analytical mechanics (e.g., Landau and Lifshitz, 1976; Goldstein, 1980) show that the quantity

$$\oint p(H, q, \lambda) \, dq,$$

where $p = \partial L / \partial \dot{q}$, is then adiabatically invariant. The integral is taken one turn around the closed orbit. The proof is not difficult, but there is no need to repeat it here. Instead, we shall show how to use this result to obtain higher-order invariants for motion of charged particles in magnetic fields.

The second and third adiabatic invariants

The adiabatic invariance of the magnetic moment arises because of the fast gyration of the particle around magnetic field lines. Frequently, the motion of the *guiding centre* is also oscillatory. In particular, this happens if the particle is trapped in a magnetic well. Familiar examples are charged particles trapped in the magnetosphere, in a magnetic mirror machine, or on the outboard side of a flux surface in a tokamak. In these cases, the guiding centre moves along the magnetic field and is reflected by the mirror force $-\mu \nabla_\parallel B$ in (6.19) when entering a region of stronger magnetic

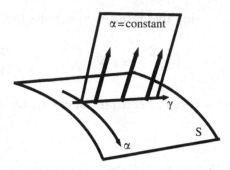

Fig. 6.2.   Construction of $\mathbf{B} = \nabla\alpha \times \nabla\beta$.

field. If the oscillatory motion along the field is more rapid than the drift across it, the guiding centre stays on approximately the same field line during each bounce period, and it may be expected that the motion has an adiabatic invariant on times scales longer than the bounce time. We may call this a higher-order or second adiabatic invariant.

To show that this is indeed the case, we first need a convenient representation of the magnetic field,

$$\mathbf{B} = \nabla\alpha \times \nabla\beta. \tag{6.26}$$

To arrive at this representation one chooses an arbitrary plane S nowhere parallel to $\mathbf{B}$ and introduces two arbitrary but well-behaved coordinates (α, γ) on S. The points on S with given α define a line on S, and the field lines of B that pass through this line define a surface, see Fig. 6.2. Let us extend the definition of $\alpha(\mathbf{r})$ to points outside S by requiring that α is constant on each such surface; we then do similarly for γ. This is simply to say that $\mathbf{B}$ is tangential to surfaces of constant α, and to surfaces of constant γ. Hence

$$\mathbf{B} \cdot \nabla\alpha = \mathbf{B} \cdot \nabla\gamma = 0,$$

and there exists some scalar λ such that

$$\mathbf{B} = \lambda\nabla\alpha \times \nabla\gamma.$$

Taking the divergence of this relation gives $\mathbf{B} \cdot \nabla\lambda = 0$, which implies that λ is expressible as a function of α and γ, i.e., $\lambda = \lambda(\alpha, \gamma)$. If we now define $\beta(\alpha, \gamma)$ through

$$\left(\frac{\partial\beta}{\partial\gamma}\right)_\alpha = \lambda(\alpha, \gamma),$$

then (6.26) follows immediately. Note that α and β are constant along $\mathbf{B}$, and can thus be regarded as field line labels, and that $\nabla \cdot \mathbf{B} = 0$ is automatically satisfied by (6.26).

It should be pointed out that this representation for the magnetic field may only hold locally. There is no guarantee that it is possible to find globally well-defined, single-valued functions α and β satisfying (6.26). For instance, as we shall see in the next chapter, the magnetic field in a tokamak or a stellarator can be expressed as

$$\mathbf{B} = \nabla(\varphi - q\theta) \times \nabla\psi, \tag{6.27}$$

which corresponds to $\alpha = \varphi - q\theta$ and $\beta = \psi$. Since the toroidal or poloidal angle, φ or θ, increases by 2π when going around the torus, the angle α is multivalued and is useful as a local, rather than global, coordinate.

Since $\mathbf{A} = \alpha\nabla\beta$ is a vector potential for $\mathbf{B}$, we can write

$$\mathbf{A} \cdot \dot{\mathbf{R}} = \alpha\dot{\mathbf{R}} \cdot \nabla\beta = \alpha\left(\dot{\beta} - \frac{\partial\beta}{\partial t}\right),$$

where we have allowed the magnetic field to vary in time. Similarly, if we let s denote the arc length along the magnetic field, so that $\partial\mathbf{R}/\partial s = \mathbf{b}$, we can write the parallel velocity as

$$\mathbf{b} \cdot \dot{\mathbf{R}} = \dot{s} + \mathbf{b} \cdot \left(\frac{\partial\mathbf{R}}{\partial\alpha}\dot{\alpha} + \frac{\partial\mathbf{R}}{\partial\beta}\dot{\beta} + \frac{\partial\mathbf{R}}{\partial t}\right).$$

Here, $\dot{s}$ dominates over the other terms on the right since the motion along the magnetic field is much faster than the drift across it (except near turning points where $\dot{s} \simeq 0$). We can now express the guiding-centre Lagrangian (6.15) in the coordinates (α, β, s) as

$$\bar{L}(\alpha, \beta, s, \dot{\alpha}, \dot{\beta}, \dot{s}, t) = \frac{m}{2}\left[\dot{s} + \mathbf{b} \cdot \left(\frac{\partial\mathbf{R}}{\partial\alpha}\dot{\alpha} + \frac{\partial\mathbf{R}}{\partial\beta}\dot{\beta} + \frac{\partial\mathbf{R}}{\partial t}\right)\right]^2$$

$$+ Ze\alpha\left(\dot{\beta} - \frac{\partial\beta}{\partial t}\right) - \mu B - Ze\Phi,$$

or approximately as

$$\bar{L} \simeq \frac{m\dot{s}^2}{2} + Ze\alpha\left(\dot{\beta} - \frac{\partial\beta}{\partial t}\right) - \mu B - Ze\Phi. \tag{6.28}$$

If the particle remains on approximately the same field line during one bounce period, then α and β are roughly constant on the time scale that $p_s = \partial\bar{L}/\partial\dot{s} = m\dot{s}$ varies, and the quantity

$$J \equiv \oint p_s ds = \oint mv_\parallel ds$$

is adiabatically invariant. It is usually called the second adiabatic invariant (μ being the first), or the longitudinal invariant. The integration is taken over one period (back and forth) of the bounce motion along the magnetic field line.

Thus, the magnetic moment μ is adiabatically invariant to perturbations that are slow on the cyclotron time scale, and the longitudinal invariant to perturbations slow on the bounce time scale. Now consider the still longer time scale on which the guiding centre drifts across field lines. Frequently, this motion is also periodic. For instance, this can be the case for particles in the magnetosphere drifting around the Earth. Their guiding centres travel along the field lines and are reflected in the regions of strong magnetic field near the poles. More slowly, magnetospheric ions drift westward (and electrons eastward) because of the grad-B and curvature drifts. (The magnetic field points north and ∇B towards Earth, so $\mathbf{B} \times \nabla B$ is west.) We may thus follow our chain of thought one step further, and take the *bounce-average*, denoted by a double overbar, of the guiding-centre Lagrangian (which we recall is the *gyroaverage* of the particle Lagrangian),

$$\overline{\overline{L}}(\alpha, \beta, \dot{\alpha}, \dot{\beta}, t; J, \mu) \simeq \frac{m\overline{\overline{\dot{s}^2}}}{2} + Ze\alpha \left(\overline{\overline{\dot{\beta}}} - \frac{\partial \overline{\overline{\beta}}}{\partial t} \right) - \mu \overline{\overline{B}} - Ze\overline{\overline{\Phi}}.$$

Since both gyromotion and bounce motion have been averaged out, the remaining motion is now the drift across the field, which is governed by $\overline{\overline{L}}$. Note that this Lagrangian does not depend on $\dot{\alpha}$, so that the momentum conjugate to α vanishes, $\partial_{\dot{\alpha}}\overline{\overline{L}}/\partial\dot{\alpha} = 0$, while that conjugate to β is $\partial_{\dot{\beta}}\overline{\overline{L}}/\partial\dot{\beta} = Ze\alpha$. The Hamiltonian for this system, $\overline{\overline{H}} = p_\alpha \dot{\alpha} + p_\beta \dot{\beta} - \overline{\overline{L}}$, therefore depends only on β, p_β, t, and the parameters μ and J. If the motion in β is periodic, and μ and J change slowly on the corresponding time scale, we are again faced with a Hamiltonian system executing rapid oscillations, and it follows that the quantity

$$\oint p_\beta d\beta = Ze \oint \alpha d\beta$$

is adiabatically invariant. This so-called third adiabatic invariant is apparently equal to the magnetic flux linked by the drift orbit (times Ze) since

$$\oint \alpha d\beta = \iint d\alpha \, d\beta = \iint \mathbf{B} \cdot d\mathbf{S}$$

where the double integral is taken over the area bounded by the orbit, the surface element is $dS = d\alpha \, d\beta / |\nabla\alpha \times \nabla\beta|$, and we have used (6.26) for the magnetic field.

Note that both the first and the third adiabatic invariants can be interpreted as magnetic fluxes linked by the orbit. These fluxes remain constant if slow perturbations are imposed, just like the magnetic flux through a loop of a perfectly conducting wire is constant by Lenz's law.

6.5 The drift kinetic equation

The kinetic equation for the distribution function of a particle species is

$$\frac{\partial f}{\partial t} + \nabla_{\mathbf{z}} \cdot (\dot{\mathbf{z}} f) = C(f), \tag{6.29}$$

where $\mathbf{z} \equiv (\mathbf{r}, \mathbf{p})$, and where $\nabla_{\mathbf{z}} = \partial/\partial \mathbf{z}$. The left-hand side of the equation is written in conservation form where $\dot{\mathbf{z}} f$ is the particle flux in six-dimensional phase space. An immediate consequence of Hamilton's equations (6.6) is that the phase-space flow velocity is divergence free since

$$\nabla_{\mathbf{z}} \cdot \dot{\mathbf{z}} = \frac{\partial}{\partial \mathbf{r}} \cdot \dot{\mathbf{r}} + \frac{\partial}{\partial \mathbf{p}} \cdot \dot{\mathbf{p}} = \frac{\partial}{\partial \mathbf{r}} \cdot \frac{\partial H}{\partial \mathbf{p}} - \frac{\partial}{\partial \mathbf{p}} \cdot \frac{\partial H}{\partial \mathbf{r}} = 0.$$

The kinetic equation (6.29) can now be written as

$$\frac{\partial f}{\partial t} + \dot{z}_k \frac{\partial f}{\partial z_k} = C(f),$$

where summation over the repeated index is understood. This form of the kinetic equation is completely independent of the choice of phase-space coordinates $\mathbf{z}$. Indeed, if we use some other set of coordinates $\mathbf{w}(\mathbf{z}, t)$, then (see Exercise 4)

$$\frac{df}{dt} = \left(\frac{\partial f}{\partial t} \right)_{\mathbf{z}} + \dot{z}_k \left(\frac{\partial f}{\partial z_k} \right)_t = \left(\frac{\partial f}{\partial t} \right)_{\mathbf{w}} + \dot{w}_k \left(\frac{\partial f}{\partial w_k} \right)_t. \tag{6.30}$$

We may, for instance, use guiding-centre variables $\mathbf{w} = (\mathbf{R}, \mathcal{E}, \mu, \vartheta)$, with $\mathcal{E}$ the energy,

$$\mathcal{E} \equiv \frac{mv^2}{2} + Ze\Phi,$$

to obtain

$$\frac{\partial f}{\partial t} + \dot{\mathbf{R}} \cdot \nabla f + \dot{\mathcal{E}} \frac{\partial f}{\partial \mathcal{E}} + \dot{\mu} \frac{\partial f}{\partial \mu} + \dot{\vartheta} \frac{\partial f}{\partial \vartheta} = C(f),$$

where the term proportional to $\dot{\mu}$ vanishes whenever the magnetic moment is conserved, $\dot{\mu} = 0$. The last term on the left describes Larmor rotation and is usually the largest term in the equation. It is larger than the drift term by a factor of $1/\delta$ defined in (6.8) and larger than the collision term by the factor $1/\Delta$ defined in (4.3). Transport theory is concerned with slow phenomena, typically $\partial/\partial t \sim \delta^2 v$, so the terms $\partial f/\partial t$ and $\dot{\mathcal{E}} \, \partial f/\partial \mathcal{E}$ are even smaller. Therefore, to the zeroth order in δ and Δ, the distribution function is independent of gyroangle,

$$\frac{\partial f_0}{\partial \vartheta} = 0.$$

An equation for $f_0(\mathbf{R}, \mathscr{E}, \mu, t)$ is obtained by taking the average over the gyroangle ϑ, with the result

$$\boxed{\frac{\partial f_0}{\partial t} + \overline{\dot{\mathscr{E}}} \frac{\partial f_0}{\partial \mathscr{E}} + \dot{\mathbf{R}} \cdot \nabla f_0 = \overline{C(f_0)}.} \tag{6.31}$$

This so-called *drift kinetic equation* is fundamental to much of plasma physics. In transport theory, where it plays a particularly prominent role, the phenomena under consideration are generally slow, in a sense which will be made precise in Chapter 8. Since the energy is then approximately conserved, the first two terms in (6.31) are small, and we have

$$v_\| \nabla_\| f_0 + \mathbf{v}_d \cdot \nabla f_0 = C(f_0).$$

Note that the lowest order distribution function f_0 depends on the *guiding-centre* position $\mathbf{R}$ rather than the particle position $\mathbf{r} = \mathbf{R} + \boldsymbol{\rho}$. The distribution function at $\mathbf{r}$ thus becomes

$$f(\mathbf{r}) \simeq f_0(\mathbf{R}) \simeq f_0(\mathbf{r}) - \boldsymbol{\rho} \cdot \nabla f_0(\mathbf{r}), \tag{6.32}$$

with $\boldsymbol{\rho} = \mathbf{b} \times \mathbf{v}/\Omega$. For instance, if $f_0(\mathbf{R})$ is Maxwellian, the distribution function at $\mathbf{r}$ will depart slightly $[O(\delta)]$ from this Maxwellian. We have seen this before, in (4.21), after solving the kinetic equation with a Lorentz operator. Now, we appreciate that it is a general result, depending only on the smallness of δ and Δ.

It is the departure $-\boldsymbol{\rho} \cdot \nabla f_0(\mathbf{r})$ from a local Maxwellian that causes the diamagnetic fluxes of particles, momentum and heat we encountered in Section 4.5. More precisely, the diamagnetic particle flux $n\mathbf{V}_\wedge = (4.64)$, the gyroviscosity $\boldsymbol{\pi}_\wedge = (4.72)$–(4.74), and the diamagnetic heat flux $\mathbf{q}_\wedge = (4.76)$ are equal to

$$\begin{pmatrix} n\mathbf{V}_\wedge \\ \boldsymbol{\pi}_\wedge \\ \mathbf{q}_\wedge \end{pmatrix} = -\int \begin{pmatrix} \mathbf{v} \\ m(\mathbf{v}'\mathbf{v}' - \mathsf{I}v'^2/3) \\ (mv'^2/2 - 5T/2)\mathbf{v}' \end{pmatrix} \boldsymbol{\rho} \cdot \nabla f_0(\mathbf{r})\, d^3v,$$

where $\mathbf{v}' = \mathbf{v} - \mathbf{V}$. Furthermore, it is the collisional relaxation of this departure from a Maxwellian that drives classical cross-field transport.

Because of the smallness of the Larmor radius in a magnetized plasma, classical transport is usually very small and is overwhelmed by neoclassical transport, which is caused by guiding-centre orbits. In the rest of this book we shall only be concerned with neoclassical transport and therefore only consider the guiding-centre distribution function f_0. Accordingly, we disregard the distinction between $\mathbf{r}$ and $\mathbf{R}$ and drop the subscript on f_0, thus regarding f as the distribution function of guiding centres.

Further reading

Old original research papers are often very readable and inspiring. The one by Alfvén (1940) on guiding-centre drifts is no exception, and Taylor's paper (1964) on plasma stability using Hamiltonian methods is also very well worth reading. Whitham's book (1974) gives a very clear and careful explanation of averaged variational principles. The book by Hazeltine and Waelbroeck (1998) contains a useful chapter on guiding-centre motion, as does the one by Goldston and Rutherford (1995), which includes topics such as magnetic ripple effects and stochastic maps. Littlejohn (1983) derived a Lagrangian for guiding-centre motion which is slightly different from our (6.15). His Lagrangian is more complicated as it operates in six-dimensional phase space, but is widely quoted in the literature. Adiabatic invariants constitute a fascinating subject in their own right and are usefully discussed by Landau and Lifshitz (1976) and Goldstein (1980), who also provide background reading on analytical mechanics. An important alternative derivation of the drift kinetic equation was developed by Hazeltine (1973) and generalized by Hazeltine and Ware (1978) to the case of a plasma with large (sonic) flow velocity.

Exercises

1. Derive Hamilton's equations of motion (6.6).

   Solution: The differential of the Lagrangian is

   $$dL = \mathbf{p} \cdot d\dot{\mathbf{r}} + \dot{\mathbf{p}} \cdot d\mathbf{r},$$

   where $\mathbf{p} = \partial L / \partial \dot{\mathbf{r}}$ and $\dot{\mathbf{p}} = \partial L / \partial \mathbf{r}$. Hamilton's equations follow immediately from the differential of the Hamiltonian (6.5),

   $$dH = \mathbf{p} \cdot d\dot{\mathbf{r}} + \dot{\mathbf{r}} \cdot d\mathbf{p} - dL = \dot{\mathbf{r}} \cdot d\mathbf{p} - \dot{\mathbf{p}} \cdot d\mathbf{r}.$$

2. Derive Hamiltonian guiding-centre equations of motion in a stationary magnetic field described by the coordinates (α, β, s).

   Solution: Start from the Lagrangian (6.28) and apply the general prescription (6.5). The generalized momenta are

   $$p_s = \frac{\partial \overline{L}}{\partial \dot{s}} = m\dot{s}, \quad p_\alpha = \frac{\partial \overline{L}}{\partial \dot{\alpha}} = 0, \quad p_\beta = \frac{\partial \overline{L}}{\partial \dot{\beta}} = Ze\alpha,$$

   and the Hamiltonian becomes,

   $$H = \frac{p_s^2}{2m} + \mu B + Ze\Phi.$$

Note that H is simply equal to the total energy. The equations of motion are

$$\dot{s} = \frac{\partial H}{\partial p_s} = p_s, \qquad\qquad \dot{p}_s = -\frac{\partial H}{\partial s} = -\mu\frac{\partial B}{\partial s} - Ze\frac{\partial \Phi}{\partial s},$$

$$\dot{\beta} = \frac{\partial H}{\partial p_\beta} = \frac{\mu}{Ze}\frac{\partial B}{\partial \alpha} + \frac{\partial \Phi}{\partial \alpha}, \qquad \dot{\alpha} = -\frac{1}{Ze}\frac{\partial H}{\partial \beta} = -\frac{\mu}{Ze}\frac{\partial B}{\partial \beta} - \frac{\partial \Phi}{\partial \beta}.$$

3. For readers familiar with analytical mechanics.

   (a) Show that the angular frequency of bounce motion executed by a particle trapped in a magnetic field is

$$\omega_b = \left(\frac{\partial H}{\partial J}\right)_{\alpha,\beta,\mu}.$$

   Solution: This follows immediately from a canonical transformation to action-angle variables $(s, p_s, \beta, p_\beta) \to (J, \theta, \beta, p_\beta)$. The Hamiltonian is then independent of the phase θ along the orbit.

   (b) Show that the drift across the magnetic field is described by

$$\dot{\alpha} = \frac{\omega_b}{Ze}\left(\frac{\partial J}{\partial \beta}\right)_{\alpha,H,\mu}.$$

   In view of (6.27), this implies that the toroidal precession frequency of trapped particles in a tokamak is

$$\omega_\varphi = \frac{\omega_b}{Ze}\left(\frac{\partial J}{\partial \psi}\right)_{H,\mu}.$$

   Solution: Express the Hamiltonian as a function of β, p_β, J, μ, and time, $H = H[\beta, p_\beta, J(\beta, p_\beta, \mu, H, t), t]$, and differentiate with respect to β, giving

$$\frac{\partial H}{\partial \beta} + \frac{\partial H}{\partial J}\frac{\partial J}{\partial \beta} = 0.$$

   Hence

$$\dot{p}_\beta = -\frac{\partial H}{\partial \beta} = \omega_b\frac{\partial J}{\partial \beta}.$$

4. Demonstrate the second equality in (6.30).

   Solution:

$$\dot{z}_k\frac{\partial f}{\partial z_k} = \frac{\partial z_k}{\partial w_l}\dot{w}_l\frac{\partial f}{\partial w_m}\frac{\partial w_m}{\partial z_k} = \delta_{lm}\dot{w}_l\frac{\partial f}{\partial w_m}$$

7

Toroidal plasmas

Tamm and Sakharov (1961), who invented the tokamak, realized that collisional transport can be very different in a torus and a cylinder. The transport depends on the *global* geometry of the magnetic field, not just on its local characteristics. The rest of the present book is devoted to this topic, so-called neoclassical transport theory. Tamm made the first outline of neoclassical transport theory in 1951. The first formal calculations were carried out by Pfirsch and Schlüter and by Galeev and Sagdeev (1968), whose work was followed by rapid progress in the United States and Europe. Like most of the literature, our presentation is focused on the tokamak, but many of the concepts have wider applicability. Mathematical tools for describing stellarator geometry are introduced in the final section of this chapter.

7.1 Magnetic field

We consider plasma in an axisymmetric magnetic field. In cylindrical coordinates (R, φ, z), all derivatives with respect to the toroidal angle φ vanish, $\partial / \partial \varphi = 0$, and

$$\mathbf{B} = \hat{\mathbf{R}} B_R + \hat{\boldsymbol{\varphi}} B_\varphi + \hat{\mathbf{z}} B_z = \nabla \times \mathbf{A},$$

where $\mathbf{A}$ is the vector potential. It is common practice in neoclassical theory to take φ to vary in the opposite sense from the usual cylindrical coordinate angle. Thus φ increases in the clockwise direction when viewed from above and $\hat{\mathbf{R}} \times \hat{\mathbf{z}} = \hat{\boldsymbol{\varphi}}$. The poloidal part of the magnetic field $\mathbf{B}_p \equiv \hat{\mathbf{R}} B_R + \hat{\mathbf{z}} B_z$ is thus

$$\mathbf{B}_p = \hat{\mathbf{R}} \frac{\partial A_\varphi}{\partial z} - \hat{\mathbf{z}} \frac{1}{R} \frac{\partial (R A_\varphi)}{\partial R} = \nabla \varphi \times \nabla \psi,$$

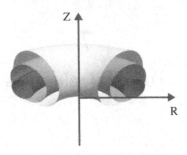

Fig. 7.1.   Nested toroidal flux surfaces.

where we have used the expression for the curl of a vector in cylindrical coordinates, and introduced the so-called *poloidal flux function*

$$\psi(R, z) \equiv -RA_\varphi(R, z).$$

It is clear that ψ is constant along magnetic field lines, $\mathbf{B} \cdot \nabla\psi = 0$. The magnetic field vector thus lies on surfaces of constant ψ. In a tokamak they usually form nested toroids and are called flux surfaces, see Fig. 7.1. The innermost flux surface, which is just a circular line, is called the magnetic axis, and for definiteness ψ can be normalized to $\psi = 0$ there. The poloidal magnetic field is produced by a toroidal current.

The toroidal part of the magnetic field is in the direction of $\nabla\varphi$ and can therefore be written as

$$\mathbf{B}_t = \hat{\boldsymbol{\varphi}}B_\varphi = I(R, z)\nabla\varphi,$$

with $\hat{\boldsymbol{\varphi}} = R\nabla\varphi$ the toroidal unit vector. The function $I(R, z)$ thus defined is related to the plasma current $\mathbf{j}$ by Ampère's law (where the displacement current is neglected)

$$\mu_0\mathbf{j} = \nabla \times \mathbf{B} = \nabla \times (I\nabla\varphi + \nabla\varphi \times \nabla\psi)$$

$$= \nabla I \times \nabla\varphi + \nabla^2\psi\nabla\varphi + (\nabla\psi \cdot \nabla)\nabla\varphi.$$

Since the vectors $\nabla^2\psi\nabla\varphi$ and $(\nabla\psi \cdot \nabla)\nabla\varphi$ are both in the toroidal direction, we see that the poloidal part of the plasma current,

$$\mathbf{j}_p = \mu_0^{-1}\nabla I \times \nabla\varphi, \tag{7.1}$$

is perpendicular to ∇I. Usually electric current cannot flow across flux surfaces, i.e., $\mathbf{j} \cdot \nabla\psi = 0$, as follows from the equilibrium condition $\mathbf{j} \times \mathbf{B} = \nabla p$ if the pressure p is a flux function. Then $\nabla I \times \nabla\psi = 0$, so the function $I(R, z)$ must be constant on flux surfaces, $I(R, z) = I(\psi)$. We are thus led to the following representation of the magnetic field:

$$\boxed{\mathbf{B} = I(\psi)\nabla\varphi + \nabla\varphi \times \nabla\psi,} \qquad (7.2)$$

which we shall use extensively. This is generally the most convenient way of writing a rotationally symmetric magnetic field. Later in this chapter more general representations will be derived which are suitable for non-axisymmetric fields, such as those in stellarators.

Let us introduce a poloidal angle coordinate θ to measure the position on a flux surface. We require θ to have a period of 2π and to vary only in the poloidal plane, $\nabla\theta \cdot \nabla\varphi = 0$, but we leave its precise definition otherwise arbitrary. Figure 7.5 shows an example of a choice of θ. The volume element becomes

$$dV = g^{1/2}\, d\psi\, d\theta\, d\varphi = d\psi\, d\theta\, d\varphi / |(\nabla\varphi \times \nabla\psi) \cdot \nabla\theta|,$$

where $g^{1/2} = 1/|(\nabla\varphi \times \nabla\psi) \cdot \nabla\theta| = 1/|\mathbf{B} \cdot \nabla\theta|$ is the Jacobian. The flux surface element (the area of an infinitesimally small rectangle with sides defined by $d\theta$ and $d\varphi$) therefore becomes

$$dS = \frac{|\nabla\psi|}{|\mathbf{B} \cdot \nabla\theta|}\, d\theta\, d\varphi, \qquad (7.3)$$

and the line element in the poloidal direction is

$$dl_p = \frac{dS}{R\, d\varphi} = \frac{B_p\, d\theta}{|\mathbf{B} \cdot \nabla\theta|}$$

since $B_p = |\nabla\psi|/R$, see Fig. 7.2. The volume inside a flux surface ψ is

$$V(\psi) = \int_0^{2\pi} d\theta \int_0^{2\pi} d\varphi \int_0^{\psi} g^{1/2}\, d\psi' = 2\pi \int_0^{\psi} d\psi' \int_0^{2\pi} \frac{d\theta}{|\mathbf{B} \cdot \nabla\theta|}. \qquad (7.4)$$

A very useful concept is that of the *flux-surface average*. Given a quantity $Q(\psi, \theta)$, its flux-surface average $\langle Q \rangle$ is defined as the volume average of Q between two neighbouring flux surfaces ψ and $\psi + d\psi$,

$$\langle Q \rangle (\psi) \equiv \int Q(\psi, \theta)\, dV \Big/ \int dV = \oint \frac{Q(\psi, \theta)}{\mathbf{B} \cdot \nabla\theta}\, d\theta \Big/ \oint \frac{d\theta}{\mathbf{B} \cdot \nabla\theta}, \qquad (7.5)$$

or, equivalently,

$$\langle Q \rangle (\psi) \equiv \oint Q \frac{dl_p}{B_p} \Big/ \oint \frac{dl_p}{B_p},$$

where $\oint$ indicates that the integration should be taken one turn around the torus in the poloidal direction. Note that the flux-surface average annihilates the operator $\mathbf{B} \cdot \nabla = g^{-1/2}\, \partial/\partial\theta$, i.e.,

$$\langle \mathbf{B} \cdot \nabla f(\psi, \theta) \rangle = 0 \qquad (7.6)$$

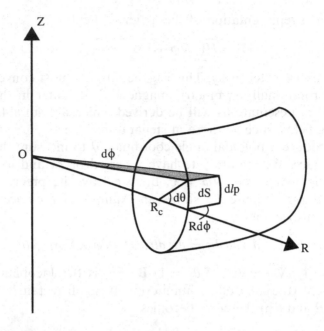

Fig. 7.2.   The length element in the poloidal direction is equal to $dl_p = B_p \, d\theta / |\mathbf{B} \cdot \nabla\theta|$, and the area of a rectangle on the flux surface defined by the infinitesimal poloidal and toroidal angles $d\theta$ and $d\varphi$ is equal to $dS = R \, dl_p \, d\varphi$.

for any periodic function $f(\psi, \theta)$. In contrast, a quantity like $\langle \mathbf{B} \cdot \nabla\theta \rangle$ does not vanish since the angle θ is not single-valued; it increases secularly by 2π when moving around the flux surface in the poloidal direction.

Using the flux-surface average, we define the *safety factor* as the average change in φ divided by the average change in θ along the magnetic field

$$q(\psi) \equiv \frac{\langle \mathbf{B} \cdot \nabla\varphi \rangle}{\langle \mathbf{B} \cdot \nabla\theta \rangle}. \tag{7.7}$$

It can be interpreted as the average number of toroidal turns the magnetic field makes in one turn around the poloidal direction since

$$q = \oint \frac{\mathbf{B} \cdot \nabla\varphi}{\mathbf{B} \cdot \nabla\theta} \, d\theta \bigg/ \oint d\theta = \oint d\varphi \bigg/ \oint d\theta,$$

where we have observed that $\mathbf{B} \cdot \nabla\varphi / \mathbf{B} \cdot \nabla\theta = d\varphi / d\theta$ along a field line. In a tokamak, the safety factor is usually of order unity or larger for stability reasons.

7.2 Magnetohydrodynamic equilibrium

Force balance

If the momentum equations (2.16) of all species in a plasma are added, one obtains

$$\mathbf{j} \times \mathbf{B} = \nabla p, \tag{7.8}$$

if inertia and viscosity are neglected. Viscosity is smaller than the pressure in any plasma close to local thermodynamic equilibrium, and the inertial term is small if the flow velocity is smaller than thermal, $(\mathbf{V}_a \cdot \nabla)\mathbf{V}_a \ll \nabla p_a$, if $V_a \ll v_{Ta}$. In (7.8), $\mathbf{j}$ denotes the total current, and p the total pressure

$$\mathbf{j} \equiv \sum_a n_a e_a \mathbf{V}_a,$$

$$p \equiv \sum_a n_a T_a.$$

Equation (7.8) describes force balance, and indicates that a plasma can be confined by a magnetic field if plasma pressure is balanced by a magnetic force. It follows that, if this is the case, no current can flow across flux surfaces, $\mathbf{j} \cdot \nabla \psi = 0$, as assumed in the previous section. Furthermore, the pressure is constant on flux surfaces, $p = p(\psi)$, since

$$\mathbf{B} \cdot \nabla p = 0.$$

It should be remembered that the time derivative was neglected in the force balance (7.8), which is thus only valid on sufficiently long time scales. We are, for instance, only considering phenomena that are slower than pressure equilibration along magnetic field lines.

The Grad–Shafranov equation

Multiplying the force balance relation (7.8) by $\nabla \psi$, and using the relation (7.2), gives

$$[I(\psi)\mathbf{j} \times \nabla \varphi - (\mathbf{j} \cdot \nabla \varphi)\nabla \psi] \cdot \nabla \psi = \nabla p \cdot \nabla \psi,$$

and by using (7.1) for the poloidal component of the current we obtain the toroidal component of the current as

$$\mathbf{j} \cdot \nabla \varphi = -\left(p' + \frac{II'}{\mu_0 R^2} \right), \tag{7.9}$$

where a prime denotes differentiation with respect to ψ. If we regard this as an equation for $I(\psi) = RB_\varphi$, we note that the variation of this quantity is driven by two sources: the pressure gradient and the plasma current. The former tends to make the plasma diamagnetic by making I'

have opposite sign from p', thus weakening the toroidal magnetic field in the centre of the discharge. The toroidal plasma current has the opposite effect, producing paramagnetism, and the balance between the two terms is described by the so-called poloidal beta parameter, which is defined and studied in an exercise at the end of this chapter.

Inserting (7.9) in the toroidal component of Ampère's law,

$$\nabla\varphi \cdot (\nabla \times \mathbf{B}) = \mu_0 \nabla \mathbf{j} \cdot \nabla\varphi,$$

recalling the vector-algebra rule $\nabla \cdot (\mathbf{B} \times \mathbf{A}) = \mathbf{A} \cdot (\nabla \times \mathbf{B}) - \mathbf{B} \cdot (\nabla \times \mathbf{A})$, and writing the left-hand side of this equation as

$$\nabla \cdot (\mathbf{B} \times \nabla\varphi) = \nabla \cdot [(\nabla\varphi \times \nabla\psi) \times \nabla\varphi] = \nabla \cdot (R^{-2}\nabla\psi),$$

gives the so-called Grad–Shafranov equation

$$\boxed{R^2 \nabla \cdot \left(\frac{\nabla\psi}{R^2}\right) = -\mu_0 R^2 \frac{dp}{d\psi} - I\frac{dI}{d\psi},} \qquad (7.10)$$

whose solutions describe the possible plasma equilibria. Whether they are stable or unstable is the subject of MHD stability theory. The operator on the left is often denoted by $\Delta^*\psi$, and is explicitly equal to

$$\Delta^*\psi = R^2 \nabla \cdot (R^{-2}\nabla\psi) = R\frac{\partial}{\partial R}\left(\frac{1}{R}\frac{\partial\psi}{\partial R}\right) + \frac{\partial^2\psi}{\partial z^2}.$$

To see what solutions to the Grad–Shafranov equation may look like, let us consider the simple case when p' and II' are constant with respect to ψ. Such an approximation frequently holds locally, e.g., near the magnetic axis. Since the operator Δ^* has the properties

$$\Delta^* 1 = \Delta^* R^2 = 0,$$
$$\Delta^* z^2 = \Delta^* \left(R^2 \ln R\right) = 2,$$
$$\Delta^* (R^2 z^2) = \Delta^* R^4/4 = 2R^2,$$

we can immediately write down the family of solutions (Zheng *et al.*, 1996)

$$\psi(R,z) = c_1 + c_2 R^2 + c_3 R^4 + c_4 z^2 R^2 + c_5 z^2 + c_6 R^2 \ln\frac{R^2}{R_0^2},$$

where the coefficients c_1 and c_2 are arbitrary, c_3 and c_4 are related to the constant p', and c_5 and c_6 are determined by II'. In order to rewrite this expression in a more geometrically instructive form, we note that one of these six coefficients controls the overall magnitude of the poloidal flux, ψ_0, say. Another one determines the radius of the magnetic axis, $R = R_0$, and a third one is fixed by our convention that ψ should vanish at the magnetic axis, $\psi(R_0, 0) = 0$. There are then three free parameters left, and we can rewrite our solution as

$$\psi(R, z) = \frac{\psi_0}{R_0^4} \left[\left(R^2 - R_0^2\right)^2 + \frac{z^2}{E^2} \left(R^2 - R_x^2\right) \right.$$

$$\left. - \tau R_0^2 \left(R^2 \ln \frac{R^2}{R_0^2} - \left(R^2 - R_0^2\right) - \frac{\left(R^2 - R_0^2\right)^2}{2R_0^2}\right) \right], \qquad (7.11)$$

where R_x, E, and τ are new constants which determine the plasma shape, as we shall see. The last term in (7.11), which is proportional to τ, is constructed so as to become small close to the magnetic axis. Indeed, if we write

$$R = R_0 \left(1 + \epsilon\right),$$

$$z = R_0 \hat{z},$$

and expand (7.11) for small ϵ and $\hat{z}$, we obtain

$$\frac{\psi(\epsilon, \hat{z})}{\psi_0} = 4(\epsilon + \delta)^2 + \frac{\hat{z}^2}{E^2} \left(1 - \frac{R_x^2}{R_0^2}\right) + O(\epsilon^4), \qquad (7.12)$$

where

$$\delta(\epsilon, \hat{z}) = \left(1 - \frac{\tau}{3}\right) \epsilon^2 + \frac{\hat{z}^2}{4E^2},$$

is of order ϵ^2. First neglecting δ, which is much smaller than ϵ, in Eq. (7.12) leads us to conclude that the flux surfaces are elliptical close to the magnetic axis, with a vertical elongation equal to

$$k = \frac{2E}{\sqrt{1 - R_x^2/R_0^2}}.$$

Specifically, the surfaces are circular if $k = 1$. The small correction caused by the presence of the term δ in (7.12) implies that flux surfaces away from the magnetic axis are not centred exactly around $R = R_0$, but are shifted by the distance $\Delta_s = \delta R_0$ toward the inboard side of the torus. This displacement is usually referred to as the Shafranov shift (Shafranov, 1966) and depends on the parameter τ. Returning to the exact solution (7.11), we note that the meaning of the parameter R_x can be understood by observing that the radial component of the magnetic field,

$$B_R = -\frac{1}{R} \frac{\partial \psi}{\partial z} = \frac{2\psi_0 z}{E^2 R_0^4} \frac{R_x^2 - R^2}{R},$$

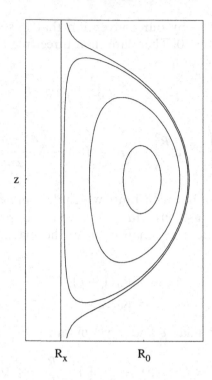

z

R_x R_0

Fig. 7.3.   Flux surfaces of the solution (7.11) to the Grad–Shafranov equation.

vanishes at $R = R_x$ if $R_x > 0$. The poloidal field is therefore vertical along this line, which forms a separatrix between open and closed flux surfaces, see Fig. 7.3. On this separatrix there are two X-points where the poloidal magnetic field vanishes, so that $RB_z = \partial\psi/\partial R = 0$.

   If the aspect ratio of the torus is very tight, as in the edge of a spherical tokamak, the field lines have a tendency to encircle the centre post many times on the inboard side of the flux surface where the toroidal field is strong, see Fig. 7.4. Indeed, the differential equation for a magnetic field line is

$$\frac{dl_\varphi}{B_\varphi} = \frac{dl_p}{B_p},$$

where the toroidal length element is $dl_\varphi = R\,d\varphi$, so that

$$d\varphi = \frac{I(\psi)\,dl_p}{R^2 B_p} = \sqrt{\left(\frac{\partial R}{\partial\theta}\right)^2_\psi + \left(\frac{\partial z}{\partial\theta}\right)^2_\psi}\,\frac{I(\psi)\,d\theta}{R|\nabla\psi|},$$

and $d\varphi/d\theta$ is thus large when R is small.

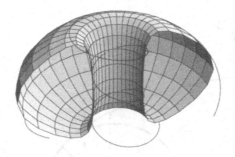

Fig. 7.4. Field line on a flux surface close to the edge of a spherical tokamak.

Large aspect ratio

In analytical transport calculations, one often simplifies the analysis by assuming that the aspect ratio $A \equiv R_0/r$ is large, so that the inverse aspect ratio is small,

$$\epsilon \equiv r/R_0 \ll 1. \tag{7.13}$$

Here, $r \equiv \sqrt{(R - R_0)^2 + z^2}$ is called the *minor radius*, as opposed to the *major radius* R. Moreover, if the outermost flux surface (at the plasma boundary) is shaped as a circle then, for small Shafranov shift, the inner flux surfaces will remain approximately circular. As we have just seen, there are indeed such solutions to the Grad–Shafranov equation in any region where the gradients of $p(\psi)$ and $I^2(\psi)$ are constant. More generally, if these gradients are allowed to vary, it can be shown that if the plasma pressure is much smaller than the magnetic field pressure, satisfying the tokamak ordering

$$\beta \equiv \frac{p}{B^2/2\mu_0} \sim \epsilon^2 \ll 1,$$

there are equilibria with nearly circular cross section, and with the safety factor (7.7) of order unity. Again, the outer magnetic surfaces are slightly shifted from the magnetic axis towards the inside of the tokamak, with a Shafranov shift of order $\Delta_s \sim \epsilon r$. (Alternatively, one can regard the inner flux surfaces as shifted toward the outside as compared with the outer flux surfaces.)

When flux surfaces are circular, the natural candidate for the poloidal angle θ, whose precise definition was left unspecified in Section 7.1, is the angle to the midplane, as in Fig. 7.5. If we let $R_c(\psi) = R_0 - \Delta_s(\psi)$ denote the centre of the magnetic surface ψ, we then have the following 'inverse' equilibrium representation

$$R = R_c(\psi) + r(\psi)\cos\theta,$$

$$z = r(\psi)\sin\theta,$$

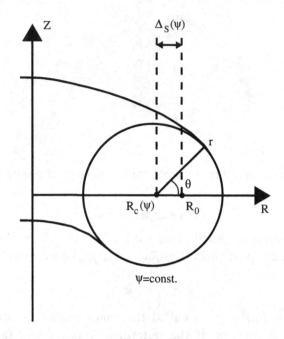

Fig. 7.5.   Flux surface with circular cross section.

whence $[R - R_c(\psi)]^2 + z^2 = r^2$. Implicit differentiation with respect to R and z, respectively, gives

$$(R - R_c)\left(1 - R_c'\frac{\partial\psi}{\partial R}\right) = rr'\frac{\partial\psi}{\partial R},$$

$$z - (R - R_c)R_c'\frac{\partial\psi}{\partial z} = rr'\frac{\partial\psi}{\partial z},$$

so that

$$\frac{\partial\psi}{\partial R} = \frac{R - R_c}{rr' + (R - R_c)R_c'},$$

$$\frac{\partial\psi}{\partial z} = \frac{z}{rr' + (R - R_c)R_c'}.$$

The poloidal magnetic field $\mathbf{B}_p = \nabla\varphi \times \nabla\psi$ thus becomes

$$\mathbf{B}_p = \hat{\boldsymbol{\theta}}B_\theta = \hat{\boldsymbol{\theta}}\frac{|\nabla\psi|}{R} = \frac{\hat{\boldsymbol{\theta}}}{R}\sqrt{\left(\frac{\partial\psi}{\partial R}\right)^2 + \left(\frac{\partial\psi}{\partial z}\right)^2}$$

$$= \hat{\boldsymbol{\theta}}\frac{r/R}{rr' + (R - R_c)R_c'} = \hat{\boldsymbol{\theta}}\frac{B_{\theta 0}(r)}{1 - \Lambda\epsilon\cos\theta},$$

with

$$B_{\theta 0}(r) = \frac{1}{R_0}\frac{d\psi}{dr},$$

$$\Lambda = -1 - \frac{R_c'}{r'} = \frac{1}{\epsilon}\frac{d\Delta_s}{dr} - 1 = O(1). \tag{7.14}$$

Thus, to lowest order in ϵ the magnetic field (7.2) in a low-beta plasma with circular flux surfaces is

$$B_\theta(r,\theta) = B_{\theta 0}(r)(1 + \Lambda\epsilon\cos\theta),$$

$$B_\varphi(r,\theta) = B_0(1 - \epsilon\cos\theta). \tag{7.15}$$

where $B_0 \equiv R_0 I(\psi)$, and the safety factor (7.7) is

$$q(r) = \frac{rB_0}{RB_{\theta 0}(r)}. \tag{7.16}$$

In this section we have not actually solved the Grad–Shafranov equation, but merely postulated solutions with nearly circular flux surfaces and large aspect ratio. The existence of such solutions is intuitively clear from the fact that cylindrical equilibria must have circular flux surfaces. Indeed, by expanding the Grad–Shafranov equation in powers of ϵ these solutions can be constructed and the Shafranov shift calculated, with the result

$$\frac{d\Delta_s}{dr} = \frac{q^2(r)}{r^3}\int_0^r \frac{r'}{R_0}\left(\frac{r'^2}{q^2(r')} - \frac{2\mu_0 R_0^2 r'}{B_0^2}\frac{dp}{dr'}\right)dr'. \tag{7.17}$$

For more information about tokamak equilibrium the reader is referred to the texts by Shafranov (1966), Freidberg (1987) and Wesson (1997).

7.3 Guiding-centre orbits in tokamaks

The motion of a charged particle in a strong, stationary, axisymmetric magnetic field is characterized by three constants of motion,

$$\mathscr{E} = \frac{mv^2}{2} + Ze\Phi,$$

$$\mu = \frac{mv_\perp^2}{2B},$$

$$p_\varphi = mRv_\varphi - Ze\psi.$$

Here the magnetic field is assumed to be strong enough to make the gyroradius $v_\perp/\Omega$ significantly smaller than the characteristic length scale of the field, so that the magnetic moment μ is conserved. The third invariant, p_φ, stems from the fact that the field is axisymmetric. The Lagrangian (6.1) does not contain φ, and a particle moving in such a field has a constant of motion

$$p_\varphi \equiv \partial L/\dot\varphi = mRv_\varphi + ZeRA_\varphi, \qquad (7.18)$$

since $\dot p_\varphi = \partial L/\partial\varphi = 0$. (Note that $v_\varphi = R\dot\varphi$ and $RA_\varphi = -\psi$.) This toroidal canonical momentum is obviously an *exact* constant of motion. Applying a similar argument to the guiding-centre Lagrangian (6.15) gives an *approximate* constant of motion, the toroidal canonical momentum of the guiding centre,

$$\bar p_\varphi \equiv \partial \bar L/\dot{\bar\varphi} = mRb_\varphi v_\parallel + ZeRA_\varphi, \qquad (7.19)$$

with $b_\varphi = B_\varphi/B = I/RB \le 1$. This is equal to the gyroaverage of Eq. (7.18) since the gyroaverage of the toroidal velocity is

$$\bar v_\varphi = \overline{\mathbf{v}\cdot\hat{\boldsymbol\varphi}} = v_\parallel \mathbf{b}\cdot\hat{\boldsymbol\varphi}.$$

Moreover, it can be shown that the canonical momenta of the particle and the guiding centre are equal to all orders in the Larmor radius (Littlejohn, 1983). For practical purposes, they are thus equivalent, and we shall make no distinction between them.

Let us estimate the relative magnitude of the two terms in p_φ. Since the poloidal magnetic field is $\mathbf{B}_p = \nabla\varphi \times \nabla\psi$ and we can estimate $\nabla\psi \sim \psi/r$, we have

$$\frac{mRv_\varphi}{Ze\psi} \sim \frac{mv_\varphi}{ZeB_p r} \sim \frac{\rho_p}{r},$$

where ρ_p is the poloidal gyroradius mv/ZeB_p. If the velocity of the particle is so small that $\rho_p \ll r$, the second term in p_φ is much larger than the first, and the particle remains close to the same flux surface ψ all the time. This condition is normally satisfied for thermal ions and electrons in tokamaks. Thus, for these particles the conservation of p_φ implies radial confinement. This conclusion is known as *Tamm's theorem*. High-energy ions sometimes have substantial poloidal gyroradii, and their orbits may deviate significantly from the flux surfaces, especially near the magnetic axis where $B_p \to 0$.

In the next chapter, we show that the electrostatic potential does not vary much over each flux surface, $\Phi = \Phi(\psi)$, if the poloidal gyroradius of most plasma particles is small and the plasma does not rotate rapidly. Since each of these particles stays close to a particular flux surface, the velocity v remains approximately constant. It is then convenient to describe the orbits by the three approximate constants of motion (v, λ, ψ), where

$$\lambda \equiv \frac{\mu B_0}{mv^2/2} = \frac{v_\perp^2 B_0}{v^2 B}, \qquad (7.20)$$

rather than the exact ones ($\mathscr{E}, \mu, p_\varphi$). Here B_0 denotes some reference field, customarily taken to be $B_0 = \langle B^2 \rangle^{1/2}$ in neoclassical theory. In the literature, this pitch-angle variable λ is sometimes defined without the normalizing B_0, so that $\lambda = v_\perp^2/v^2 B$.

Now let us consider in more detail orbits of particles moving on a flux surface where the magnetic field varies between $B_{\min}$ on the outboard side and $B_{\max}$ on the inboard side of the surface. In general, a particle with a given energy $\mathscr{E}$ and magnetic moment μ can never enter a region where the magnetic field is so strong that $\mu B > \mathscr{E} - Ze\Phi$, since then $v_\perp > v$ and $v_\parallel^2 = 2(\mathscr{E} - Ze\Phi - \mu B)/m < 0$, which is unphysical. If Φ is constant along the orbit, so that v is conserved, this means that λ always stays below B_0/B. Thus, while all particles must satisfy

$$0 \leq \lambda \leq \frac{B_0}{B_{\min}},$$

the ones with

$$\frac{B_0}{B_{\max}} < \lambda \leq \frac{B_0}{B_{\min}}$$

can only be on the outboard side of the flux surface. When one of the latter particles moves along a field line towards the inside, it is reflected by the mirror force $F_\parallel = -\mu \nabla_\parallel B$, cf. (6.19), so that $v_\parallel$ changes sign. Such particles are referred to as 'trapped' (on the outside of the torus). The phenomenon of particle trapping also occurs in the magnetosphere, where charged particles are reflected by the strong magnetic field near the poles. Particles in a tokamak with $0 < \lambda < B_0/B_{\max}$ are free to move along field lines, and explore the entire flux surface. The mirror force slows them down as they approach the strong magnetic field on the inside, but is not strong enough to reflect them. Such particles are referred to as 'passing', 'circulating' (around the torus), or 'untrapped'.

The parallel velocity can be expressed in terms of constants of motion by $v_\parallel = \sigma v(1 - \lambda B/B_0)^{1/2}$, where $\sigma = \pm 1$ stays constant for circulating particles and changes sign for trapped ones. Thus, strictly speaking, a passing particle orbit is characterized by the *four* constants of motion $(v, \lambda, \psi, \sigma)$. Expressed in these variables, the volume element in velocity-space is equal to

$$d^3v = 2\pi v_\perp \, dv_\perp \, dv_\parallel = \sum_\sigma \frac{\pi v^3 B}{|v_\parallel| B_0} \, dv \, d\lambda. \qquad (7.21)$$

Large aspect ratio

In a large-aspect-ratio tokamak with circular flux surfaces and $q \sim O(1)$, the magnetic field (7.15) is dominated by its toroidal component, so that

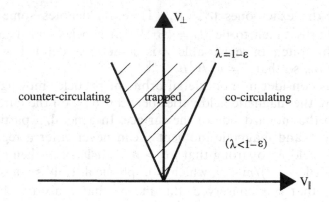

Fig. 7.6.   Regions of trapped and circulating particles.

$B \simeq B_\varphi$. Its strength varies between $B_{\min} = B_0/(1 + \epsilon)$ on the outside and $B_{\max} = B_0/(1 - \epsilon)$ on the inside of the torus, and the particle orbits can thus be classified as

$$\text{circulating if } 0 \le \lambda < 1 - \epsilon,$$

$$\text{trapped if } 1 - \epsilon < \lambda < 1 + \epsilon.$$

In view of (7.21), we see that the trapped particles constitute a small fraction, $O(\sqrt{\epsilon}) \ll 1$, of the total number of particles if $\epsilon \ll 1$.

The orbits typically look as in Fig. 7.7. Because of their shape, the trapped orbits are also called banana orbits. The parallel velocity of a particle is

$$v_\parallel = \sigma v\sqrt{1 - \lambda B/B_0} \simeq \sigma v\sqrt{1 - \lambda(1 - \epsilon\cos\theta)}. \tag{7.22}$$

For trapped orbits, $v_\parallel$ is thus largest in the midplane $\theta = 0$, and is of the order $v_\parallel \sim \epsilon^{1/2}v$. Since $p_\varphi \simeq mRv_\parallel - Ze\psi$ is conserved, $v_\parallel$ is positive on the outside and negative on the inside of a trapped orbit. The banana width is

$$\Delta r = \frac{\Delta\psi}{RB_p} = \frac{\Delta v_\parallel}{\Omega_p} \sim \epsilon^{1/2}\rho_p. \tag{7.23}$$

On the other hand, for passing orbits (except the ones very close to the trapped/passing boundary $\lambda = 1 - \epsilon$) $v_\parallel$ does not change much over the orbit, $\Delta v_\parallel \sim \epsilon v$, so the circulating-particle excursion from the flux surface is relatively small,

$$\Delta r = \frac{\Delta v_\parallel}{R\Omega_p} \sim \frac{qv}{\Omega}.$$

The reason why a particle orbit departs from a flux surface is, of course, the perpendicular drift (6.20). In a large-aspect-ratio torus it is in the

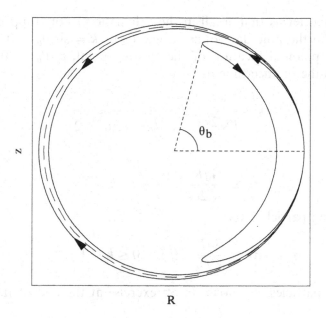

Fig. 7.7. Poloidal projection of a flux surface (dashed) and particle orbits far from the magnetic axis in a tokamak.

vertical direction

$$\mathbf{v}_d = -\frac{v^2 + v_\parallel^2}{2\Omega_\varphi R}\hat{\mathbf{z}},$$

(7.24)

where $\Omega_\varphi = ZeB_\varphi/m$. Ions drift downward and electrons drift upward if $B_\varphi > 0$, and vice versa if the toroidal field is reversed.

The time required for a particle to complete its poloidal orbit (often called the bounce time) is equal to

$$\tau_h = \oint \frac{d\theta}{v_\parallel \mathbf{b} \cdot \nabla \theta} \simeq \frac{B}{B_\theta} \int \frac{r\, d\theta}{\sigma v \sqrt{1 - \lambda[1 - \epsilon(1 - 2\sin^2\theta/2)]}},$$

where we have used the identity $\cos\theta = 1 - 2\sin^2\theta/2$ in (7.22), and the integral is to be taken one poloidal turn around the orbit. For circulating particles, this means that the integral is taken around the flux surface, $0 \le \theta \le 2\pi$, while for trapped particles it follows a banana orbit. It is useful to introduce the trapping parameter

$$k^2 \equiv \frac{1 - \lambda(1 - \epsilon)}{2\epsilon\lambda},$$

(7.25)

so that $0 < k < 1$ for trapped particles and $k > 1$ for circulating ones. (This trapping condition differs from that given above by $O(\epsilon^2)$, since we have neglected terms that small in the calculation.) For trapped particles k is related to the poloidal bounce angle θ_b by $k = \sin \theta_b/2$. The bounce angle is the poloidal location of the turning point, $v_\parallel(\theta_b) = 0$. We may now express the bounce time as

$$\tau_b = \frac{qR}{v\sqrt{2\epsilon\lambda}} \oint \frac{d\theta}{\sigma\sqrt{k^2 - \sin^2\theta/2}}, \tag{7.26}$$

i.e.,

$$\tau_b \simeq \frac{4qR}{v\sqrt{2\epsilon\lambda}} \frac{K(k^{-1})}{k}, \quad k > 1,$$

for circulating particles, and

$$\tau_b \simeq \frac{8qR}{v\sqrt{2\epsilon}} K(k), \quad 0 \le k < 1 \tag{7.27}$$

for trapped particles, as shown in an exercise at the end of the chapter. Here

$$K(k) \equiv \int_0^{\pi/2} \frac{dx}{\sqrt{1 - k^2 \sin^2 x}}, \quad 0 \le k < 1,$$

is the complete elliptic integral of the first kind. It has the properties

$$K(k) = \frac{\pi}{2} \left(1 + \frac{k^2}{4} + O(k^4)\right), \quad k \to 0,$$

$$K(k) \to \ln \frac{4}{\sqrt{1 - k^2}}, \quad k \to 1.$$

The limit $k \to 0$ describes deeply trapped particles on the outboard side of the torus, and $k \to 1$ corresponds to barely trapped particles, with turning points on the inside of the torus. The bounce time for these particles approaches infinity as $k \to 1$ because as $\lambda \to 1 - \epsilon$ it takes increasingly longer for a barely trapped banana particle to 'climb up' to the top of the magnetic well. In the limit $k \to \infty$, the velocity vector is almost parallel to the magnetic field, $v \simeq v_\parallel$, and the bounce time becomes $\tau_b \to 2\pi qR/v_\parallel$, which is the time it takes to travel q times around the torus in the toroidal direction.

Particle motion in a large-aspect-ratio tokamak is entirely analogous to the motion of an ordinary pendulum. The angle of the pendulum to the vertical can be identified with the poloidal angle θ. Trapped motion corresponds to the pendulum executing finite oscillations, and circulating motion corresponds to the pendulum swinging all the way over the top around its pivot. The expression for the bounce time derived above is similar to the period of a pendulum. In particular, it tends to infinity if the turning point of the pendulum approaches the upper, unstable, equilibrium point. Note that the number of particles with very long bounce times is exponentially small since the bounce time approaches infinity logarithmically at the trapped–passing boundary.

Potato orbits

The general picture of tokamak particle orbits just presented relies on the assumption that these orbits stay close to a magnetic surface, so that r is nearly constant. This is true for thermal ions and electrons in most of the plasma, but must fail close to the magnetic axis where the minor radius r becomes as small as the banana width (7.23). Since the poloidal gyroradius is $\rho_p = q\rho/\epsilon$ this occurs when $q\rho/\epsilon^{1/2} \sim r$, i.e., inside the potato-orbit region

$$r \sim r_p \equiv (q^2 \rho^2 R)^{1/3}. \tag{7.28}$$

At such short distances from the magnetic axis, trapped orbits no longer resemble thin bananas, and are instead called 'potato orbits', see Fig. 7.8. Neoclassical transport theory breaks down when r is comparable to, or smaller than, the potato width r_p. For high-energy ions r_p can be comparable to the radius of the plasma column, so that a considerable fraction of these particles follow potato orbits.

In order to analyse guiding-centre orbits without assuming $r \simeq$ constant, it is convenient to introduce the dimensionless variables

$$\Psi = \frac{Ze\psi}{mR_0v},$$

$$J = -\frac{p_\varphi}{mR_0v} = \Psi - \frac{R}{R_0}\xi,$$

with $\xi = v_\parallel/v$. Thus, Ψ and J are normalized versions of ψ and p_φ. In the second equality we have used $B \simeq B_0/R$, which is true near the magnetic axis because $B_\theta/B_\varphi = \epsilon/q \ll 1$. Since then also $\lambda = (1-\xi^2)R/R_0$, we have

$$\left(\frac{R}{R_0}\right)^2 - \frac{\lambda R}{R_0} = (\Psi - J)^2.$$

This equation determines the shape of orbits in coordinates (R, Ψ), and is valid for any flux-surface geometry as long as $B_p/B_t \ll 1$. If we confine our attention to a region close to the magnetic axis where the safety factor q is nearly constant and assume that the flux surfaces are circular, then $d\psi/dr = B_0r/q$, so that

$$\psi = \frac{B_0r^2}{2q},$$

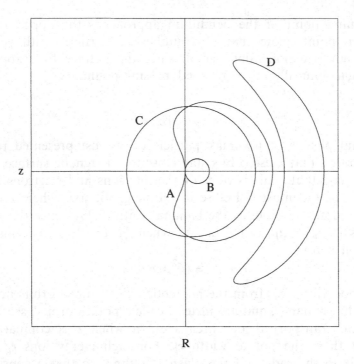

Fig. 7.8. Orbits close to the magnetic axis. Trapped orbits that pass through the axis (A), also known as potato orbits, have relatively large widths. Circulating orbits passing through the axis (B) stay within a few Larmor radii of the axis. Most orbits in the near-axis region are circulating (C) and stay close to a particular flux surface. At a distance of only one potato width from the axis, the trapped orbits resemble bananas (D).

and thus $\Psi = R_0\epsilon^2/q\rho_0$, where $\rho_0 = 2v/\Omega_0$ and $\Omega_0 = eB_0/m$. Thus, we can write down the following quartic equation for the locations $\epsilon = 1 - R/R_0$ where orbits intersect the midplane, $\theta = 0$,

$$(1+\epsilon)^2 - \lambda(1+\epsilon) = \left(\frac{R_0}{q\rho_0}\epsilon^2 - J\right)^2. \qquad (7.29)$$

If we write this equation as

$$\epsilon = \epsilon_0 \left(1 \pm \frac{q\rho_0}{\epsilon_0^2 R_0}\sqrt{(1+\epsilon)(1+\epsilon-\lambda)}\right)^{1/2},$$

where $\epsilon_0 \equiv (q\rho_0 J/R_0)^{1/2}$, and note that $\rho_0/R_0 \ll 1$, it becomes apparent that for most solutions the term involving the square root is small, so that $\epsilon \simeq \epsilon_0$. This corresponds to the thin-orbit-width approximation used previously. It is also clear that this approximation fails when

$$\frac{q\rho_0}{\epsilon_0^2 R_0}\sqrt{1+\epsilon-\lambda} \sim 1,$$

i.e., when $1-\lambda = O(\epsilon)$ and $\epsilon \lesssim (q\rho/R_0)^{2/3}$ simultaneously. This corresponds to the potato width estimate (7.28) above.

To see what happens to orbits which certainly cannot remain on magnetic flux surfaces, let us analyse the orbits that pass through the magnetic axis, $\epsilon = 0$. For such orbits the constant term in the quartic equation (7.29) must vanish, $1-\lambda-J^2 = 0$, so that $\epsilon = 0$ is a solution. By finding the other solutions we can determine the other point where such an orbit intersects the midplane and thus the orbit width. These solutions satisfy

$$\epsilon + 2 - \lambda = \left(\frac{R_0}{q\rho_0}\right)^2 \epsilon^3 - \frac{2R_0 J}{q\rho_0}\epsilon.$$

If we denote the value of $\xi = v_\parallel/v$ at the magnetic axis by ξ_0, then we have $J = -\xi_0$ and $\lambda = 1-\xi_0^2$, and we can write this equation for the orbit width as

$$\epsilon^3 + \frac{2q\rho_0}{R_0}\xi_0\epsilon - \frac{q^2\rho_0^2}{R_0^2}\left(1+\xi_0^2+\epsilon\right) = 0. \tag{7.30}$$

There are two types of solutions to this equation. For most orbits $\xi_0 = O(1)$ and the term ϵ^3 is small, so that the solution becomes

$$\epsilon \simeq \frac{1+\xi_0^2}{\xi_0}\frac{q\rho_0}{2R_0}.$$

Thus, these orbits stay within about q gyroradii from the magnetic axis. The parallel velocity remains nearly constant along such an orbit since the first term in $J = \Psi - R\xi/R_0$ is smaller than the second one,

$$\Psi = \frac{R_0\epsilon^2}{q\rho_0} = O\left(\frac{q\rho_0}{R_0}\right) \ll \frac{\xi R}{R_0} = O(1).$$

This implies that ξ cannot change sign, and that the particle circulates around the torus. Just like farther away from the magnetic axis, most particle orbits are circulating and only make small radial excursions of order $O(q\rho_0)$. On the other hand, if ξ_0 is very small, $\xi_0 \sim (q\rho_0/R_0)^{1/3}$, then the cubic term in (7.30) is no longer negligible and the orbit intersects the midplane at a position

$$\epsilon \sim \left(\frac{q\rho_0}{R_0}\right)^{2/3},$$

which again corresponds to the potato width estimate (7.28). For these particles the parallel velocity may change sign somewhere along the orbit, which is then toroidally trapped.

In summary, most orbits stay within a distance of the order of the Larmor radius multiplied by the safety factor q from a particular flux surface. This is true everywhere in the plasma. A small proportion of the particles are trapped (in the toroidal direction) and have wider orbits. These particles are characterized by small values of $\xi = v_\parallel / v$. Far from the magnetic axis, trapped orbits have $\xi \sim \epsilon^{1/2}$ and ϵ is nearly constant over the orbits. In the region close to the centre, as defined by (7.28), the orbit width is comparable to r, so that ϵ varies significantly. The trapping criterion can still be expressed as $\xi \sim \epsilon^{1/2}$ if the potato width (7.28) is used in $\epsilon = r/R$, and then becomes $\xi \sim (q\rho_0/R_0)^{1/3}$.

7.4 Non-axisymmetric systems

The description of magnetic field and particle orbits in a non-axisymmetric system such as a stellarator is more complicated than in a tokamak. The existence of nested flux surfaces is not guaranteed for three-dimensional systems without symmetry (Grad, 1967). In general, such systems have regions with magnetic islands and chaotic magnetic fields, but modern stellarators are carefully designed to minimize the volume of such regions. To a good approximation, the magnetic field thus has nested flux surfaces with toroidal topology. Unlike the tokamak, however, the stellarator configuration is not axisymmetric, so the poloidal cross section varies with the toroidal angle. Since the magnetic field plays such a dominant role in determining particle orbits and neoclassical transport, it is advantageous to describe the system in terms of coordinates that are aligned with magnetic surfaces. In this section, we derive a special set of magnetic coordinates, known as Boozer coordinates (Boozer, 1981), which are widely used in stellarator analysis, and formulate the guiding-centre equations of motion in these coordinates.

Magnetic coordinates

We begin by recalling some results from classical vector analysis. Given a set of coordinates (x^1, x^2, x^3) in three-dimensional space, a vector field $\mathbf{B(r)}$ can be represented in two fundamental ways,

$$\mathbf{B(r)} = B^i(\mathbf{r})\mathbf{e}_i(\mathbf{r}) = B_i(\mathbf{r})\mathbf{e}^i(\mathbf{r}),$$

as linear combinations of the co- and contravariant basis vectors

$$\mathbf{e}_i = \frac{\partial \mathbf{r}}{\partial x^i},$$

$$\mathbf{e}^i = \nabla x^i,$$

respectively, where the vector components are

$$B^i = \mathbf{B} \cdot \mathbf{e}^i,$$

$$B_i = \mathbf{B} \cdot \mathbf{e}_i.$$

Scalar quantities with raised indices are called contravariant, and those with lowered indices are called covariant. Note that the basis vectors are not of unit length generally. As usual, summation over repeated indices is understood. However, in this section our notation differs from that in the rest of the book regarding the vector components. Elsewhere, we have followed the usual convention in physics and written vectors as $\mathbf{B} = B_i \hat{\mathbf{x}}_i$, where $\hat{\mathbf{x}}_i = \mathbf{e}_i / |\mathbf{e}_i|$, but in this section B_i instead denotes the covariant component $B_i = \mathbf{B} \cdot \mathbf{e}_i$.

The Jacobian associated with the coordinate system is

$$g^{1/2} = \left| \frac{\partial \mathbf{r}}{\partial x^i} \right| = \frac{1}{(\nabla x^1 \times \nabla x^2) \cdot \nabla x^3},$$

and the co- and contravariant basis vectors are related to one another by

$$\mathbf{e}^i = g^{-1/2} \mathbf{e}_j \times \mathbf{e}_k,$$
$$\mathbf{e}_i = g^{1/2} \mathbf{e}^j \times \mathbf{e}^k, \tag{7.31}$$

where (i, j, k) is a cyclic permutation of $(1, 2, 3)$. Finally, the divergence and curl of a vector are given by

$$\nabla \cdot \mathbf{B} = g^{-1/2} \frac{\partial}{\partial x^i} \left(g^{1/2} B^i \right), \tag{7.32}$$

$$\nabla \times \mathbf{B} = g^{-1/2} \epsilon_{ijk} \frac{\partial B_j}{\partial x^i} \mathbf{e}_k, \tag{7.33}$$

where ϵ_{ijk} is the completely antisymmetric tensor.

Armed with these results, let us now consider the equilibrium of a plasma in a non-axisymmetric magnetic field,

$$\mathbf{j} \times \mathbf{B} = \nabla p,$$

and let us assume that surfaces of constant pressure have toroidal topology, as in the non-chaotic regions in a stellarator. We can then introduce a coordinate system

$$(x^1, x^2, x^3) = (\psi_p, \theta, \varphi),$$

such that pressure only depends on the first coordinate, $p = p(\psi_p)$, and the other coordinates have the character of poloidal and toroidal angles, respectively. The exact definitions of the coordinates are somewhat arbitrary at this point. We only require that surfaces of constant ψ_p should coincide with isobars (surfaces of constant pressure), and that θ and φ should be periodic in the poloidal and toroidal directions, respectively, each with period 2π on the torus. Since $\mathbf{B} \cdot \nabla p = 0$, the magnetic field vector lies in surfaces of constant ψ_p if the pressure gradient does not vanish. Because of the requirement $\nabla \cdot \mathbf{B} = 0$, or using Eq. (7.32),

$$\frac{\partial}{\partial \theta} \left(g^{1/2} B^\theta \right) + \frac{\partial}{\partial \varphi} \left(g^{1/2} B^\varphi \right) = 0,$$

there must exist a 'stream function' $\eta(\psi_p, \theta, \varphi)$ such that

$$B^\theta = -\frac{\partial \eta}{\partial \varphi}, \tag{7.34}$$

$$B^\varphi = \frac{\partial \eta}{\partial \theta}. \tag{7.35}$$

The function η can be chosen in the following form to yield a single-valued magnetic field,

$$\eta(\psi_p, \theta, \varphi) = \tilde{\eta}(\psi_p, \theta, \varphi) + f_1(\psi_p)\theta - f_2(\psi_p)\varphi,$$

where the functions f_1 and f_2 are to be chosen shortly, and the magnetic field then becomes

$$\begin{aligned}
\mathbf{B} &= B^\theta \mathbf{e}_\theta + B^\varphi \mathbf{e}_\varphi = g^{1/2} \left(B^\theta \nabla \varphi \times \nabla \psi_p + B^\varphi \nabla \psi_p \times \nabla \theta \right) \\
&= f_1 \nabla \psi_p \times \nabla \theta + f_2 \nabla \varphi \times \nabla \psi_p + \nabla \psi_p \times \nabla \tilde{\eta},
\end{aligned} \tag{7.36}$$

where we have used Eqs. (7.31), (7.34) and (7.35).

Our next task is to choose the functions ψ_p, f_1 and f_2. We choose ψ_p to be the poloidal flux divided by 2π,

$$\psi_p = \frac{1}{2\pi} \int_0^{\psi_p} d\psi_p' \int_0^{2\pi} g^{1/2} \mathbf{B} \cdot \nabla \theta \, d\varphi,$$

and introduce the toroidal flux

$$\psi_t(\psi_p) = \int_0^{\psi_p} d\psi_p' \int_0^{2\pi} g^{1/2} \mathbf{B} \cdot \nabla \varphi \, d\theta.$$

Elsewhere in this book, we have denoted the poloidal flux by just ψ, but in this section we instead use the notation ψ_p to distinguish from the toroidal flux ψ_t. The safety factor is defined for a general non-axisymmetric system as

$$q(\psi_p) = \frac{d\psi_t}{d\psi_p},$$

and reduces to our earlier definition (7.7) in a tokamak. The function f_1 is chosen to be equal to unity and f_2 to be equal to $q(\psi_p)$. The magnetic field (7.36) then becomes

$$\mathbf{B} = \nabla\varphi \times \nabla\psi_p + q(\psi_p)\nabla\psi_p \times \nabla\theta + \nabla\psi_p \times \nabla\tilde{\eta}, \qquad (7.37)$$

where the last term can be eliminated by redefining the poloidal angle by $\theta \to \theta + \tilde{\eta}/q(\psi_p)$. This concludes our definition of the coordinate system $(\psi_p, \theta, \varphi)$. Note that the magnetic field is straight in these coordinates since in the direction along a field line

$$\frac{d\varphi}{d\theta} = \frac{\mathbf{B} \cdot \nabla\varphi}{\mathbf{B} \cdot \nabla\theta} = \text{constant}.$$

Coordinates that are aligned with the magnetic field in this way are called *magnetic coordinates*. Clearly, q measures the number of toroidal revolutions that a magnetic field line makes in one poloidal turn around the flux surface. In stellarators, it is conventional to use the rotational transform $\iota = 1/q$. Rational values of q or ι correspond to surfaces about which magnetic islands can form.

Let us now turn to the current. To arrive at the representation (7.36) of the magnetic field we used the two equations $\nabla \cdot \mathbf{B} = 0$ and $\mathbf{B} \cdot \nabla p = 0$, which both follow from the equilibrium condition $\mathbf{j} \times \mathbf{B} = \nabla p$. Since the same relations hold for the current, $\nabla \cdot \mathbf{j} = 0$ and $\mathbf{j} \cdot \nabla p = 0$, this must have a similiar representation:

$$\mathbf{j} = j^\theta \mathbf{e}_\theta + j^\varphi \mathbf{e}_\varphi = F_1'(\psi_p)\nabla\psi_p \times \nabla\theta + F_2'(\psi_p)\nabla\varphi \times \nabla\psi_p + \nabla\psi_p \times \nabla v(\psi_p, \theta, \varphi),$$

where $F_1'(\psi_p)$ and $F_2'(\psi_p)$ are arbitrary, and

$$g^{1/2} j^\theta = F_2' - \frac{\partial v}{\partial\varphi},$$

$$g^{1/2} j^\varphi = F_1' + \frac{\partial v}{\partial\theta}.$$

By using the expression (7.33) in Ampère's law, $\nabla \times \mathbf{B} = \mu_0 \mathbf{j}$, we now obtain the equations

$$\frac{\partial B_\theta}{\partial\psi_p} - \frac{\partial B_{\psi_p}}{\partial\theta} = \mu_0 \left(\frac{dF_1}{d\psi_p} + \frac{\partial v}{\partial\theta} \right),$$

$$\frac{\partial B_{\psi_p}}{\partial\varphi} - \frac{\partial B_\varphi}{\partial\psi_p} = \mu_0 \left(\frac{dF_2}{d\psi_p} - \frac{\partial v}{\partial\varphi} \right),$$

which are easily integrated to give

$$B_{\psi_p} = \mu_0 \left(-v + \frac{dV}{d\psi_p} \right),$$

$$B_\theta = \mu_0 \left(F_1 + \frac{dV}{d\theta} \right),$$

$$B_\varphi = \mu_0 \left(-F_2 + \frac{dV}{d\varphi} \right),$$

where $V(\psi_p, \theta, \varphi)$ is an arbitrary function (essentially an integration constant) corresponding to a current-free magnetic field, so that in the absence of currents $\mathbf{B} = \nabla V$. This term can be neglected if we account for all the relevant currents in the functions F_1 and F_2. This leaves us with the following very useful expressions for the magnetic field,

$$\mathbf{B} = \nabla\varphi \times \nabla\psi_p + q(\psi_p)\nabla\psi_p \times \nabla\theta$$

$$= B_{\psi_p}(\psi_p, \theta, \varphi)\nabla\psi_p + B_\theta(\psi_p)\nabla\theta + B_\varphi(\psi_p)\nabla\varphi, \qquad (7.38)$$

where the first line is our earlier contravariant representation (7.37), and the second line is the new covariant representation. The important feature of this expression is that $B_\theta = \mu_0 F_1(\psi_p)$ and $B_\varphi = -\mu_0 F_2(\psi_p)$ are flux functions. Magnetic coordinates with this property are called *Boozer coordinates* (Boozer, 1981).

It is perhaps surprising that a term proportional to $\nabla\psi_p$ should appear in the covariant representation (7.38) since this vector is normal to the flux surface. The reason for this is that Boozer coordinates are not orthogonal, so that $\nabla\theta$ and $\nabla\varphi$ are not tangential to the flux surface. The term $B_{\psi_p}\nabla\psi_p$ is needed in Eq. (7.38) to cancel the perpendicular components of the other two terms. Finally, from the scalar product of the two representations of $\mathbf{B}$ in Eq. (7.38) it follows that the Boozer coordinate Jacobian is inversely proportional to B^2 on each flux surface,

$$g^{1/2} = \frac{B_\theta + qB_\varphi}{B^2},$$

since the numerator is a flux function. There are also other magnetic coordinates used extensively in the analysis of stellarators. For example, Hamada coordinates differ from Boozer coordinates only through the Jacobian (which is constant in Hamada coordinates), and are useful for certain kinetic transport calculations.

Guiding-centre motion

We now use Boozer coordinates to formulate guiding-centre equations of motion in a stellarator. It follows from Eq. (7.38) that the parallel velocity is equal to

$$v_\parallel = \frac{\mathbf{v} \cdot \mathbf{B}}{B} = \frac{1}{B} \left(B_{\psi_p} \dot{\psi}_p + B_\theta \dot{\theta} + B_\varphi \dot{\varphi} \right). \qquad (7.39)$$

In the small-gyroradius expansion underlying the drift equations the perpendicular velocity is smaller than the parallel one. This implies that the first term in (7.39) is smaller than the other two, e.g.,

$$\frac{B_{\psi_p} \dot{\psi}_p}{B_\varphi \dot{\varphi}} \sim \frac{v_d}{v_\parallel} \ll 1,$$

except very near turning points where $v_\parallel \to 0$. To the accuracy that guiding-centre equations are usually derived, it is therefore justified to neglect this term. Recalling that $q = d\psi_t / d\psi_p$, we conclude from the first line in Eq. (7.38) that the vector potential for the magnetic field is $\mathbf{A} = \psi_t \nabla \theta - \psi_p \nabla \varphi$. The guiding-centre Lagrangian (6.15) thus becomes

$$\bar{L} = \frac{m}{2B^2} \left(B_\theta \dot{\theta} + B_\varphi \dot{\varphi} \right)^2 + Ze \left(\psi_t \dot{\theta} - \psi_p \dot{\varphi} \right) - \mu B - Ze\Phi. \qquad (7.40)$$

The canonical momenta conjugate to θ and φ are

$$p_\theta = \frac{\partial \bar{L}}{\partial \dot{\theta}} = \frac{m v_\parallel B_\theta}{B} + Ze\psi_t,$$

$$p_\varphi = \frac{\partial \bar{L}}{\partial \dot{\varphi}} = \frac{m v_\parallel B_\varphi}{B} - Ze\psi_p,$$

with $v_\parallel = (B_\theta \dot{\theta} + B_\varphi \dot{\varphi})/B$, while p_{ψ_p} vanishes since $\bar{L}$ does not depend on ψ_p.

The guiding-centre Hamiltonian is equal to

$$H = p_\theta \dot{\theta} + p_\varphi \dot{\varphi} - \bar{L} = \frac{m v_\parallel^2}{2} + \mu B + Ze\Phi,$$

and is to be regarded as a function of the phase-space coordinates $(\theta, \varphi, p_\theta, p_\varphi)$ and time. For instance, the radial variation of $B(\psi_p, \theta, \varphi)$ should be expressed in terms of p_θ and p_φ. Note that the guiding centre is described by *four* coordinates in phase space, while *six* are needed for particles in general. The gyroaveraging procedure eliminated the gyroangle and led to the conclusion that the magnetic moment μ is constant, which therefore only appears as a parameter in the guiding-centre equations.

Further reading

There are several good texts discussing MHD in toroidal plasmas, e.g., Shafranov (1966), Freidberg (1987), White (1989), Biskamp (1993) and Wesson (1997). The topics covered in this chapter are also very carefully discussed by Hazeltine and Meiss (1992). In much of the neoclassical transport literature, the underlying MHD-equilibrium is assumed to be known. The following lucidly written papers can fill this gap: Callen and Dory (1972), Weening (1997), McCarthy (1999), and for rotating plasmas the classic paper by Maschke and Perrin (1980). In a notable paper by Cowley *et al.* (1991), analytical solutions to the Grad–Shafranov equation are constructed in the limit of large plasma pressure, $\epsilon \beta_p \gg 1$, and large aspect ratio. Conventional neoclassical theory is only applicable far (many banana widths) from the magnetic axis. Recently a number of attempts have been made to extend the theory to cover the region within a potato orbit width from the plasma centre. The reader is referred to Helander (2000) for a critical survey. The special formalism of Hamiltonian guiding-centre motion in toroidal magnetic fields is developed in depth in the books by White (1989), and Hazeltine and Waelbroeck (1998). A simple introduction to stellarator magnetic fields was given by Boozer (1998), and much more material on this topic can be found in the book by Wakatani (1998). Our derivation of Boozer coordinates follows unpublished work by Hirshman (1982).

Exercises

1. The *poloidal beta* can be defined in different ways; one is

$$\beta_p(\psi) \equiv \frac{8\pi}{\mu_0 I_p^2} \int p \, dS,$$

where the integral is taken over the area inside the flux surface ψ in a poloidal cross section of the torus, and I_p is the total toroidal plasma current inside this flux surface.

(a) Show that in a large-aspect-ratio tokamak with circular cross section

$$\beta_p(\psi) = \frac{\bar{p}}{B_\theta^2(a)/2\mu_0},$$

where $\bar{p}$ is the average pressure inside r.

Solution: This follows directly from

$$\int p \, dS = \pi r^2 \bar{p},$$

$$B_\theta = \frac{\mu_0 I_p}{2\pi r}.$$

(b) Derive the equilibrium condition

$$\frac{d}{dr}\left(p + \frac{B_\varphi^2}{2\mu_0}\right) + \frac{B_\theta}{\mu_0 r}\frac{d(rB_\theta)}{dr} = 0.$$

Solution: The force balance $\mathbf{j} \times \mathbf{B} = \nabla p$ is to lowest order in ϵ equivalent to $p' = j_\theta B_\varphi - j_\varphi B_\theta$. The desired relation follows after eliminating the currents by using

$$\mu_0 j_\theta = -\frac{dB_\varphi}{dr}, \quad \mu_0 j_\varphi = \frac{1}{r}\frac{d(rB_\theta)}{dr}$$

from Ampere's law $\nabla \times \mathbf{B} = \mu_0 \mathbf{j}$ in cylindrical coordinates.

(c) Show that

$$\beta_p = 1 + \frac{1}{(rB_\theta)^2}\int_0^r r'^2 \frac{dB_\varphi^2}{dr'}\,dr',$$

and note the significance of this result.

Solution: This expression follows immediately from (b). If $\beta_p > 1$ the toroidal field strength has a minimum at the magnetic axis because the current j_θ set up in the plasma produces a field that opposes the external field. The plasma is then diamagnetic. If $\beta_p < 1$ the situation is the opposite and the plasma is paramagnetic.

2. Show that at the plasma boundary, $r = a$, the radial derivative of the Shafranov shift is equal to

$$\frac{d\Delta_s}{dr}\bigg|_{r=a} = \epsilon\left(\beta_p + \frac{l_i}{2}\right),$$

and that (7.14) satisfies

$$\Lambda(a) = \beta_p + \frac{l_i}{2} - 1,$$

where

$$l_i = \frac{2}{a^2 B_p^2(a)}\int_0^a B_p^2(r)r\,dr$$

is the internal inductance of the plasma.

Solution: This follows from evaluating (7.14) and (7.17) at $r = a$, assuming that the edge pressure vanishes, $p(a) = 0$.

3. The tokamak safety factor.

(a) Show that the tokamak safety factor can be expressed as

$$q(\psi) = \frac{I(\psi)V'(\psi)}{4\pi^2}\left\langle\frac{1}{R^2}\right\rangle,$$

where $V(\psi)$ is the volume inside the flux surface ψ.

Solution: This follows immediately from the expression (7.4) for the volume $V(\psi)$, and definitions (7.5) and (7.7).

(b) For an equilibrium with circular cross section and large aspect ratio, q is given by (7.16). How is this formula modified if the cross section is elliptical?

Solution: From the preceding exercise

$$q = \frac{V'(\psi)B_\varphi}{4\pi^2 R},$$

at large aspect ratio, regardless of the shape of the cross section. If this is elliptical, we have $\psi = \psi(r)$, with

$$r^2 = z^2/k^2 + x^2,$$

where k is the vertical elongation. Then $V/2\pi R = \pi k r^2$ is the area of the elliptical cross section, and it follows that

$$q(r) = \frac{krB_\varphi}{\psi'(r)},$$

which differs from the circular expression (7.16) by a factor k.

4. Evaluate the bounce time (7.26) for trapped particles.

   Solution: The integral is taken around the banana orbit, i.e., first from $-\theta_b$ to θ_b and then back again. This means that

   $$\tau_b = (7.26) = \frac{4qR}{v\sqrt{2\epsilon\lambda}} \int_0^{\theta_b} \frac{d\theta}{\sqrt{k^2 - \sin^2 \theta/2}}$$

   $$= (y = \theta/2) \simeq \frac{8qR}{v\sqrt{2\epsilon}} \int_0^{\theta_b/2} \frac{dy}{\sqrt{k^2 - \sin^2 y}},$$

   where we recall that $k = \sin \theta_b/2$ and $\lambda = 1 + O(\epsilon)$ for trapped particles. To evaluate this integral, introduce the new variable x defined by $k \sin x = \sin y$, so that

   $$\cos y\, dy = k \cos x\, dx = k\, dx\sqrt{1 - \sin^2 x} = dx\sqrt{k^2 - \sin^2 y}.$$

   The bounce time (7.27) now follows.

5. As discussed in Section 6.4, trapped orbits drift slowly across the magnetic field. Without the drift a trapped particle would remain on the same field line all the time, but due to the magnetic drift the orbit does not quite close on itself, so that the field line label $\varphi_0 = \varphi - q\theta$ varies in time. This implies that trapped orbits undergo slow toroidal precession. Calculate the precession frequency in a large-aspect-ratio tokamak with circular cross section. Solution: The change in φ_0 caused by the magnetic drift is given by

   $$\dot\varphi_0 = \mathbf{v}_d \cdot \nabla(\varphi - q\theta) \simeq -\mathbf{v}_d \cdot \nabla(q\theta) \simeq \frac{v^2}{2\Omega_\varphi R}\left(\frac{q(r)\cos\theta}{r} + q'(r)\theta\sin\theta\right),$$

since Eq. (7.24) implies $\mathbf{v}_d \cdot \nabla r = -v_d \sin\theta$ and $\mathbf{v}_d \cdot \nabla\theta = -(v_d/r)\cos\theta$. Evaluating the bounce average, here denoted by an overbar, note that

$$
\overline{\cos\theta} = \frac{4}{\tau_b} \int_0^{\theta_b} \cos\theta \frac{d\theta}{\dot\theta} = \frac{4qR}{\tau_b v} \int_0^{\theta_b} \frac{\cos\theta\, d\theta}{\sqrt{1 - \lambda(1 - \epsilon\cos\theta)}}
$$

$$
= \frac{1}{K(k)} \int_0^{\theta_b/2} \frac{1 - 2\sin^2 y}{\sqrt{k^2 - \sin^2 y}}\, dy
$$

$$
= \frac{1}{K(k)} \int_0^{\pi/2} \frac{1 - 2k^2\sin^2 x}{\sqrt{1 - k^2\sin^2 x}}\, dx = \frac{2E(k)}{K(k)} - 1,
$$

and

$$
\overline{\theta\sin\theta} = \frac{4}{K(k)} \int_0^{\theta_b/2} \frac{y\sin y\cos y\, dy}{\sqrt{k^2 - \sin^2 y}} = \text{(partial integration)}
$$

$$
= \frac{4}{K(k)} \int_0^{\theta_b/2} \sqrt{k^2 - \sin^2 y}\, dy = \frac{4k^2}{K(k)} \int_0^{\pi/2} \frac{\cos x\, dx}{\sqrt{1 - k^2\sin^2 x}}
$$

$$
= 4\left(\frac{E(k)}{K(k)} + k^2 - 1\right).
$$

The toroidal precession thus becomes

$$
\overline{\dot\varphi}_0 = \frac{q^2}{\Omega_\varphi Rr}\left[\frac{E(k)}{K(k)} - \frac{1}{2} + \frac{2rq'}{q}\left(\frac{E(k)}{K(k)} + k^2 - 1\right)\right].
$$

This result can also be obtained by differentiating the second adiabatic invariant J, as described in an exercise at the end of Chapter 6. Note that the toroidal precession can be either positive or negative depending on the bounce angle θ_b, and is small in the sense that $\overline{\dot\varphi}_0 \tau_b \sim q\epsilon^{1/2}\rho_\theta/r \ll 1$. A trapped particle thus remains on approximately the same field line during a bounce period. Particles can also drift toroidally as a result of a radial electric field. As discussed in Section 13.2, this causes the entire plasma to rotate with the velocity $V_\varphi = E_r/B_\theta$.

8

Transport in toroidal plasmas

This chapter is is devoted to general results underpinning transport theory in an axisymmetric torus. The details of the theory that depend sensitively on the collision frequency are considered separately in the following chapters. Here, we outline some general features that are common to all collisionality regimes.

8.1 Transport ordering

Transport theory relies on a particular ordering of the relative importance of various physical processes in the plasma. The fundamental expansion parameter is that of the smallness of the Larmor radius ρ compared with the macroscopic scale length L,

$$\delta \equiv \rho/L \ll 1. \tag{8.1}$$

The requirement that this parameter be small must hold for all particle species; it is most severe for ions. We shall expand the distribution function in powers of δ, $f_a = f_{a0} + f_{a1} + \cdots$. Gradients (of density, temperature, magnetic field etc.) are thus assumed to be small on the scale of ρ. The task of transport theory is to describe the gradual relaxation of these gradients towards global thermodynamic equilibrium. As long as the step length in the random walk is smaller than the gradient length scale this process is diffusive, so the associated time derivative is expected to be of order

$$\frac{\partial}{\partial t} \sim \frac{D}{L^2} \sim \delta^2 v, \tag{8.2}$$

where v is the collision frequency and $D \sim v\rho^2$ the diffusion coefficient.

As usual, we shall also assume that the plasma is magnetized in the sense that the gyrofrequency exceeds the collision frequency,

$$\Delta = \frac{v}{\Omega} \ll 1,$$

a requirement that is practically always satisfied with a wide margin. Otherwise the magnetic field would be too weak to influence the confinement of the plasma. We note that we can also express Δ as the ratio between the Larmor radius $\rho = v_T/\Omega$ and the mean-free path $\lambda = v_T/v$,

$$\Delta = \frac{\rho}{\lambda},$$

so the condition $\Delta \ll 1$ again expresses the smallness of the Larmor radius, and

$$\Delta = O(\delta)$$

if the mean-free path is comparable to the macroscopic scale length L. When considering the issue of collisionality, we discuss various *subsidiary* orderings of $\delta/\Delta = \lambda/L$. Thus, while δ is considered to be the smallest parameter present in the problem, following the basic expansion in the smallness of δ, a secondary expansion in the smallness or largeness of λ/L is sometimes carried out. The requirement $\Delta = O(\delta)$ implies that the time derivative (8.2) is considerably smaller than the transit frequency,

$$\frac{\partial}{\partial t} \sim \delta^2 \frac{v_T}{L}. \tag{8.3}$$

Finally, it is convenient to take the flow velocities to be smaller than the thermal speed,

$$V \sim \delta v_T, \tag{8.4}$$

so that the drift kinetic equation can be used in the form derived in Chapter 6. This is usually justified experimentally. The case of rapidly rotating plasmas is discussed in Chapter 13.

A few words should be said about the large-aspect-ratio limit $\epsilon \ll 1$, where $\epsilon = r/R$ is the ratio of the minor to major radius of the toroidal plasma. We have already seen that magnetohydrodynamic equilibrium and particle dynamics are particularly simple in this limit. This is true also for transport theory. However, we develop the theory as far as possible without invoking $\epsilon \ll 1$. The macroscopic scale length L can thus be identified with either the major or minor radius of the torus; they are comparable. Occasionally, we show how the results are simplified when $\epsilon \ll 1$. In these cases, the requirement (8.1) that the plasma be magnetized must be interpreted in the stronger sense that the *poloidal* Larmor radius $\rho_p = mv_\perp/ZeB_p$ is smaller than the *minor* radius, $\rho_p/r \ll 1$.

Electric field

The electric field is related to potentials by

$$\mathbf{E} = -\nabla\Phi - \partial\mathbf{A}/\partial t \equiv -\nabla\Phi + \mathbf{E}^{(A)},$$

where the first term is the electrostatic field and the second term the inductive electric field. The transport ordering (8.3) implies that the first term dominates since

$$\frac{\partial\mathbf{A}/\partial t}{\nabla\Phi} \sim \frac{BL}{T/eL}\frac{\partial}{\partial t} \sim \delta$$

if the electrostatic potential is of order $\Phi \sim T/e$. This is consistent with the requirement (8.4) that the $\mathbf{E} \times \mathbf{B}$ drift should be smaller than the thermal speed. In the toroidal direction, $\nabla\Phi$ vanishes because of axisymmetry, leaving the small induction term,

$$E_\varphi = E_\varphi^{(A)} = -\partial\mathbf{A}_\varphi/\partial t, \tag{8.5}$$

which is normally used in a tokamak to drive Ohmic current.

It should be noted that the existence of this induced electric field implies that the magnetic equilibrium evolves in the sense that the poloidal flux, $\psi = -RA_\varphi$, varies in time, $\partial\psi/\partial t = RE_\varphi$. The flux surface labelled by ψ therefore moves (inward) with a velocity $\mathbf{V}_\psi$ satisfying

$$\frac{\partial\psi}{\partial t} + \mathbf{V}_\psi \cdot \nabla\psi = 0. \tag{8.6}$$

Although in general this implies that flux surfaces change shape, nothing observable need actually change at all in the plasma. For instance, if $RE_\varphi = c$ is constant, so that $\partial\psi/\partial t = c$, then we may simply introduce $\Psi = \psi - ct$. This modified flux function does not change in time, and the magnetic field is equal to $\mathbf{B} = I\nabla\Psi + \nabla\varphi \times \nabla\Psi$. On the other hand, if the loop voltage $V_{\text{loop}} = 2\pi RE_\varphi$ varies from point to point in the plasma, then the poloidal magnetic field must necessarily be non-stationary.

8.2 Collisionality

The physics of neoclassical transport depends on the relative magnitude of the collision frequency v and the transit frequency $\omega_t = v_T/qR$, the so-called collisionality. It is similar for electrons and ions since $v_{ee}/v_{ii} \sim v_{Te}/v_{Ti}$, but may differ among ion species with very disparate masses. (Highly charged impurities are more collisional than bulk ions and electrons.) If the collisionality is large,

$$\frac{L}{\lambda} \sim \frac{v}{v_T/qR} \gg 1, \tag{8.7}$$

the mean-free path $\lambda = v_T/v$ is shorter than the parallel distance around a flux surface $L \sim qR$, and the Braginskii fluid equations may be applied for the analysis. The particle orbits discussed in the preceding chapter are then not fully completed by a typical thermal particle since its motion is disturbed by collisions before an orbit has been completed. This high-collisionality regime is called the *Pfirsch–Schlüter*, or fluid, regime.

In the opposite limit,

$$\frac{v}{v_T/qR} \ll 1,$$

referred to as the *banana–plateau* regime, orbits are completed and short-mean-free-path closure of the fluid equations is inapplicable. The core of a tokamak is usually in this regime. For instance, if $n_e = 10^{20}$ m^{-3}, $T_e = 10$ keV, $R = 3$ m, and $q = 1.5$, we have

$$\frac{v_{ee}}{v_{Te}/qR} \sim 4 \cdot 10^{-4}.$$

If the aspect ratio is large, $\epsilon \ll 1$, the banana–plateau regime is subdivided into two regimes: the plateau regime

$$\epsilon^{3/2} \ll \frac{v}{v_T/qR} \ll 1, \tag{8.8}$$

and the banana regime

$$\frac{v}{v_T/qR} \ll \epsilon^{3/2}. \tag{8.9}$$

In the former, most circulating particle orbits are completed but trapped orbits are destroyed by collisions before completion since the effective collision frequency, $v_{\text{eff}} = v/\epsilon$, required to scatter a trapped particle out of its magnetic well, $\Delta B/B \sim \epsilon$, is larger than the bounce frequency $\omega_b \sim \sqrt{\epsilon} v_T/qR$, i.e.,

$$v_* \equiv \frac{v/\epsilon}{\omega_b} = \frac{v/\epsilon^{3/2}}{v_T/qR} \gg 1.$$

It is important to note that the effective collision frequency for scattering the velocity vector by an angle $\Delta\vartheta$ is $v/(\Delta\vartheta)^2$, as follows for instance from the form of the pitch-angle scattering operator (3.29). Trapped particles occupy the region $v_\parallel/v \sim \sqrt{\epsilon}$ in velocity space, so it is appropriate to take $\Delta\vartheta = \sqrt{\epsilon}$. Finally, in the banana regime, $v_* \ll 1$, the particle dynamics is virtually collisionless, and both types of orbits (trapped and circulating) can be completed.

It should be noted that since the collision frequency for an individual particle is sensitive to its energy, different particles on the same flux surface may have different collisionalities. The orderings we have discussed only refer to thermal particles.

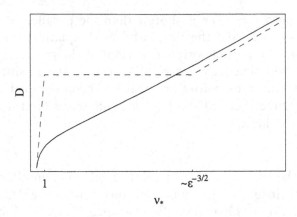

Fig. 8.1. Dependence of neoclassical diffusion coefficient on collisionality at large aspect ratio. The solid curve represents the asymptotic limit $\varepsilon \to 0$, while the dashed curve represents $\varepsilon = 0.2$.

Since $\epsilon \to 0$ on the magnetic axis, the central region of a sufficiently hot plasma (so that $vqR/v_T \ll 1$) is always formally in the plateau regime. However, if the collisionality is low this region may be very small. If it is narrower than an orbit width (potato width) it is of little importance for transport theory, which fundamentally only describes length scales much larger than the orbits. For instance, transport is only diffusive on length scales much larger than the random-walk step size.

At large aspect ratio, transport is quite different in different collisionality regimes. As we shall see in the next three chapters, the diffusivity of particles and heat is proportional to the collision frequency in the banana and Pfirsch–Schlüter regimes (with different proportionality constants) but is independent of collisionality (forms a 'plateau') in the plateau regime, see Fig. 8.1. While this is strictly true in the limit $\epsilon \to 0$, at realistic aspect ratios the distinction between the regimes is blurred, and the plateau is difficult to discern.

8.3 Distribution function

In any plasma, the distribution function of each particle species is approximately (to lowest order in δ) Maxwellian if the mean-free path is shorter than the macroscopic length scale (R in a torus), i.e., in the Pfirsch–Schlüter regime. As we have just seen, in itself this condition is not satisfied in most regions of a typical tokamak plasma, nor does it hold in most astrophysical plasmas.

However, in a system with closed magnetic flux surfaces, like a tokamak, the distribution function is nevertheless approximately Maxwellian, regardless of the collisionality, as long as the Larmor radius is small. To demonstrate this result, we turn to the drift kinetic equation (6.31). In the lowest order (in δ) we may neglect the slow guiding-centre drift across the magnetic field as well as the time derivative. For each species we then have

$$v_\parallel \nabla_\parallel f_0 = C(f_0). \tag{8.10}$$

We multiply this equation by $\ln f_0$, integrate over velocity space, and apply the flux-surface average (7.5). This procedure annihilates the left-hand side since

$$\left\langle \int (\ln f_0) v_\parallel \nabla_\parallel f_0 \, d^3v \right\rangle = \sum_\sigma \frac{2\pi B_0}{m^2} \sigma \left\langle \mathbf{B} \cdot \nabla \int f_0(\ln f_0 - 1) d\lambda \, \mathscr{E} \, d\mathscr{E} \right\rangle = 0,$$

where we have used (7.6),

$$d^3v = \sum_\sigma \frac{2\pi B}{m^2 v_\parallel B_0} \mathscr{E} \, d\mathscr{E} \, d\lambda,$$

and the relation $(\ln f_0)\nabla f_0 = \nabla(f_0 \ln f_0 - f_0)$. Hence, on flux-surface average

$$\left\langle \int (\ln f_{a0}) C_a(f_{a0}) d^3v \right\rangle = 0$$

for each particle species. For the ions in a simple plasma, collisions with electrons are weak, $C_{ie}/C_{ii} \sim (m_e/m_i)^{1/2}$, so $C_i \simeq C_{ii}$. But according to the H-theorem (Exercise 1 in Chapter 3)

$$\int (\ln f_a) C_{aa}(f_a) d^3v \leq 0,$$

with equality if and only if f_a is Maxwellian. We conclude that to lowest order in δ the ion distribution must be Maxwellian,

$$f_{i0} = f_{Mi}(\psi) = N_i(\psi) \left[\frac{m_i}{2\pi T_i(\psi)} \right]^{3/2} \exp\left[-\frac{m_i v^2/2 + Ze\Phi}{T_i(\psi)} \right]. \tag{8.11}$$

For the electrons $C_e = C_{ee} + C_{ei}$, and in the ion rest frame

$$\int (\ln f_{e0}) C_{ee}(f_{e0}) d^3v \leq 0,$$

$$\int (\ln f_{e0}) C_{ei}(f_{e0}) d^3v = S_L \left[f_{e0}/f_{Me}, \ln f_{e0} \right] \leq 0,$$

in the notation of (3.65). The electrons are therefore Maxwellian with the same flow velocity as the ions in lowest order,

$$f_{e0} = f_{Me}(\psi) = N_e(\psi) \left[\frac{m_e}{2\pi T_e(\psi)} \right]^{3/2} \exp\left[-\frac{m_e v^2/2 - e\Phi}{T_e(\psi)} \right].$$

Note that the electron–ion collision operator vanishes only if the electron Maxwellian is stationary in the ion rest frame. Quasineutrality forces the densities to be the same,

$$n_e = N_e(\psi)\exp(e\Phi/T_e) = Zn_i = ZN_i(\psi)\exp(-Ze\Phi/T_i), \qquad (8.12)$$

but the temperatures may well differ, $T_e \neq T_i$, since C_{ei} does not involve energy exchange, except as a small $O(m_e/m_i)$ correction to the Lorentz operator. Dividing the two expressions in (8.12) for the density shows that the electrostatic potential Φ is also approximately constant on flux surfaces, $\Phi \simeq \Phi_0(\psi)$. The poloidal variation of electrostatic potential, $\tilde{\Phi}$, is therefore usually weak in tokamaks, i.e., much smaller than the temperature, $e\tilde{\Phi}/T_e = O(\delta) \ll 1$. In rapidly rotating plasmas, however, stronger variation may arise because of the centrifugal force. Anomalous transport is probably caused by the $\mathbf{E} \times \mathbf{B}$ drift associated with potential fluctuations of plasma turbulence.

We have thus proved that the lowest-order drift kinetic equation (8.10) only has Maxwellian solutions that are constant on flux surfaces. In the case of low collisionality (higher bounce frequency than collision frequency), the underlying physics is the following. The rapid parallel streaming described by the left-hand side of (8.10) establishes a distribution that is constant on each flux surface, $\nabla_\parallel f_0 = 0$, in a bounce time, while collisions drive this distribution towards a Maxwellian in a collision time. On the other hand, if the collisionality is high, local thermodynamic equilibrium is first established at each point in the plasma in a collision time. Then, on the longer time scale associated with parallel collisional transport, which is of order $(L/\lambda)^2/\nu$ since this transport turns out to be diffusive, the density and temperature equilibrate on each flux surface. These processes are faster than allowed by the transport ordering (8.3). Therefore, in transport theory the relaxation to Maxwellian distributions has already occurred within each flux surface separately, and the discussion focuses on the slower relaxation of cross-field gradients.

The first-order [i.e., one order higher than (8.10)] drift kinetic equation is

$$\boxed{v_\parallel \nabla f_{a1} + \mathbf{v}_d \cdot \nabla f_{a0} + e_a v_\parallel E_\parallel^{(A)} \frac{\partial f_{a0}}{\partial \mathscr{E}} = C_a(f_{a1}).} \qquad (8.13)$$

It determines the correction f_{a1} to the lowest-order Maxwellian guiding centre distribution f_{a0}. Here and throughout the rest of the book C_a denotes the *linearized* collision operator of species a. The term $\mathbf{v}_d \cdot \nabla f_{a0} = (\mathbf{v}_d \cdot \nabla\psi)\,\partial f_{a0}/\partial\psi$ represents the effect of the cross-field drift, and by using (6.25) we can write

$$\dot{\mathbf{R}} \cdot \nabla\psi = v_\parallel(\mathbf{b} \times \nabla\psi) \cdot \nabla\left(\frac{v_\parallel}{\Omega_a}\right) = Iv_\parallel \nabla_\parallel\left(\frac{v_\parallel}{\Omega_a}\right), \qquad (8.14)$$

with the derivative $\nabla_\parallel$ taken at constant $\mathscr{E}$ and μ. To arrive at Eq. (8.14) we have used $RB_\varphi = I$ and

$$\boxed{\frac{\mathbf{B} \times \nabla\psi}{B^2} = \frac{I}{B}\mathbf{b} - R\hat{\boldsymbol{\varphi}},} \tag{8.15}$$

where $\hat{\boldsymbol{\varphi}} = R\nabla\varphi$ is the unit vector in the toroidal direction. This very useful relation follows from $\mathbf{B} \times \nabla\psi = I(\psi)\nabla\varphi \times \nabla\psi - |\nabla\psi|^2\nabla\varphi$ and $R^2B^2 = I^2 + |\nabla\psi|^2$. Hence the first-order drift kinetic equation becomes

$$v_\parallel \nabla_\parallel \left(f_{a1} + \frac{Iv_\parallel}{\Omega_a}\frac{\partial f_{a0}}{\partial\psi} \right) - \frac{e_a E_\parallel^{(A)}}{T_a} v_\parallel f_{a0} = C_a(f_{a1}). \tag{8.16}$$

The task of neoclassical theory is essentially to solve this equation in various collisionality regimes.

8.4 Current

Plasma confinement is caused by a current which together with the magnetic field produces a $\mathbf{j} \times \mathbf{B}$ force opposing the pressure gradient, according to the force balance equation (7.8). Taking the vectorial product of this equation with $\mathbf{B}$ gives the perpendicular (diamagnetic) current,

$$\mathbf{j}_\perp = \frac{\mathbf{B} \times \nabla p}{B^2} = \left(\frac{I\mathbf{B}}{B^2} - R\hat{\boldsymbol{\varphi}} \right) \frac{dp}{d\psi},$$

where we have used (8.15) in the second equality. Incidentally, this expression for $\mathbf{j}_\perp$ can also be obtained by taking the sum $\mathbf{j}_\perp = \sum_a n_a e_a \mathbf{V}_{a\perp}$ of the diamagnetic flows (4.64) of all species.

In a torus, this perpendicular current is generally not divergence free since

$$\nabla \cdot \mathbf{j}_\perp = \mathbf{B} \cdot \nabla \left(\frac{I}{B^2}\frac{dp}{d\psi} \right) \neq 0$$

if the magnetic field strength varies along the field lines. (Here we have used the relation $\nabla \cdot (A\mathbf{B}) = \mathbf{B} \cdot \nabla A$, which holds for any scalar A since $\nabla \cdot \mathbf{B} = 0$.) On the other hand, the total current must be divergence free to ensure quasineutrality, $\nabla \cdot (\mathbf{j}_\perp + \mathbf{j}_\parallel) = 0$. The existence of a parallel current is necessary to satisfy this requirement. Since

$$0 = \nabla \cdot \left(\frac{j_\parallel \mathbf{B}}{B} + \mathbf{j}_\perp \right) = \mathbf{B} \cdot \nabla \left(\frac{j_\parallel}{B} + \frac{I}{B^2}\frac{dp}{d\psi} \right)$$

it follows that the quantity within the parentheses must be constant of flux surfaces, and the parallel current becomes

$$j_\parallel = -\frac{I}{B}\frac{dp}{d\psi} + K(\psi)B,$$

where $K(\psi)$ is an arbitrary integration constant. This is the 'return current' necessary to close the diamagnetic current. Clearly, this is a geometrical effect associated with toroidicity; no return current is necessary in a cylinder since $\nabla \cdot \mathbf{j}_\perp = 0$ there. The sum of the parallel and perpendicular components of the current is

$$\mathbf{j} = -\frac{dp}{d\psi} R\hat{\varphi} + K(\psi)\mathbf{B}. \tag{8.17}$$

By relating $K(\psi)$ to the flux-surface average $\langle j_\| B \rangle$, we can write the parallel current as

$$j_\| = \frac{B}{\langle B^2 \rangle} \langle j_\| B \rangle - \frac{I(\psi)}{B} \frac{dp}{d\psi} \left(1 - \frac{B^2}{\langle B^2 \rangle} \right) \equiv j_{OH} + j_{PS}, \tag{8.18}$$

where the two terms are governed by essentially different physics. In a collisional plasma the first term is usually driven by the parallel electric field, and is then called the Ohmic current. We have therefore denoted it by j_{OH} here. (It can sometimes also be driven by neutral beams or plasma waves.) However, in a collisionless plasma this term may have a substantial contribution from the so-called 'bootstrap current', which is driven by perpendicular gradients. The second term is the *Pfirsch–Schlüter current*, and is driven by the pressure gradient. Note that we have invoked nothing but force balance (7.8) to arrive at the result (8.18), which is therefore generally valid in all collisionality regimes.

To see that the Ohmic current arises in response to the induced electric field in a collisional plasma, we use Braginskii's electron momentum equation (2.16) multiplied by $\mathbf{B}/n_e$ to obtain

$$0 = -e\mathbf{B} \cdot \mathbf{E} - \frac{\mathbf{B} \cdot \nabla p_e}{n_e} - \frac{0.51 m_e}{\tau_e} (V_{e\|} - V_{i\|})B - 0.71\mathbf{B} \cdot \nabla T_e.$$

Here, (4.36) has been used for the friction, and electron inertia has been neglected. Viscosity is also neglected as it is small in the collisional regime, see (4.50)–(4.55). The Ohmic current is now obtained from the flux-surface average, yielding

$$\langle j_\| B \rangle = \sigma \langle \mathbf{B} \cdot \mathbf{E} \rangle, \tag{8.19}$$

where the electrical conductivity is $\sigma = n_e e^2 \tau_e / 0.51 m_e$. We have used (7.6) and the fact that the temperature is nearly constant on the flux surface, so that $\langle (\mathbf{B}/n_e) \cdot \nabla p_e \rangle \simeq T_e \langle \mathbf{B} \cdot \nabla \ln p_e \rangle = 0$.

The appearance of the parallel Pfirsch–Schlüter current in (8.18) is perhaps surprising. Apparently, it arises in response to the pressure gradient, and vanishes on a flux-surface average in the sense that $\langle B j_{PS} \rangle = 0$.

(Note however that $\langle j_{PS} \rangle \neq 0$.) Since the magnetic field decreases with increasing R, the Pfirsch–Schlüter current has the same sign as $I(\psi)$ on the outboard side of the torus and the opposite sign on the inboard side (if $dp/d\psi < 0$). It therefore reinforces the Ohmic current on the outboard side and opposes it on the inboard side, which has the effect of increasing the poloidal magnetic field on the outboard side.

Evidently, the Pfirsch–Schlüter current is associated with the curvature of the magnetic field. (If the field had been straight, the diamagnetic current would have been divergence free.) In a sense, it can be said to be the fluid manifestation of the guiding-centre drift in the curved field. Electron and ion guiding centres drift in opposite directions according to (7.24), leading to a vertical current. The Pfirsch–Schlüter return current along the magnetic field closes the circuit.

In a large-aspect-ratio tokamak, the Ohmic current can be estimated as

$$j_{OH} \sim B_p/\mu_0 r, \tag{8.20}$$

from $\nabla \times \mathbf{B} = \mu_0 \mathbf{j}$, and the Pfirsch–Schlüter current in (8.18) as

$$j_{PS} \sim \epsilon p/r B_p,$$

since $d\psi \simeq R B_p \, dr$, $B \simeq I/R$, and $1 - B^2/\langle B^2 \rangle \sim \epsilon$. Their ratio is thus

$$j_{PS}/j_{OH} \sim \epsilon \beta_p,$$

where $\beta_p \equiv p/(B_\theta^2/2\mu_0)$ is the poloidal beta, which is of order unity in the standard tokamak ordering, $\beta = O(\epsilon^2)$, see Exercise 7.1. The Pfirsch–Schlüter current is thus smaller than the Ohmic current unless $\epsilon \beta_p \sim 1$, which is the tokamak equilibrium limit. It is also smaller than the bootstrap current, which we shall calculate in Chapter 11, by a factor of $\epsilon^{1/2}$. Moreover, the bootstrap current does not vanish on a flux-surface average.

8.5 Parallel particle and heat fluxes

We have just seen that a parallel return current is necessary to close the diamagnetic current in a toroidal magnetic field. A similar argument can be made for particle and heat fluxes, again independently of collisionality.

In Section 8.3, it was shown that if the transport orderings are satisfied, the distribution function for each species is nearly Maxwellian with a flow velocity we assumed to be small (8.4). Therefore the perpendicular component of the flow in lowest order equals the diamagnetic flow (4.64), which we write as

$$n_a \mathbf{V}_{a\perp} = n_a(\psi) \omega_a(\psi) \left(R\hat{\varphi} - \frac{I}{B}\mathbf{b} \right), \tag{8.21}$$

$$\omega_a(\psi) \equiv -\frac{d\Phi}{d\psi} - \frac{1}{n_a e_a}\frac{dp_a}{d\psi}, \tag{8.22}$$

using (8.15). Note that to lowest order p_a and Φ are functions of ψ only, so the flow is perpendicular to $\nabla\psi$ and thus tangential to the flux surface. Like all diamagnetic flows, it does not relax the gradients that drive it, and does not lead directly to any transport. Incidentally, it is often convenient in neoclassical theory to decompose vectors in their toroidal and parallel components, as has been done here. At large aspect ratio, these directions are nearly, but not exactly, parallel. Nevertheless, the plasma flow components in these directions are governed by fundamentally different physics, namely rapid flow along the field versus axisymmetry in the toroidal direction.

Equation (8.21) represents the perpendicular component of the flow velocity within the flux surface and is of order $V \sim \delta v_T$. The parallel component can be determined by observing that the total (perpendicular + parallel) flow within the flux surface must be divergence free to zeroth order in δ. Otherwise the continuity equation

$$\frac{\partial n}{\partial t} + \nabla \cdot (n\mathbf{V}) = 0,$$

would imply larger time derivatives,

$$\frac{\partial \ln n}{\partial t} \sim \frac{V}{L} \sim \delta \frac{v_T}{L},$$

than allowed by the transport ordering (8.3). However, even in an axisymmetric field ($\partial/\partial\varphi = 0$), the diamagnetic flow (8.21) is not itself divergence free. The general formula for the divergence of a vector $\mathbf{B}$ in an arbitrary coordinate system is given by (7.32), where the Jacobian becomes $g^{1/2} = (\mathbf{B} \cdot \nabla\theta)^{-1}$ if the coordinates are (ψ, θ, φ). The divergence of the perpendicular particle flux is thus

$$\nabla \cdot (n_a \mathbf{V}_{a\perp}) = \frac{1}{g^{1/2}} \frac{\partial}{\partial\theta} \left(g^{1/2} n_a \mathbf{V}_{a\perp} \cdot \nabla\theta\right) = -\frac{1}{g^{1/2}} \frac{\partial}{\partial\theta} \left(\frac{\omega_a n_a I}{B^2}\right) \neq 0,$$

and must therefore be balanced by a parallel 'return flow' $V_{a\parallel}$ with

$$\nabla \cdot (n_a \mathbf{V}_{a\parallel}) = -\nabla \cdot (n_a \mathbf{V}_{a\perp}).$$

Using

$$\nabla \cdot (n_a \mathbf{V}_{a\parallel}) = \frac{1}{g^{1/2}} \frac{\partial}{\partial\theta} \left(g^{1/2} n_a V_{a\parallel} \mathbf{b} \cdot \nabla\theta\right) = \frac{1}{g^{1/2}} \frac{\partial}{\partial\theta} \left(\frac{n_a V_{a\parallel}}{B}\right),$$

we conclude that parallel flow is

$$n_a V_{a\parallel} = \frac{I(\psi) n_a(\psi) \omega_a(\psi)}{B} + K_a(\psi) B, \tag{8.23}$$

where $K_a(\psi)$ is an integration constant. The total flow (8.21)+(8.23) within the flux surface thus becomes

$$\boxed{n_a \mathbf{V}_a = \omega_a(\psi) n_a(\psi) R \hat{\varphi} + K_a(\psi) \mathbf{B},} \tag{8.24}$$

where it is apparent that $\nabla \cdot (n_a \mathbf{V}_a) = 0$. In fact, this is the most general form for an incompressible flow that is tangential to flux surfaces. The first term in (8.24) describes purely toroidal rotation of the species a. The angular frequency ω_a may differ from surface to surface and from species to species, but within every flux surface each species fluid rotates as a rigid body. The *poloidal* rotation is entirely contained in the second term, which is parallel to the magnetic field. Here, the velocity varies within the flux surface in proportion to the field strength. The parallel velocity is thus larger on the inboard side of the torus than on the outboard side. The reason is that the area dS of the cross section of a magnetic flux tube is inversely proportional to field strength B to keep the magnetic flux $B\,dS$ constant. But the particle flux $n_a V_{a\parallel}\,dS$ must also be constant, so when the flux tube contracts on the inboard side of the torus the velocity $V_{a\parallel}$ increases correspondingly.

The current is obtained by multiplying (4.24) by the charge e_a, summing over species, and using charge neutrality, $\sum n_a e_a = 0$,

$$\mathbf{j} = \sum_a n_a e_a \mathbf{V}_a = -\frac{dp}{d\psi} R \hat{\varphi} + \mathbf{B} \sum_a e_a K_a(\psi).$$

This expression reproduces (8.17) if one identifies $\sum_a e_a K_a$ with K.

Finally, the heat flux can be treated in a completely analogous way. The perpendicular heat flux of each species is approximately diamagnetic (4.76),

$$\mathbf{q}_{a\perp} = \frac{5 p_a}{2 m_a \Omega_a} \mathbf{b} \times \nabla T_a.$$

If the total heat flux of each species is divergence free, there must be a parallel return flow,

$$q_{a\parallel} = -\frac{5 n_a T_a I}{2 e_a B} \frac{dT_a}{d\psi} + L_a(\psi) B, \tag{8.25}$$

where $L_a(\psi)$ is an arbitrary integration constant analogous to the quantity $K_a(\psi)$ in (8.24). The total heat flux (the sum of parallel and perpendicular components) can be written as

$$\mathbf{q}_a = -\frac{5 p_a}{2 e_a B} \frac{dT_a}{d\psi} R \hat{\varphi} + L_a(\psi) \mathbf{B}. \tag{8.26}$$

In the next chapter, we shall derive this result more carefully (for a pure plasma) by scrutinizing the energy equation and convincing ourselves that indeed $\nabla \cdot \mathbf{q}_i = 0$ since the other (convective) terms are small. Here, we note that (8.25) is not completely general since even if the convective terms in the energy equation are negligible, we can only say that the total heat flux is divergence free, $\sum_a \nabla \cdot \mathbf{q}_a = 0$. In taking the heat flux for *each* particle species to be divergence free, we have implicitly assumed that there is very little energy exchange between different species. This is true for particle species with widely disparate masses (as in a plasma consisting of electrons, ions, and heavy impurities) since the energy transfer term is small for collisions between such particles, see (3.56). For relatively light impurities, such as carbon, the situation is more complicated.

8.6 Flow across flux surfaces

Applying Gauss's law to the continuity equation (2.9) gives

$$-\frac{\partial}{\partial t} \int_V n_a d^3 r = \int_{\partial V} n_a \mathbf{V}_a \cdot d\mathbf{S}.$$

Let us take the first integral over the volume V inside some flux surface ψ. Since the surface element is $d\mathbf{S} = \nabla \psi \, d\theta \, d\varphi / \mathbf{B} \cdot \nabla \theta$ by (7.3), we conclude that the particle flux across the flux surface is

$$-\frac{\partial}{\partial t} \int_V n_a d^3 r = \langle n_a \mathbf{V}_a \cdot \nabla \psi \rangle \oint \frac{d\theta}{\mathbf{B} \cdot \nabla \theta}.$$

One aim of transport theory is to calculate this quantity, which apart from a geometric factor thus equals $\langle n_a \mathbf{V}_a \cdot \nabla \psi \rangle$, or

$$\langle \mathbf{\Gamma}_a \cdot \nabla \psi \rangle = \langle R \hat{\boldsymbol{\varphi}} \cdot (n_a \mathbf{V}_a \times \mathbf{B}) \rangle ,$$

where $\mathbf{\Gamma}_a \equiv n_a \mathbf{V}_a$, and we have used the representation (7.2) for the magnetic field. This form of the cross-field flux suggests that we use the flux-surface averaged $R\hat{\boldsymbol{\varphi}}$-projection of the momentum equation (2.10) to write

$$\langle e_a \mathbf{\Gamma}_a \cdot \nabla \psi \rangle = \left\langle R \hat{\boldsymbol{\varphi}} \cdot \frac{\partial m_a n_a \mathbf{V}_a}{\partial t} \right\rangle + \langle R \hat{\boldsymbol{\varphi}} \cdot \nabla \cdot \mathbf{\Pi} \rangle - \left\langle n_a e_a R E_\varphi^{(A)} \right\rangle - \langle R F_{a\varphi} \rangle ,$$

where we have recalled (8.5) and now denote the force by $\mathbf{F}$ rather than $\mathbf{R}$ as in Chapter 2. Note that the pressure gradient does not appear since $R\hat{\boldsymbol{\varphi}} \cdot \nabla p_a = \partial p_a / \partial \varphi = 0$ by axisymmetry. As shown in the exercises, the two first terms on the right are small in transport theory so we have

$$\boxed{\langle e_a \mathbf{\Gamma}_a \cdot \nabla \psi \rangle = -\left\langle n_a e_a R E_\varphi^{(A)} \right\rangle - \langle R F_{a\varphi} \rangle .} \qquad (8.27)$$

That the $\langle R\hat{\boldsymbol{\varphi}} \cdot \nabla \cdot \mathbf{\Pi} \rangle$ component of the viscosity is small implies weak radial transport of angular momentum, and has important consequences for the evolution of the radial electric field. This was discussed in Chapter 5 in the simpler context of transport in a cylinder. The first term on the right in Eq. (8.27) has a simple interpretation. In view of Eq. (8.6), this term can be written as $\langle n_a e_a \mathbf{V}_\psi \cdot \nabla \psi \rangle$ and thus indicates that the plasma has a tendency to move with the flux surface labelled by ψ.

It is useful to decompose the cross-field flux (8.27) into three terms,

$$\langle \mathbf{\Gamma}_a \cdot \nabla\psi \rangle = \langle \mathbf{\Gamma}_a \cdot \nabla\psi \rangle^{\mathrm{cl}} + \left\langle n_a \frac{\mathbf{E}^{(A)} \times \mathbf{B}}{B^2} \cdot \nabla\psi \right\rangle + \langle \mathbf{\Gamma}_a \cdot \nabla\psi \rangle^{\mathrm{neo}} .$$

The first is the classical flux,

$$\langle \mathbf{\Gamma}_a \cdot \nabla\psi \rangle^{\mathrm{cl}} \equiv \langle R F_{a\perp\varphi}/e_a \rangle, \qquad (8.28)$$

which is driven by perpendicular friction as we saw in Chapter 5. The second term is the $\mathbf{E} \times \mathbf{B}$ drift across the flux surface, and the remainder is defined to be the neoclassical flux. It can be written as

$$\langle \mathbf{\Gamma}_a \cdot \nabla\psi \rangle^{\mathrm{neo}} \equiv -I \left\langle \frac{F_{a\parallel} + n_a e_a E_\parallel^{(A)}}{e_a B} \right\rangle, \qquad (8.29)$$

by using (8.15). Thus, while the the classical flux is associated with *perpendicular* friction, the neoclassical flux has to do with *parallel* friction. This reflects the different mechanisms of these transport processes. Classical transport is caused by frictional relaxation of the diamagnetic flow associated with gyromotion, which is perpendicular to the magnetic field. Neoclassical transport is caused by friction acting on guiding centres, which predominantly move along the field. In a torus with large aspect ratio the neoclassical flux is usually much larger than the classical flux. Both the classical and the neoclassical particle flux are *automatically ambipolar*. That is, regardless of any radial electric field, we always have

$$\sum_a e_a \langle \mathbf{\Gamma}_a \cdot \nabla\psi \rangle = 0, \qquad (8.30)$$

because of momentum conservation, $\sum_a \mathbf{F}_a = 0$, and quasineutrality, $\sum_a n_a e_a = 0$. This ambipolarity property is a consequence of the transport ordering we have adopted. As we saw in the context of classical transport in Section 5.2, non-ambipolar transport is small in the Larmor radius. Non-ambipolar transport can also occur on time scales shorter than allowed by the transport ordering (8.2).

It is instructive to also construct the neoclassical flux (8.29) kinetically from

$$\langle \mathbf{\Gamma}_a \cdot \nabla\psi \rangle^{\mathrm{neo}} = \left\langle \int f_a \mathbf{v}_d \cdot \nabla\psi \, d^3v \right\rangle, \qquad (8.31)$$

which is done in Exercise 3. Neoclassical transport is thus driven by the deviation $f_{a1}(\mathbf{R})$ of the guiding-centre distribution from a Maxwellian. Classical transport is caused by the difference (6.32) between the distribution function of particles, $f_a(\mathbf{r})$, and guiding centres, $f_a(\mathbf{R})$, i.e., by the departure from a Maxwellian caused by finite Larmor radius.

The conductive heat flux across flux surfaces,

$$\langle \mathbf{q}_a \cdot \nabla\psi \rangle = \langle \mathbf{Q}_a \cdot \nabla\psi \rangle - \frac{5T_a}{2}\Gamma_a,$$

(conductive $=$ total$-$convective) is obtained in a similar way to the particle flux. Its neoclassical part is

$$\langle \mathbf{q}_a \cdot \nabla\psi \rangle^{\mathrm{neo}} = \left\langle \int f_a \left(\frac{m_a v^2}{2} - \frac{5T_a}{2} \right) \mathbf{v}_d \cdot \nabla\psi \, d^3v \right\rangle$$

$$= -\frac{I}{e_a} \left\langle \int \left(\frac{m_a v^2}{2} - \frac{5T_a}{2} \right) m_a v_\| C_a(f_a) \frac{d^3v}{B} \right\rangle. \qquad (8.32)$$

These fluxes can be related to the rate of collisional entropy production on a flux surface

$$\dot{S}_a = -\left\langle \int \frac{f_{a1}}{f_{a0}} C_a(f_{a1}) \, d^3v \right\rangle, \qquad (8.33)$$

by using the kinetic equation (8.13) to write

$$\dot{S}_a = -\left\langle \int \frac{f_{a1}}{f_{a0}} \left(v_\| \nabla f_{a1} + \mathbf{v}_d \cdot \nabla f_{a0} + e_a v_\| E_\|^{(A)} \frac{\partial f_{a0}}{\partial \mathscr{E}} \right) d^3v \right\rangle.$$

The first term in this expression vanishes because

$$\left\langle \int v_\| \nabla_\| \left(\frac{f_{a1}^2}{2f_{a0}} \right) d^3v \right\rangle = \sum_\sigma \sigma \left\langle \int B\nabla_\| \left(\frac{f_{a1}^2}{2f_{a0}} \right) \frac{\pi v^3}{B_0} dv \, d\lambda \right\rangle = 0$$

since $\langle B\nabla_\| A \rangle = 0$ for any A, and the flux-surface averaged entropy production associated with species a thus becomes

$$\dot{S}_a = \frac{\langle j_{a\|}E_\| \rangle}{T_a} - \left(\frac{d\ln p_a}{d\psi} + \frac{e_a}{T_a}\frac{d\Phi_a}{d\psi} \right) \langle \Gamma_a \cdot \nabla\psi \rangle^{\mathrm{neo}} - \frac{d\ln T_a}{d\psi}\frac{\langle \mathbf{q}_a \cdot \nabla\psi \rangle^{\mathrm{neo}}}{T_a},$$

where we have used expressions (8.31) and (8.32) for the neoclassical transport. The meaning of this result is clear: transport tends to relax gradients and so produces entropy. For instance, transport of heat down a temperature gradient produces entropy $(\mathbf{q} \cdot \nabla T)/T^2$ in accordance with the general thermodynamic relation $dq = T \, dS$.

8.7 Confinement time

The quality of confinement is usually expressed in terms of the *energy confinement time*, which is defined by

$$\tau_E \equiv \frac{W}{P_{\text{loss}}}, \tag{8.34}$$

where W is the total energy of the plasma and P_{loss} is the energy loss rate. Roughly speaking, this is the time scale on which plasma energy would decay if the heating were switched off and if this did not affect energy transport (which it usually does, however). In a steady state, P_{loss} is equal to the heating power, which is usually well known, but if the plasma is evolving some care is required in evaluating the confinement time. It is the purpose of this section to derive an expression for τ_E which remains valid when the plasma moves around or changes shape, and when the currents in the plasma and the coils vary rapidly.

There are several reasons why a plasma evolves during a discharge. The plasma may be heated, or lose energy, so rapidly that the stored energy changes on the time scale of τ_E. Alternatively, the currents in the plasma and field coils may vary, thus changing the magnetic field structure and plasma shape. In the plasma core, the magnetic field usually changes quite slowly because of high electrical conductivity, so that in the diffusion equation for the magnetic field,

$$\frac{\partial \mathbf{B}}{\partial t} = \frac{1}{\mu_0 \sigma} \nabla^2 \mathbf{B},$$

the magnetic diffusivity $1/\mu_0\sigma$ is small. This equation is obtained by using $\mathbf{j} = \sigma\mathbf{E}$ in Faraday's law, $\nabla \times \mathbf{E} = -\partial\mathbf{B}/\partial t$. (Even if this simple Ohm's law is not strictly accurate (due to, e.g., the existence of a bootstrap current) we may still use it for the present order-of-magnitude estimate.) Comparing the 'skin time', $\tau_{\text{skin}} \sim a^2\mu_0\sigma$ associated with magnetic field evolution with a typical neoclassical (ion) energy confinement time,

$$\tau_E \sim \frac{a^2}{\epsilon^{1/2}\rho_{i\theta}^2/\tau_i},$$

see Eq. (1.10), gives

$$\frac{\tau_{\text{skin}}}{\tau_E} \sim \beta_p \sqrt{\frac{\epsilon m_i}{m_e}}, \tag{8.35}$$

which is usually large. Here we have used the Spitzer value, Eq. (4.27), for the conductivity σ, and recalled the poloidal beta from Chapter 7, Exercise 1. Thus, even if confinement is neoclassical, the energy confinement time tends to be shorter than the resistive skin time in the plasma core. However, in low-pressure discharges magnetic field evolution can be important.

To analyse the energy balance in an evolving discharge, we recall the energy conservation equation (2.11) summed over all species in the plasma,

$$\frac{\partial}{\partial t}\left(\frac{3p}{2} + \frac{\rho V^2}{2}\right) + \nabla \cdot \mathbf{Q} = \mathbf{j} \cdot \mathbf{E} + P,$$

where $\rho = \sum m_a n_a$ is the mass density, $\mathbf{V}$ the mean flow velocity, $p = \sum n_a T_a$ the total pressure, and $\mathbf{Q}$ the total (conductive + convective) heat flux. We have added a term P representing the heating power density by neutral-beam injection, radio-frequency waves etc., on the right-hand side. Ohmic heating is described by $\mathbf{j} \cdot \mathbf{E}$ and should not be included in P.

We also need Poynting's theorem (Jackson, 1975),

$$\frac{\partial}{\partial t}\left(\frac{B^2}{2\mu_0} + \frac{\epsilon_0 E^2}{2}\right) + \nabla \cdot \mathbf{s} = -\mathbf{j} \cdot \mathbf{E},$$

where $\mathbf{s} = \mathbf{E} \times \mathbf{B}/\mu_0$ is the Poynting vector and where the electric field energy term is negligible since $\epsilon_0 \mu_0 E^2/B^2 = (E/cB)^2 \ll 1$ if the $\mathbf{E} \times \mathbf{B}$ velocity is subluminal. Adding these two equations gives a conservation equation for total energy,

$$\frac{\partial}{\partial t}\left(\frac{3p}{2} + \frac{\rho V^2}{2} + \frac{B^2}{2\mu_0}\right) + \nabla \cdot (\mathbf{Q} + \mathbf{s}) = P. \tag{8.36}$$

In order to allow the plasma to move and to change shape, we assume that it occupies a region $U(t)$ which varies smoothly, but otherwise arbitrarily, in time. The boundary of this region, which we denote by ∂U, thus moves with some speed $\mathbf{u}$, and is assumed to form a flux surface at each t. In a magnetic field with a poloidal divertor, ∂U could typically be the separatrix. As shown in an exercise at the end of this chapter, the volume integral $J(t)$ over $U(t)$ of any quantity $A(\mathbf{r}, t)$,

$$J(t) = \int_{U(t)} A(\mathbf{r}, t)\, dV$$

varies in time according to

$$\frac{dJ}{dt} = \int_{U(t)} \frac{\partial A}{\partial t} dV + \int_{\partial U(t)} A\mathbf{u} \cdot d\mathbf{S}, \tag{8.37}$$

where the second integral is an area integral over the plasma boundary. This result can be used together with Eq. (8.36) to calculate the rate of change of the thermal energy

$$W \equiv \int_U \left(\frac{3p}{2} + \frac{\rho V^2}{2}\right) dV$$

as

$$\frac{dW}{dt} = P_{\text{aux}} - \int_U \frac{\partial}{\partial t} \left(\frac{B^2}{2\mu_0} \right) dV - \int_{\partial U} \left[-\left(\frac{3p}{2} + \frac{\rho V^2}{2} \right) \mathbf{u} + \mathbf{Q} + \mathbf{s} \right] \cdot d\mathbf{S};$$

(8.38)

where $P_{\text{aux}} = \int P \, dV$ is the total auxiliary heating power and we have used Gauss's theorem. It is reasonable to assume that the energy density is very small at the plasma edge, so that

$$\frac{3p}{2} + \frac{\rho V^2}{2} \simeq 0 \quad \text{on } \partial U$$

can be neglected. The surface integral of the Poynting vector appearing in (8.38) is equal to

$$\int_{\partial U} \mathbf{s} \cdot d\mathbf{S} = \mu_0^{-1} \int (E_p B_\varphi - E_\varphi B_p) \, 2\pi R \, dl_p$$

$$= \mu_0^{-1} \left(2\pi R B_\varphi \oint E_p dl_p - V_{\text{loop}} \oint B_p \, dl_p \right),$$

where the line integrals are taken one poloidal turn around the flux surface (with dl_p the line element), and where $V_{\text{loop}} = 2\pi R E_\varphi$ is the loop voltage at the plasma edge. By Ampère's and Faraday's laws these integrals are equal to

$$\oint B_p dl_p = \mu_0 I_p,$$

$$\oint E_p dl_p = -\int \frac{\partial B_\varphi}{\partial t} dS_p,$$

where I_p is the toroidal plasma current and the surface integral is taken over the poloidal cross section of the plasma. In the remaining integral in (8.38),

$$\int_U \frac{\partial}{\partial t} \left(\frac{B^2}{2\mu_0} \right) dV = \int_U \frac{B_p \dot{B}_p + B_\varphi \dot{B}_\varphi}{\mu_0} dV,$$

it is useful to decompose the toroidal field into two parts,

$$B_\varphi = B_\varphi^{(\text{vacuum})} + B_\varphi^{(\text{plasma})},$$

where the first term represents the 'vacuum' field produced by the toroidal field coils (and any other poloidal currents outside the plasma), and the second term is the toroidal field generated by currents inside the plasma (which is paramagnetic or diamagnetic depending on β_p). The former term then cancels one of the terms from the Poynting flux, and Eq. (8.38) can be written as

$$\frac{dW}{dt} = P_{\text{aux}} + I_p V_{\text{loop}} - X - \int_{\partial U} \mathbf{Q} \cdot d\mathbf{S},$$

where

$$X = \mu_0^{-1} \int_U \left[B_p \dot{B}_p + B_\varphi^{(\text{plasma})} \left(\dot{B}_\varphi^{(\text{vacuum})} + \dot{B}_\varphi^{(\text{plasma})} \right) \right] dV. \tag{8.39}$$

The confinement time is generally defined as the stored thermal energy divided by the loss power,

$$\tau_E = W \bigg/ \int_{\partial U} \mathbf{Q} \cdot d\mathbf{S},$$

and we are now in a position to write down an expression for this quantity in a plasma that evolves arbitrarily,

$$\tau_E = \frac{W}{P_{\text{aux}} + V_{\text{loop}} I_p - \dot{W} - X}, \tag{8.40}$$

which is our final result. Broadly speaking, the first two terms in the denominator represent the auxiliary and Ohmic heating power, and the last two terms represent the power expended in changing the plasma and magnetic field, so that the sum of all the terms in the denominator is equal to the loss power. However, it should be noted that the term $V_{\text{loop}} I_p$ denotes the flux of electromagnetic energy associated with the induced toroidal electric field into the plasma and is actually not equal to the Ohmic power unless the plasma is in a steady state. Also, the term X, which is defined by Eq. (8.39), contains the rate of change of the poloidal magnetic field energy inside the plasma, and a part (but not all) of the toroidal field energy.

Further reading

Collisional transport in toroidal plasmas has been the subject of several widely quoted review articles by Hinton and Hazeltine (1976), Galeev and Sagdeev (1979), and Hirshman and Sigmar (1981), and books (Balescu, 1988), (Fussmann, 2001). Hazeltine and Meiss (1992) also discuss the basic concepts very carefully.

Exercises

1. Demonstrate that the neglected inertia term in (8.27) is small in comparison with the friction.

Solution: The friction is $F_\varphi \sim mnVv$, and for the inertial term use (8.2) to find

$$\frac{mnR\, dV/dt}{RF_\varphi} \sim \frac{dV/dt}{vV} \sim \delta^2.$$

2. Show that the viscosity term, which was neglected in (8.27), is small by going through the following steps.

   (a) Show that $R\hat{\varphi} \cdot \nabla \cdot \boldsymbol{\pi} = \nabla \cdot (R\hat{\varphi} \cdot \boldsymbol{\pi})$.

   Solution: It is clear that $R\hat{\varphi}\cdot\nabla\cdot\boldsymbol{\pi} = \nabla\cdot(R\hat{\varphi}\cdot\boldsymbol{\pi}) - \boldsymbol{\pi} : \nabla(R\hat{\varphi})$, so we need to show that $\boldsymbol{\pi} : \nabla(R\hat{\varphi}) = 0$. (The notation means $\mathsf{A} : \mathsf{B} = \sum_{jk} A_{jk} B_{kj}$ for any two tensors.) This follows from the fact that $\nabla(R\hat{\varphi}) = \hat{\mathbf{R}}\hat{\varphi} - \hat{\varphi}\hat{\mathbf{R}}$ is anti-symmetric and $\boldsymbol{\pi}$ is symmetric, $\pi_{jk} = \pi_{kj}$.

   (b) Verify that for any vector $\mathbf{A}$

   $$\langle \nabla \cdot \mathbf{A} \rangle = \oint \frac{\partial(g^{1/2}\mathbf{A}\cdot\nabla\psi)}{\partial\psi} d\theta \bigg/ \oint g^{1/2} d\theta = \frac{1}{V'}\frac{\partial}{\partial\psi}\left(V'\langle \mathbf{A}\cdot\nabla\psi\rangle\right),$$

   where $V' \equiv 2\pi \oint g^{1/2} d\theta$ and $g^{-1/2} = \mathbf{B}\cdot\nabla\theta$.

   (c) By inspecting (4.68), show that the off-diagonal elements in the viscosity tensor are $O(\delta^2 p)$.

   Solution: The largest term is the one proportional to Ω. By using the notation (4.70), we can write Eq. (4.68) as

   $$\mathsf{K}(\mathsf{P}) = O(\delta^2 p).$$

   This is understood from inspecting the magnitude of one of the terms in (4.69), e.g., $(\partial V_l/\partial x_i) \sim \delta v_T/L$, recalling that the flow velocity is small, $V \sim \delta v_T$. It follows that the off-diagonal elements of P, and hence $\boldsymbol{\pi}$, are at most of order $O(\delta^2 p)$.

   (d) Use this information to compare viscosity with friction.

   Solution: Since $R\hat{\varphi} \cdot \boldsymbol{\pi} \cdot \nabla\psi = 0$ for any diagonal tensor $\boldsymbol{\pi}$, we have

   $$\frac{R\hat{\varphi} \cdot \nabla \cdot \boldsymbol{\pi}}{RF_\varphi} \sim \frac{\delta^2 n T}{Lmn\,\delta v_T\, v} \ll 1.$$

3. Prove (8.29) from (8.31).

   Solution: First we employ (7.21) and (8.14) to write

   $$\langle \boldsymbol{\Gamma}_a \cdot \nabla\psi \rangle^{\text{neo}} = \oint \frac{d\theta}{\mathbf{B}\cdot\nabla\theta} \sum_\sigma \sigma \int f_a I \nabla_\parallel \left(\frac{v_\parallel}{\Omega_a}\right) \frac{\pi B}{B_0} v^3\, dv\, d\lambda \bigg/ \oint \frac{d\theta}{\mathbf{B}\cdot\nabla\theta}$$

   $$= -\sum_\sigma \sigma \int \frac{\pi v^3}{B_0} dv\, d\lambda \oint \frac{I v_\parallel}{\Omega_a} \frac{\partial f_a}{\partial\theta} d\theta \bigg/ \oint \frac{d\theta}{\mathbf{B}\cdot\nabla\theta},$$

where we have integrated by parts in θ. Next, using $B\nabla_\| f_a = (\mathbf{B}\cdot\nabla\theta)\, \partial f_a/\partial\theta$ and (8.16) gives

$$\langle \mathbf{\Gamma}_a \cdot \nabla\psi \rangle^{\text{neo}} = -\left\langle \frac{I}{\Omega_a} \int v_\|^2 \nabla_\| f_{a1} d^3 v \right\rangle$$

$$= -I \left\langle \int \left[m_a v_\| C_a(f_{a1}) + \frac{m_a v_\|^2}{T_a} e_a E_\|^{(A)} f_{a0} \right] \frac{d^3 v}{e_a B} \right\rangle = (8.29).$$

4. Prove Eq. (8.37).

Solution: From the definition of the derivative we have

$$\frac{dJ}{dt} = \lim_{\Delta t \to 0} \frac{1}{\Delta t} \left(\int_{U(t+\Delta t)} A(\mathbf{r}, t+\Delta t) dV - \int_{U(t)} A(\mathbf{r}, t) dV \right)$$

$$= \int_{U(t)} \frac{\partial A}{\partial t} dV + \lim_{\Delta t \to 0} \frac{1}{\Delta t} \left(\int_{U(t+\Delta t)} A(\mathbf{r}, t) dV - \int_{U(t)} A(\mathbf{r}, t) dV \right).$$

The difference between the regions $U(t+\Delta t)$ and $U(t)$ is the volume spanned by $\Delta t \mathbf{u} \cdot d\mathbf{S}$.

9

Transport in the Pfirsch–Schlüter regime

With the exception of (8.19) the results in the previous chapter are valid in all collisionality regimes. We now turn to the specific transport properties of the collisional regime (8.7). The relatively cold edge plasma of a tokamak is often in this regime, which is named after the two German physicists who first analysed it.

All neoclassical transport, including Pfirsch–Schlüter transport, occurs as a consequence of the movement of guiding centres under the combined action of drifts and collisions. In the banana regime, the resulting diffusion can be understood relatively easily in terms of banana orbits and effective collision frequencies according to the random-walk argument presented in Chapter 1. In the Pfirsch–Schlüter regime, guiding-centre orbits are constantly interrupted by collisions, which must be taken into account when making a random-walk estimate of the transport. Since the collision frequency exceeds the transit frequency, parallel particle motion is diffusive, with a diffusion coefficient

$$D_\parallel \sim \lambda^2/\tau \sim v_T^2/\nu,$$

where $\lambda = v_T/\nu$ is the mean-free path. Thus, the time it takes for a particle to move around a flux surface is of the order

$$\Delta t \sim \frac{(qR)^2}{D_\parallel} \sim \nu \left(\frac{qR}{v_T}\right)^2,$$

since the parallel distance is of the order qR. The cross-field transport is caused by the guiding-centre drift (7.24), which is inward if the particle is above the midplane and outward if it is below the midplane (if $\Omega > 0$). As the particle diffuses in the parallel direction, the cross-field drift is sometimes outward and sometimes inward. This leads to a random walk in the radial direction, with a step length

$$\Delta r \sim v_d \Delta t \sim \frac{\rho v_T \Delta t}{R},$$

and a step time Δt. The resulting diffusion coefficient is thus

$$D_\perp \sim \frac{(\Delta r)^2}{\Delta t} \sim v q^2 \rho^2, \tag{9.1}$$

which is larger than the classical diffusion coefficient by a factor q^2. In the tokamak edge, which is often collisional, the diffusion coefficient is thus an order of magnitude larger than suggested by classical transport theory.

9.1 Ion heat flux

We begin by studying ion energy transport in a pure plasma. Since the collisionality is high, the plasma obeys Braginskii's fluid equations, and we need not solve the kinetic problem (8.16). Instead, we can simply use the ion version of the energy equation (2.17),

$$\frac{3n_i}{2}\left(\frac{\partial}{\partial t} + \mathbf{V}_i \cdot \nabla\right) T_i + p_i \nabla \cdot \mathbf{V}_i = -\nabla \cdot \mathbf{q}_i - \boldsymbol{\pi}_i : \nabla \mathbf{V}_i + Q_i. \tag{9.2}$$

The conductive heat flux is given by (4.45),

$$\mathbf{q}_i = -\kappa_\parallel^i \nabla_\parallel T_i + \kappa_\wedge^i \mathbf{b} \times \nabla T_i - \kappa_\perp^i \nabla_\perp T_i, \tag{9.3}$$

with $\kappa_\parallel^i : \kappa_\wedge^i : \kappa_\perp^i \sim 1 : \Delta : \Delta^2$ and $\Delta = (\Omega_i \tau_i)^{-1} \ll 1$. The largest term in the energy equation (9.2) is therefore the parallel heat flux, which operates on a time scale

$$\left(\frac{\partial}{\partial t}\right)_\parallel = \frac{\nabla_\parallel(\kappa_\parallel^i \nabla_\parallel T_i)}{n_i T_i} \sim \frac{T_i \tau_i}{m_i L^2} \sim \frac{\lambda}{L}\frac{v_{Ti}}{L}, \tag{9.4}$$

since $\kappa_\parallel^i \sim n_i T_i \tau_i / m_i$. As usual, we denote the macroscopic scale length by L. The convective terms operate on the longer time scale

$$\left(\frac{\partial}{\partial t}\right)_{\mathbf{V}_i} = \frac{n_i(\mathbf{V}_i \cdot \nabla)T_i}{n_i T_i} \sim \frac{p_i \nabla \cdot \mathbf{V}_i}{n_i T_i} \sim \frac{V_i}{L} \sim \delta \frac{v_{Ti}}{L},$$

and the other terms in (9.2) are still smaller. Both these time scales are faster than the time scale (8.3) associated with cross-field transport, which we are primarily interested in.

Therefore, if we expand the temperature,

$$T_i = T_{i0} + T_{i1} + \cdots,$$

then to lowest order, $\nabla_\|(\kappa^i_\| \nabla_\| T_{i0}) = 0$, so

$$T_{i0} = T_{i0}(\psi).$$

Thus, on the fastest time scale (9.4) parallel heat conduction makes the temperature uniform on flux surfaces. This result is not new; we have already derived it from kinetic theory in (8.11).

In next order, we have

$$\nabla \cdot \left(\kappa^i_\| \nabla_\| T_{i1} - \kappa^i_\wedge \mathbf{b} \times \nabla T_{i0} \right) = 0,$$

since the convective heat flux associated with T_{i0} vanishes. Hence

$$\frac{1}{g^{1/2}} \frac{\partial}{\partial \theta} \left[g^{1/2} \nabla \theta \cdot \left(\kappa^i_\| \mathbf{b} \nabla_\| T_{i1} - \kappa^i_\wedge \mathbf{b} \times \nabla T_{i0} \right) \right]$$

$$= \frac{1}{g^{1/2}} \frac{\partial}{\partial \theta} \left(\frac{\kappa^i_\|}{B} \nabla_\| T_{i1} - \frac{I \kappa^i_\wedge}{B} \frac{dT_{i0}}{d\psi} \right) = 0,$$

where we have used $\mathbf{B} \cdot \nabla \theta = 1/g^{1/2}$ from (7.2), and $(\mathbf{B} \times \nabla \psi) \cdot \nabla \theta = I/g^{1/2}$ from (8.15). Integrating in θ gives

$$\nabla_\| T_{i1} = \frac{I \kappa^i_\wedge}{\kappa^i_\|} \frac{dT_{i0}}{d\psi} + L_i(\psi) B,$$

where the integration constant $L_i(\psi)$ can be determined from the relation $\langle B \nabla_\| T_{i1} \rangle = 0$, implied by (7.6). Since $\kappa^i_\wedge B$ is constant over the flux surface, it follows that

$$\nabla_\| T_{i1} = \frac{I \kappa^i_\wedge}{\kappa^i_\|} \left(1 - \frac{B^2}{\langle B^2 \rangle} \right) \frac{dT_{i0}}{d\psi}. \tag{9.5}$$

We see that T_{i1} varies over the flux surface. The mechanism is the same as that giving rise to the Pfirsch–Schlüter current. The diamagnetic heat flux $\kappa^i_\wedge \mathbf{b} \times \nabla T_{i0}$ is not divergence free, and must therefore be balanced by a parallel return flow, which, in turn, implies a small but essential (for transport) parallel temperature gradient. Since the latter (9.5) is positive on the inside of the torus and negative on the outside (if $dT_{i0}/d\psi < 0$), the temperature is up–down asymmetric. The magnitude of the temperature variation is

$$T_1/T_0 \sim \nu/\Omega = \rho/\lambda = \Delta, \tag{9.6}$$

where we have used $\kappa_\wedge/\kappa_\| \sim \Delta = \nu/\Omega$. This small temperature variation in the flux surface is the essential driving force for radial Pfirsch–Schlüter heat transport. In the estimate (9.6) we neglected several geometrical factors, which are formally of order unity in the expansion procedure but which may be important in practice. For instance, the length scale that enters on the left of (9.5) is the connection length along the magnetic field, $L_\parallel \sim qR$, while that on the right is the radial scale length $L_\perp$ associated with the temperature gradient ∇T_0. The latter is normally much shorter than the former. A more careful estimate of the relative temperature variation is

$$T_1/T_0 \sim \frac{L_\parallel B \Delta}{L_\perp B_\theta} = \frac{\rho}{L_\perp} \frac{L_\parallel}{\lambda} \frac{q}{\epsilon}. \tag{9.7}$$

This quantity is required to be small in neoclassical theory (since $\delta = \rho/L_\perp$ is formally the smallest parameter), but can quite easily be of order unity or larger in the tokamak edge, where the radial scale length $L_\perp$ is short.

Knowing the temperature variation across and along the field, T_{i0} and T_{i1} respectively, it is straightforward to construct the heat flux across the flux surface from (9.3) and (9.5),

$$\begin{aligned}
\mathbf{q}_i \cdot \nabla \psi &= \left(-\kappa_\perp^i \nabla_\perp T_{i0} + \kappa_\wedge^i \mathbf{b} \times \nabla T_{i1} \right) \cdot \nabla \psi \\
&= -\left[\kappa_\perp^i |\nabla \psi|^2 + \frac{(I \kappa_\wedge^i)^2}{\kappa_\parallel^i} \left(1 - \frac{B^2}{\langle B^2 \rangle} \right) \right] \frac{dT_{i0}}{d\psi} \\
&\equiv \left(\mathbf{q}_i^{\text{cl}} + \mathbf{q}_i^{PS} \right) \cdot \nabla \psi.
\end{aligned} \tag{9.8}$$

The first term is the classical cross-field heat flux, and the second term is the neoclassical Pfirsch–Schlüter heat flux, which arises entirely because of toroidicity. Both heat fluxes are ultimately driven by the radial temperature gradient, but the neoclassical heat flux is also fundamentally associated with a parallel gradient.

Noting that $\kappa_\wedge^i \propto B^{-1}$ while $\kappa_\parallel^i$ is constant on flux surfaces, we write the Pfirsch–Schlüter heat flux as

$$\mathbf{q}_i^{PS} \cdot \nabla \psi = -\frac{(I B \kappa_\wedge^i)^2}{\kappa_\parallel^i} \left(\frac{1}{B^2} - \frac{1}{\langle B^2 \rangle} \right) \frac{dT_{i0}}{d\psi}.$$

Its flux-surface average is always in the direction opposite to the temperature gradient since for any flux-surface geometry

$$\left\langle \frac{1}{B^2} - \frac{1}{\langle B^2 \rangle} \right\rangle \geq 0$$

as shown in an exercise at the end of this chapter. To determine the relative importance of the classical and neoclassical heat fluxes, it is instructive to study the case of large aspect ratio, $\epsilon \ll 1$. Then $B = B_0[1 - \epsilon \cos \theta + O(\epsilon^2)]$, and thus

$$\frac{B_0^2}{B^2} = 1 + 2\epsilon \cos \theta + O(\epsilon^2),$$

$$1 - \frac{B^2}{\langle B^2 \rangle} = 2\epsilon \cos \theta + O(\epsilon^2), \tag{9.9}$$

where the second expression vanishes exactly on a flux-surface average. Hence

$$\left\langle \frac{1}{B^2}\left(1 - \frac{B^2}{\langle B^2 \rangle}\right)\right\rangle \simeq \frac{1}{2\pi B_0^2}\int_0^{2\pi}(1 + 2\epsilon\cos\theta)\,2\epsilon\cos\theta\,d\theta = \frac{2\epsilon^2}{B_0^2},$$

and it follows from (9.8) that

$$\left\langle \mathbf{q}_i^{PS}\cdot\nabla\psi\right\rangle \simeq -\frac{2(\kappa_\wedge^i r B)^2}{\kappa_\parallel^i}\frac{dT_{i0}}{d\psi},$$

where we have used $B_0 \simeq I/R$ and $q = \epsilon B_0/B_\theta$. Finally, since $\nabla\psi \simeq RB_\theta\hat{\mathbf{r}}$, and $(\kappa_\wedge^i)^2/\kappa_\parallel^i\kappa_\perp^i \simeq 0.8$ for $Z = 1$ (see (4.46)–(4.48)), we can write the total flux-surface averaged heat flux (9.8) as

$$\boxed{\langle q_{ir}\rangle = \frac{\langle \mathbf{q}_i\cdot\nabla\psi\rangle}{|\nabla\psi|} = -\kappa_\perp^i\left(1 + 1.6q^2\right)\frac{dT_{i0}}{dr},} \qquad (9.10)$$

where the first term is the classical and the second term the neoclassical contribution. Thus, on the slow transport time scale (8.2), heat flows across the flux surfaces as if enhanced over the classical flux by a factor $(1 + 1.6q^2)$. This number can be substantial, especially in the tokamak edge where q is large, and agrees with the estimate (9.1) at the beginning of the chapter. The exact value of the factor 1.6 depends on the effective ion charge of the plasma $Z_{\text{eff}} = \sum_a n_a Z_a^2/\sum_a n_a Z_a$ (where the sums are taken over all ion species); here we have assumed $Z_{\text{eff}} = 1$.

It is interesting also to consider the Pfirsch–Schlüter heat flux *before* flux-surface averaging. By (9.8) and (9.9) it is

$$q_{ir}^{PS} = -1.6q^2\left[\epsilon^{-1}\cos\theta + O(1)\right]\kappa_\perp^i\frac{dT_{i0}}{dr}.$$

The local heat flux is thus much larger than the flux-surface averaged one. It is in the inward direction on the inboard side of the torus and outward on the outboard side, see Fig. 9.1. It is only the small $O(\epsilon)$ difference between these two net flows that survive the flux-surface average. It nevertheless yields the substantial 'neoclassical' enhancement over the classical heat flux. In a rotating plasma, where the centrifugal force makes heavy impurities accumulate on the outboard side of each flux surface, the fine balance between inward and outward Pfirsch–Schlüter fluxes is tilted, and transport can be enhanced by a factor of order ϵ^{-1}, see Fülöp and Helander (1999).

9.2 Several species

We have just calculated the ion heat flux from Braginskii's equations. In doing so we only had to deal with one particle species since electrons

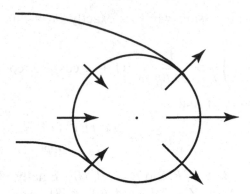

Fig. 9.1.   Pfirsch–Schlüter heat flux in a torus with large aspect ratio.

do not influence ion energy transport much. When calculating particle transport, it is necessary to consider several species (e.g., electrons and ions, or electrons and several ion species) since collisions within one species do not lead to any particle transport. The whole problem then becomes more complex, and it is appropriate to treat particle and energy transport from a single point of view, which is the purpose of the present section.

First, we recall the expressions for the return flows of particles (8.23) and heat (8.25),

$$V_{a\|} = -\frac{I}{n_a e_a B}\left(n_a e_a \frac{d\Phi}{d\psi} + \frac{dp_a}{d\psi}\right)\left(1 - \frac{B^2}{\langle B^2 \rangle}\right) + \frac{B\left\langle B V_{a\|}\right\rangle}{\langle B^2 \rangle}, \qquad (9.11)$$

$$q_{a\|} = -\frac{5 n_a T_a I}{2 e_a B}\frac{d T_a}{d\psi}\left(1 - \frac{B^2}{\langle B^2 \rangle}\right) + \frac{B\left\langle B q_{a\|}\right\rangle}{\langle B^2 \rangle}. \qquad (9.12)$$

These relations are valid for all collisionalities (not only the Pfirsch–Schlüter regime), but the last terms on the right depend very differently on v in different collisionality regimes. Equation (9.12) assumes that the masses of different species are sufficiently disparate, so that the energy exchange between them can be neglected.

Our next task is to relate these parallel flows to the neoclassical cross-field flows (8.29), (8.32),

$$\langle \mathbf{\Gamma}_a \cdot \nabla\psi \rangle^{\text{neo}} = -\left\langle \frac{I(R_{a\|} + n_a e_a E_\|^{(A)})}{e_a B}\right\rangle, \qquad (9.13)$$

$$\frac{\langle \mathbf{q}_a \cdot \nabla\psi \rangle^{\text{neo}}}{T_a} = -\left\langle \frac{I H_{a\|}}{e_a B}\right\rangle, \qquad (9.14)$$

where the friction and 'heat friction' are defined by

$$\begin{pmatrix} R_{a\parallel} \\ H_{a\parallel} \end{pmatrix} = \int \begin{pmatrix} 1 \\ m_a v^2/2T_a - 5/2 \end{pmatrix} m_a v_\parallel C_a(f_a)\, d^3v.$$

Thus, what is required is to find the relation between these parallel forces and the parallel flows $V_{a\parallel}$ and $q_{a\parallel}$. In general, this must be done by using kinetic theory, proceeding from the drift kinetic equation (8.13)

$$v_\parallel \nabla f_{a1} + \mathbf{v}_d \cdot \nabla f_{a0} + e_a v_\parallel E_\parallel^{(A)} \frac{\partial f_{a0}}{\partial \mathscr{E}} = C_a(f_{a1}). \tag{9.15}$$

We recall that this equation is of first order in the fundamental expansion parameter δ. In this chapter, we are interested in the case of high collisionality

$$\lambda/L = \delta/\Delta \ll 1,$$

and can therefore simplify the problem further by following Hazeltine (1974) and making a subsidiary expansion in λ/L, writing

$$f_{a1} = f_{a1}^{(-1)} + f_{a1}^{(0)} + \cdots.$$

The reason why we have chosen the expansion to start at the order -1 is that we have already seen that the temperature variation (9.6) within a flux surface is of order $O\left[\delta(\lambda/L)^{-1}\right]$. In this order the drift kinetic equation only includes the collision term

$$C_a(f_{a1}^{(-1)}) = 0,$$

which implies that each $f_{a1}^{(-1)} = f_{Ma}(\psi, \theta)$ is Maxwellian. The density and temperature moments of $f_{a1}^{(-1)}$ are as yet unknown and may vary with the poloidal angle θ (in contrast to f_{a0}, which is constant on flux surfaces). It is important to remember that the collision operator describes collisions with all species. In next order then, we have the multi-species Spitzer problem, cf. (4.9),

$$\left[\nabla_\parallel \ln p_a - \frac{e_a E_\parallel}{T_a} + \left(\frac{m_a v^2}{2T_a} - \frac{5}{2} \right) \nabla_\parallel T_a \right] v_\parallel f_{a1}^{(-1)} = C_a(f_{a1}^{(0)}). \tag{9.16}$$

The term in (9.15) containing the inductive electric field $E_\parallel^{(A)}$ is ordered to be $O(\delta\Delta)$ and is thus comparable to the other driving term, $\mathbf{v}_d \cdot \nabla f_{a0}$.

Both can be neglected in (9.16). Taking the number moment and the $(m_a v^2/2T_a - 5/2)$-moment of this equation shows that the parallel forces are proportional to parallel gradients,

$$R_{a\parallel} = \nabla_\parallel \ln p_a - n_a e_a E_\parallel, \tag{9.17}$$

$$H_{a\parallel} = \frac{5n_a}{2} \nabla_\parallel T_a. \tag{9.18}$$

On the other hand, solving the kinetic equation for $f_{a1}^{(0)}$ (which is a major undertaking) gives the relation between driving forces and flows. Since the equation is linear we can write the result as

$$R_{a\parallel} = \sum_b \left(l_{11}^{ab} V_{b\parallel} + l_{12}^{ab} \frac{2q_{b\parallel}}{5p_b} \right), \tag{9.19}$$

$$H_{a\parallel} = \sum_b \left(l_{21}^{ab} V_{b\parallel} + l_{22}^{ab} \frac{2q_{b\parallel}}{5p_b} \right), \tag{9.20}$$

where the solutions of the kinetic equations (9.16) determine the coefficients l_{jk}^{ab}. This system of equations expresses the forces in terms of flows and is thus the inverse of (4.22), (4.23).

We now have all information necessary to construct the cross-field fluxes. The heat flux, for instance, is obtained by combining (9.14) and (9.20),

$$\frac{\langle \mathbf{q}_a \cdot \nabla \psi \rangle^{\mathrm{neo}}}{T_a} = -\frac{I}{e_a} \sum_b \left\langle \left(l_{21}^{ab} V_{b\parallel} + l_{22}^{ab} \frac{2q_{b\parallel}}{5p_b} \right) \frac{1}{B} \left(1 - \frac{B^2}{\langle B^2 \rangle} \right) \right\rangle,$$

where we have also used $\langle B H_{a\parallel} \rangle = 0$, which follows from (9.18) and (7.6). The end result is finally obtained by using the expressions (9.11) and (9.12) for the parallel fluxes $V_{b\parallel}$ and $q_{b\parallel}$. This procedure gives the desired radial Pfirsch–Schlüter transport laws,

$$\langle \mathbf{\Gamma}_a \cdot \nabla \psi \rangle^{\mathrm{neo}} = I^2 \sum_b \frac{T_b}{e_a e_b} \left(l_{11}^{ab} \frac{d \ln p_b}{d\psi} + l_{12}^{ab} \frac{d \ln T_b}{d\psi} \right) \left\langle \frac{1}{B^2} - \frac{1}{\langle B^2 \rangle} \right\rangle$$

$$- I n_a \left\langle \frac{E_\parallel^{(A)}}{B} \left(1 - \frac{B^2}{\langle B^2 \rangle} \right) \right\rangle,$$

$$\frac{\langle \mathbf{q}_a \cdot \nabla \psi \rangle^{\mathrm{neo}}}{T_a} = I^2 \sum_b \frac{T_b}{e_a e_b} \left(l_{21}^{ab} \frac{d \ln p_b}{d\psi} + l_{22}^{ab} \frac{d \ln T_b}{d\psi} \right) \left\langle \frac{1}{B^2} - \frac{1}{\langle B^2 \rangle} \right\rangle.$$

Electron transport in a pure plasma

For electrons in a pure hydrogenic plasma, $Z_{\text{eff}} = 1$, we need not go through the trouble of solving the Spitzer problem (9.16). Instead, the transport coefficients can be extracted from Braginskii's laws of parallel electron transport (4.36)–(4.42),

$$R_{e\parallel} = -0.51\frac{m_e n_e}{\tau_e}u_\parallel - 0.71 n_e \nabla_\parallel T_e,$$

$$q_{e\parallel} = 0.71 n_e T_e u_\parallel - 3.16\frac{n_e T_e \tau_e}{m_e}\nabla_\parallel T_e,$$

where $u_\parallel = V_{e\parallel} - V_{i\parallel}$. We rewrite these relations in the form (9.19)–(9.20), using (9.18) to express the parallel temperature gradient in terms of $H_{e\parallel}$,

$$R_{e\parallel} = -\frac{m_e n_e}{\tau_e}\left[\left(0.51 + \frac{0.71^2}{3.16}\right)u_\parallel - \frac{0.71}{3.16}\frac{q_{e\parallel}}{p_e}\right],$$

$$H_{e\parallel} = \frac{5n_e}{2}\nabla_\parallel T_e = \frac{5m_e n_e}{2\cdot 3.16\tau_e}\left(0.71 u_\parallel - \frac{q_{e\parallel}}{p_e}\right).$$

Comparing these expressions with (9.19) and (9.20), it is now evident that $l_{11}^{ee} = -l_{11}^{ei} = -\left(0.51 + 0.71^2/3.2\right)m_e n_e/\tau_e$, $l_{12}^{ee} = l_{21}^{ee} = -l_{21}^{ei} = (5 \cdot 0.71/2 \cdot 3.16)m_e n_e/\tau_e$, $l_{22}^{ee} = -(5^2/2^2 \cdot 3.16)m_e n_e/\tau_e$, and the other transport coefficients vanish. The cross-field fluxes become

$$\langle \mathbf{\Gamma}_e \cdot \nabla\psi\rangle^{\text{neo}} = -\left\langle\frac{1}{B^2} - \frac{1}{\langle B^2\rangle}\right\rangle\frac{I^2 m_e}{e^2\tau_e}\left(0.66\frac{dp}{d\psi} - 0.55 n_e\frac{dT_e}{d\psi}\right)$$

$$- I n_a\left\langle\frac{E_\parallel^{(A)}}{B}\left(1 - \frac{B^2}{\langle B^2\rangle}\right)\right\rangle,$$

$$\frac{\langle \mathbf{q}_e \cdot \nabla\psi\rangle^{\text{neo}}}{T_e} = -\left\langle\frac{1}{B^2} - \frac{1}{\langle B^2\rangle}\right\rangle\frac{I^2 m_e}{e^2\tau_e}\left(1.95 n_e\frac{dT_e}{d\psi} - 0.55\frac{dp}{d\psi}\right),$$

where $p = p_e + p_i = n_e(T_e + T_i)$ is the total pressure.

These formulas are valid for arbitrary aspect ratio and flux-surface geometry. If the flux surfaces are circular and $\epsilon \ll 1$, then the term containing the electric field is small, and

$$\frac{I^2 m_e}{e^2\tau_e}\left\langle\frac{1}{B^2} - \frac{1}{\langle B^2\rangle}\right\rangle\frac{T_e}{|\nabla\psi|^2} \simeq 2q^2 D_e,$$

with

$$D_e \equiv \frac{T_e}{m_e\Omega_e^2\tau_e} = \frac{\rho_e^2}{\tau_e},$$

the classical diffusion coefficient, cf. (5.7). Hence

$$\langle \mathbf{\Gamma}_e \cdot \nabla r \rangle \simeq \frac{\langle \mathbf{\Gamma}_e \cdot \nabla \psi \rangle^{\mathrm{neo}}}{|\nabla \psi|} \simeq -2q^2 D_e \left(\frac{0.66}{T_e} \frac{dp}{dr} - \frac{0.55 n_e}{T_e} \frac{dT_e}{dr} \right), \qquad (9.21)$$

$$\langle \mathbf{q}_e \cdot \nabla r \rangle \simeq \frac{\langle \mathbf{q}_e \cdot \nabla \psi \rangle^{\mathrm{neo}}}{|\nabla \psi|} \simeq -2q^2 D_e \left(1.95 n_e \frac{dT_e}{dr} - 0.55 \frac{dp}{dr} \right). \qquad (9.22)$$

The Pfirsch–Schlüter transport is, again, $O(q^2)$ faster than classical diffusion. Because of ambipolarity (itself a consequence of momentum conservation $R_{ei} = R_{ie}$) the ion particle flux is equal to the electron flux, $\Gamma_i = \Gamma_e$, and is therefore $O(\sqrt{m_e/m_i})$ smaller than the ion energy flux in a clean plasma.

Impurity transport

As noted already in Chapter 5, the ion particle flux increases dramatically if impurities are introduced and then becomes comparable to the ion heat flux. For simplicity, consider a plasma with a main ion species (i) and a single impurity (Z). For bulk ions, the collision operator is

$$C_i = C_{ie} + C_{ii} + C_{iZ},$$

where $C_{ie}/C_{ii} \sim \sqrt{m_e/m_i} \ll 1$, and

$$\frac{C_{iZ}}{C_{ii}} \sim \frac{n_Z e_Z^2}{n_i e_i^2} \equiv \alpha, \qquad (9.23)$$

so C_{ie} is negligible in comparison with C_{iZ} whenever $\alpha \gg \sqrt{m_e/m_i}$, which is practically always the case, cf. (5.8).

To construct Pfirsch–Schlüter transport laws in an impure plasma, we need to solve the Spitzer problem (9.16) and find the transport coefficients l_{jk}^{ab}. If the impurity is heavy, $m_Z \gg m_i$ the ion–impurity collision operator is given by (3.32), which is *identical* to the usual electron–ion collision operator (3.28) apart from numerical coefficients. The ion–ion and electron–electron collision operators are also similar in the same sense. Therefore, the Spitzer problem for bulk ions in an impure plasma

$$v_\parallel \nabla_\parallel f_{Mi} = C_{ii}(f_{i1}) + C_{iZ}(f_{i1})$$

is mathematically equivalent to the usual electron problem

$$v_\parallel \nabla_\parallel f_{Me} = C_{ee}(f_{e1}) + C_{ei}(f_{e1}),$$

if the ratio between the two collision operators is similar. In the former equation the ratio is (9.23), and in the latter equation it is simply equal to the ion charge, $C_{ei}/C_{ee} \sim n_i e_i^2/n_e e^2 = e_i/e$. (Recall that $n_i e_i = n_e$.) The two equations are thus identical if $\alpha = e_i/e$.

For instance, Braginskii's transport coefficients for $e_i/e = 1$, which we have just used to construct electron transport laws, can now be applied to the problem of ion transport in an impure plasma with $\alpha = 1$, i.e.,

$$Z_{\text{eff}} = \frac{n_i e_i^2 + n_Z e_Z^2}{n_e e^2} = 2$$

if the main ions are hydrogenic, $e_i = e$. In analogy with the preceding subsection, the non-vanishing transport coefficients are $l_{11}^{ii} = -l_{11}^{iZ} = -0.66 m_i n_i/\tau_{iZ}$, $l_{12}^{ii} = l_{21}^{ii} = -l_{21}^{iZ} = 0.55 m_i n_i/\tau_{iZ}$, and $l_{22}^{ii} = -1.95 m_i n_i/\tau_{iZ}$. Then, in a large-aspect-ratio torus,

$$\langle \mathbf{\Gamma}_i \cdot \nabla r \rangle \simeq -2q^2 D_i n_i \left[0.66 \left(\frac{d \ln p_i}{dr} - \frac{T_Z}{Z T_i} \frac{d \ln p_Z}{dr} \right) - \frac{0.55 n_i}{T_i} \frac{dT_i}{dr} \right],$$

$$\langle \mathbf{q}_i \cdot \nabla r \rangle \simeq -2q^2 D_i n_i T_i \left[1.95 \frac{d \ln T_i}{dr} + 0.55 \left(\frac{d \ln p_i}{dr} - \frac{T_Z}{Z T_i} \frac{d \ln p_Z}{dr} \right) \right].$$

The classical diffusion coefficient here is

$$D_i \equiv \frac{T_i}{m_i \Omega_i^2 \tau_{iZ}}$$

with the ion–impurity collision time

$$\tau_{iZ} = \frac{12\pi^{3/2}}{2^{1/2}} \frac{m_i^{1/2} T_i^{3/2} \epsilon_0^2}{n_Z e_i^2 e_Z^2 \ln \Lambda}$$

in analogy with (1.4). For other values of α, the transport coefficients are numerically different, and can be looked up in the literature (Braginskii, 1965; Rutherford, 1974; Hirshman and Sigmar, 1981).

Since $D_e/D_i \sim \sqrt{m_e/m_i} \ll 1$, collisional electron diffusion is much slower than the ion diffusion. To maintain ambipolarity, the ion and impurity fluxes must therefore be opposite and carry equal charge,

$$\Gamma_i \simeq -Z\Gamma_Z.$$

If the main ions diffuse outward, the impurities move inward, just as we found for classical diffusion in (5.10). Indeed, in those tokamak plasmas where anomalous, fluctuation-driven diffusion is reduced (e.g., in H-mode plasmas and reversed magnetic shear plasmas), impurity accumulation in the plasma core is frequently observed experimentally.

Finally, note that the transport is primarily driven by gradients in the bulk (light) ion parameters. The term proportional to the impurity pressure gradient is multiplied by the small factor $1/Z$, and is therefore relatively unimportant. The radial flux of a heavy impurity is therefore practically independent of the impurity pressure profile!

Further reading

The original paper by Pfirsch and Schlüter was never published. Of course, the reviews by Hinton and Hazeltine (1976), Hirshman and Sigmar (1981), and Balescu (1988) all discuss Pfirsch–Schlüter transport. Our presentation is influenced by the seminal papers by Hirshman (1977) and, for impurity transport, by Rutherford (1974). A synoptic paper by Sugama and Horton (1996) summarizes this collisional regime (as well as the plateau and banana regimes).

Exercise

1. Show that

$$\left\langle \frac{1}{B^2} \right\rangle - \frac{1}{\langle B^2 \rangle} \geq 0.$$

Solution: This follows immediately from

$$\left\langle \left(\frac{1}{B} - \frac{B}{\langle B^2 \rangle} \right)^2 \right\rangle \geq 0.$$

10

Transport in the plateau regime

The plateau regime of collisionality was defined in (8.8) and only exists at large aspect ratio, $\epsilon \ll 1$. In this regime, passing particle orbits are collisionless, $\nu \ll v_T/qR$, so we must solve the kinetic equation (8.16) rather than Braginskii's equations. On the other hand, trapped orbits are typically interrupted by collisions,

$$\omega_b \ll \nu_{\text{eff}},$$

where the bounce frequency is of the order $\omega_b \sim \epsilon^{1/2} v_T/qR$, and the effective collision frequency is $\nu_{\text{eff}} = \nu/\epsilon$. This ordering will allow us to simplify the kinetic problem by essentially ignoring the trapped population. In the plateau regime it is circulating particles that dominate radial transport, which, although caused by collisions, turns out to be independent of the collision frequency.

10.1 Physical picture

To understand the physics of plateau transport, we first study a simplified version of the drift kinetic equation (8.13). Recalling the general form (6.30) for kinetic equations, we have

$$\dot\theta \frac{\partial f_1}{\partial \theta} + \dot v_\parallel \frac{\partial f_1}{\partial v_\parallel} + \dot v \frac{\partial f_1}{\partial v} + \mathbf{v}_d \cdot \nabla f_0 = C(f_1), \tag{10.1}$$

where the distribution function is expressed in the variables $(\psi, \theta, v_\parallel, v)$. As usual, these variables refer to the guiding centre rather than the particle itself. Note that we use the variables $(v_\parallel, v)$ rather than the more conventional $(\mathscr{E}, \mu)$. Of course, the equation (10.1) can also be derived from (6.31) by transforming from the velocity-space coordinates $(\mathscr{E}, \mu)$ to

179

$(v_\parallel, v)$. The latter variables are not conserved quantities, but their variation is relatively weak,

$$\dot{v}_\parallel = -\frac{\mu \nabla_\parallel B}{m} = -v_\perp^2 \frac{\epsilon \sin \theta}{2qR} \sim \epsilon \frac{v}{qR},$$

(10.2)

$$\dot{v} = \frac{e v_\parallel E_\parallel}{mv} \sim \delta \frac{v^2}{qR},$$

which enables us to regard them as constants, which will be justified in more detail later in this chapter. Here we have used (6.19) in the first equation and $eE_\parallel/T \sim \delta/qR$ in the second one. We restrict our attention to passing particles since, clearly, the neglect of the mirror force is not justified for trapped ones. Furthermore, we replace the full collision operator by a simple Krook operator,

$$C(f_1) \rightarrow -\nu_{\text{eff}} f_1,$$

$$\nu_{\text{eff}} = \left(\frac{v}{v_\parallel}\right)^2 \nu,$$

(10.3)

where the form for the effective collision frequency ν_{eff} is motivated by the fact that small-angle scattering, by an angle $\Delta \vartheta \sim v_\parallel/v \ll 1$ turns out to play the dominant part in plateau transport. As usual, this means that the collision frequency is enhanced by a factor $1/(\Delta \vartheta)^2$. Since the drift velocity (7.24) is vertical and $\dot{\theta} = v_\parallel/qR$ in a torus with large aspect ratio and circular cross section, we obtain

$$\frac{v_\parallel}{qR} \frac{\partial f_1}{\partial \theta} + \nu_{\text{eff}} f_1 = -v_d \frac{\partial f_0}{\partial r} \sin \theta.$$

(10.4)

This equation, which embodies the crucial physics of plateau transport, is conveniently solved by writing $f_1 = f_+ \cos \theta + f_- \sin \theta$, giving rise to the system

$$\omega_t f_- + \nu_{\text{eff}} f_+ = 0,$$

$$-\omega_t f_+ + \nu_{\text{eff}} f_- = -v_d \frac{\partial f_0}{\partial r},$$

where the transit frequency is $\omega_t \equiv v_\parallel/qR$. The solution is

$$f_- = -G(\omega_t) v_d \frac{\partial f_0}{\partial r},$$

$$G(\omega_t) \equiv \frac{\nu_{\text{eff}}}{\nu_{\text{eff}}^2 + \omega_t^2}.$$

Now recall that most passing particles are nearly collisionless, $v_{\text{eff}} \ll \omega_t$. This means that our solution f_- is highly peaked around $v_\parallel \simeq 0$. Indeed, $G(\omega_t)$ approaches a delta function in this limit,

$$G(\omega_t) \to \pi \delta(\omega_t), \quad v_{\text{eff}}/\omega_t \to 0,$$

where the factor π comes from the circumstance that

$$\int_{-\infty}^{\infty} G \, d\omega_t = \pi.$$

The mathematical reason why f_- has this localized structure is the cancellation between the two terms on the left of (10.4). Physically, it signifies a *resonance* between the collision frequency and the transit frequency. The resonance occurs only for a small subset of the circulating particles, namely, those for which $v_{\text{eff}} \sim \omega_t$ (since this is where the function $G(\omega_t)$ peaks), i.e.,

$$\left(\frac{v_\parallel}{v}\right)^3 \sim \frac{v}{v/qR} \ll 1. \tag{10.5}$$

It is the requirement that these particles be circulating, $v_\parallel/v \gg \epsilon^{1/2}$, while most passing particles are collisionless, that underlies the plateau ordering (8.8).

Only the part of f_1 that is odd in θ, i.e., f_-, contributes to the radial particle and heat fluxes

$$\langle \mathbf{\Gamma} \cdot \nabla r \rangle = \left\langle \int f_1 v_d \sin \theta \, d^3 v \right\rangle = -\pi \left\langle \int \delta \left(\frac{v_\parallel}{qR}\right) \frac{\partial f_0}{\partial r} (v_d \sin \theta)^2 \, d^3 v \right\rangle$$

$$= -\hat{\omega}_t \rho^2 n \frac{q^2 \pi^{1/2}}{4} \left(\frac{d \ln n}{dr} + \frac{3}{2} \frac{d \ln T}{dr}\right),$$

$$\langle \mathbf{q} \cdot \nabla r \rangle = \left\langle \int f_1 \left(\frac{mv^2}{2} - \frac{5T}{2}\right) v_d \sin \theta \, d^3 v \right\rangle$$

$$= -\hat{\omega}_t \rho^2 n T \frac{q^2 \pi^{1/2}}{8} \left(\frac{d \ln n}{dr} + \frac{15}{2} \frac{d \ln T}{dr}\right),$$

where $\rho = v_T/\Omega$ is the gyroradius and $\hat{\omega}_t = v_T/qR$ the transit frequency of a particle moving at the thermal speed $v_T^2 = 2T/m$, and we have used $\delta(v_\parallel/qR) = qR\delta(v_\parallel)$. Note that the flux is independent of the collision frequency and, in fact, of all the details of the collision operator. The latter is only needed to resolve the singularity at $v_\parallel = 0$, which justifies its replacement by a Krook operator.

The scaling of the diffusion coefficient associated with the plateau fluxes,

$$D \sim \hat{\omega}_t (q\rho)^2, \tag{10.6}$$

can be understood in the following way. We would expect to be able to write D as

$$D \sim \frac{n_{\text{res}}}{n} \cdot v_{\text{eff}} \cdot (\Delta r)^2,$$

where n_{res} is the number of contributing (resonant) particles, Δr is the radial step size, and v_{eff} the frequency at which these steps are taken in the random walk. The step size Δr is equal to the distance a particle has time to drift across a flux surface before suffering a collision, and we have

$$\frac{n_{\text{res}}}{n} \sim \frac{v_\parallel}{v},$$

$$v_{\text{eff}} \sim \left(\frac{v}{v_\parallel}\right)^2 v,$$

$$\Delta r \sim \frac{v_d}{v_{\text{eff}}}.$$

Using these estimates together with (10.5) to determine the typical parallel velocity, $v_\parallel$, of the resonant particles gives (10.6).

In summary, while the banana regime relies on an expansion in the smallness of the collision frequency parameter $v_{\text{eff}}/\hat\omega_t \ll 1$, and the Pfirsch–Schlüter regime on the opposite expansion, the plateau regime results from a resonance between the collision frequency and the transit frequency. The exact nature of the collisions then becomes unimportant, and the transport becomes independent of the collision frequency.

10.2 Transport laws

Guided by the physical picture just considered, let us now more rigorously derive the plateau transport laws for a tokamak with large aspect ratio and circular cross section. We still neglect the parallel electric field, but now include the mirror force (10.2) in the drift kinetic equation, which we write as

$$\frac{v}{qR}\left(\xi \frac{\partial f_{a1}}{\partial \theta} - \epsilon \sin\theta \frac{1-\xi^2}{2} \frac{\partial f_{a1}}{\partial \xi}\right) - C_a(f_{a1}) = -v_d \frac{\partial f_{a0}}{\partial r}\sin\theta, \quad (10.7)$$

where $\xi = v_\parallel/v$ is the cosine of the pitch angle. We remind ourselves that f_{a0} is Maxwellian,

$$f_{a0} = n_a \left(\frac{m_a}{2\pi T_a}\right)^{3/2} \exp\left(-\frac{\mathscr{E} - e_a\Phi}{T_a}\right),$$

and that the radial derivative on the right-hand side of the drift kinetic equation is to be taken at fixed energy, see (8.16), so that

$$\frac{\partial f_{a0}}{\partial r} = \left[\frac{d\ln p_a}{dr} + \frac{e_a}{T_a}\frac{d\Phi}{dr} + \left(\frac{m_a v^2}{2T_a} - \frac{5}{2}\right)\frac{d\ln T_a}{dr}\right]f_{a0}.$$

In the previous section we neglected the radial electric field $d\Phi/dr$, without justification.

The essential feature of the plateau distribution function is that it is localized around $\xi = 0$. In general, f_{a1} itself does not have this property since there may be a parallel flow associated with f_{a1}. To allow for this possibility, we write

$$f_{a1} = \frac{m_a v_\parallel V_\parallel}{T_a} f_{a0} + h_a,$$

and require h_a to be localized. If so, pitch-angle scattering of h_a dominates in the collision operator,

$$C_a(f_{a1}) \simeq v_a \mathcal{L}(h_a) \simeq \frac{v_a}{2} \frac{\partial^2 h_a}{\partial \xi^2}.$$

Here, v_a is the total collision frequency (for instance, $v_e = v_{ee} + v_{ei}$), and we have neglected $\xi^2 \ll 1$ in the Lorentz operator,

$$\mathcal{L} = \frac{1}{2} \frac{\partial}{\partial \xi} (1 - \xi^2) \frac{\partial}{\partial \xi}.$$

Incidentally, it is the scattering character of the collision operator, with its two ξ-derivatives, that motivated the factor $(v/v_\parallel)^2$ in front of the effective collision frequency (10.3). With this simplification, the drift kinetic equation (10.7) becomes, for $\xi \ll 1$,

$$\xi \frac{\partial h_a}{\partial \theta} - \frac{\epsilon \sin \theta}{2} \frac{\partial h_a}{\partial \xi} - \frac{v_a q R}{2v} \frac{\partial^2 h_a}{\partial \xi^2} = s_a(v, r) \sin \theta, \qquad (10.8)$$

where the source term on the right is

$$s_a \equiv -\frac{v_d q R}{v} \frac{\partial f_{a0}}{\partial r} + \frac{m_a v V_\parallel \epsilon}{2 T_a} f_{a0}.$$

Since we expect h_a to be localized around small ξ in the sense quantified by (10.5), it is convenient to introduce a stretched pitch-angle variable,

$$\eta \equiv \xi \hat{v}^{-1/3},$$

where $\hat{v} = v_a/(2v/qR) \ll 1$ is the collision frequency normalized to the transit frequency for well circulating particles. The drift kinetic equation now becomes

$$\eta \frac{\partial h_a}{\partial \theta} - \frac{\epsilon}{\hat{v}^{2/3}} \sin \theta \frac{\partial h_a}{\partial \eta} - \frac{\partial^2 h_a}{\partial \eta^2} = \frac{s_a}{\hat{v}^{1/3}} \sin \theta. \qquad (10.9)$$

The second term on the left represents the mirror force, and can be neglected because of the plateau ordering (8.8), which implies $\epsilon^{3/2} \ll \hat{v}$, and thus requires trapped particles to be collisional. As worked out in Exercise 1, the remaining equation has the solution

$$h_a = \frac{s_a}{\hat{v}^{1/3}} \int_0^\infty e^{-\tau^3/3} \sin(\theta - \eta\tau) d\tau, \tag{10.10}$$

which is a function peaked around $\eta \leq 1$. Only the part of h_a that is even in ξ contributes to the fluxes. As also shown in the exercise, this function is peaked around $\xi \leq \hat{v}^{1/3}$, and in the limit $\hat{v} \to 0$

$$h_a^{\text{even}}(\xi) \to \pi s_a \delta(\xi) \sin \theta.$$

Knowing h_a, we can now calculate the particle flux,

$$\langle \mathbf{\Gamma}_a \cdot \nabla r \rangle = \left\langle \int_0^\infty 2\pi v^2 \, dv \int_{-1}^1 v_d \sin^2 \theta \pi s_a \delta(\xi) \, d\xi \right\rangle$$

$$= -\hat{\omega}_{ta} \rho_a^2 n_a \frac{\pi^{1/2} q^2}{4} \left[\frac{d \ln n_a}{dr} + \frac{3}{2} \frac{d \ln T_a}{dr} + \frac{e_a}{T_a} \left(\frac{d\Phi}{dr} + V_\parallel B_\theta \right) \right],$$

where again $\hat{\omega}_{ta} = v_{Ta}/qR$ and $\rho_a = v_{Ta}/\Omega_a$. The parallel velocity $V_\parallel$ can be determined by requiring that the total flux be ambipolar, see (8.30). In a simple plasma consisting of electrons and a single ion species of charge Z, the radial ion particle flux must be equal to the electron flux divided by Z. Thus $\Gamma_{ir} = \Gamma_{er}/Z$, and the ambipolarity condition requires

$$\boxed{V_\parallel = -\frac{T_i}{ZeB_\theta} \left(\frac{d \ln n_i}{dr} + \frac{3}{2} \frac{d \ln T_i}{dr} + \frac{Ze}{T_i} \frac{d\Phi}{dr} \right).}$$

With this information we can finally calculate the plateau particle flux

$$\boxed{\begin{aligned} \langle \mathbf{\Gamma}_e \cdot \nabla r \rangle &= Z \langle \mathbf{\Gamma}_i \cdot \nabla r \rangle \\ &= -\hat{\omega}_{te} \rho_e^2 n_e \frac{\pi^{1/2} q^2}{4} \left[\left(1 + \frac{T_i}{ZT_e} \right) \frac{d \ln n_e}{dr} + \frac{3}{2} \frac{d \ln T_e}{dr} + \frac{3}{2ZT_e} \frac{d \ln T_i}{dr} \right], \end{aligned}}$$

and, completely analogously, the heat fluxes,

$$\boxed{\begin{aligned} \langle \mathbf{q}_e \cdot \nabla r \rangle &= -\hat{\omega}_{te} \rho_e^2 n_e \frac{\pi^{1/2} q^2}{8} \left[\left(1 + \frac{T_i}{ZT_e} \right) \frac{d \ln n_e}{dr} \right. \\ &\qquad\qquad \left. + \frac{15}{2} \frac{d \ln T_e}{dr} + \frac{3}{2ZT_e} \frac{d \ln T_i}{dr} \right], \\ \langle \mathbf{q}_i \cdot \nabla r \rangle &= -\hat{\omega}_{ti} \rho_i^2 n_i \frac{3\pi^{1/2} q^2}{4} \frac{dT_i}{dr}. \end{aligned}} \tag{10.11}$$

As usual, the ion heat flux is $O(\sqrt{m_i/m_e})$ larger than the electron heat flux, which is comparable to the particle fluxes. In multi-species plasma, which we consider in Chapter 12, the ion particle flux is larger and is comparable to the heat flux.

Further reading

Galeev and Sagdeev (1968) discovered the phenomenon of plateau transport, and the subject is reviewed in the standard texts by Hinton and Hazeltine (1976), Hirshman and Sigmar (1981), and Balescu (1988). In Hirshman (1978c) an important earlier conundrum was resolved demonstrating in detail how intrinsic ambipolarity of transport arises in the plateau regime. This paper also contains a lucid derivation of the time evolution of the toroidal and poloidal plasma rotation and the radial electric field on the corresponding multiple time scales. By definition, the plateau regime only exists at large aspect ratio, but it does not require circular flux surfaces. Solano and Hazeltine (1994) show how to derive plateau transport coefficients for arbitrarily shaped flux surfaces.

Exercise

1. The resonant distribution function.

(a) Solve (10.9), ignoring the second term on the left.
Solution: It is clear that h_a is equal to $s_a \hat{v}^{-1/3}$ multiplied by the solution to

$$\eta \frac{\partial \bar{h}}{\partial \theta} - \frac{\partial^2 \bar{h}}{\partial \eta^2} = \sin\theta.$$

If we write $\bar{h} = \text{Im}\,[H(\eta)e^{i\theta}]$, then H must satisfy

$$i\eta H - \frac{\partial^2 H}{\partial \eta^2} = 1.$$

Verify that

$$H(\eta) = \int_0^\infty \exp(-i\eta\tau - \tau^3/3)\,d\tau$$

is a solution to this equation, and note that $H(\eta) \to 0$ as $\eta \to \pm\infty$ by Riemann–Lebesgue's lemma. Hence $h_a = s_a v^{-1/3}\bar{h} = (10.10)$.

(b) Show that in the limit $\hat{v} \to 0$, the part of h_a that is even in $\xi = \eta\hat{v}^{1/3}$,

$$h_a^{\text{even}} = \frac{s_a}{\hat{v}^{1/3}} \sin\theta \int_0^\infty e^{-\tau^3/3} \cos\eta\tau\,d\tau,$$

approaches a delta function, $h_a^{\text{even}}(\xi) \to \pi s_a \delta(\xi) \sin\theta$.

Solution: This is equivalent to proving that

$$\int_{-1}^{1} h_a^{\text{even}} d\xi \rightarrow \pi s_a \sin \theta,$$

and

$$h_a^{\text{even}}(\xi) \rightarrow 0, \quad \xi \neq 0,$$

as $\hat{v} \rightarrow 0$. The first of these statements follows from

$$\int_{-\infty}^{\infty} H(\eta) \, d\eta = \lim_{N \to \infty} \int_{-N}^{N} H(\eta) \, d\eta$$

$$= \lim_{N \to \infty} \int_{0}^{\infty} e^{-\tau^3/3} d\tau \int_{-N}^{N} e^{-i\eta\tau} \, d\eta$$

$$= 2 \int_{0}^{\infty} e^{-\tau^3/3} \frac{\sin N\tau}{\tau} \, d\tau = \pi,$$

where we have used

$$\int_{0}^{\infty} \frac{\sin x}{x} \, dx = \frac{\pi}{2}.$$

Integrating twice by parts shows that

$$\int_{0}^{\infty} e^{-\tau^3/3} \cos \eta\tau \, d\tau = O(\eta^{-2}), \quad \eta \rightarrow \infty,$$

and it follows that the asymptotic behaviour of h_a^{even} is

$$h_a^{\text{even}}(\xi) = O(\hat{v}^{1/3}), \quad \hat{v} \rightarrow 0$$

away from $\xi = 0$.

11
Transport in the banana regime

11.1 Drift kinetic equation

In the banana regime (8.9), the effective collision frequency for scattering of trapped particles, $\nu_{\text{eff}} = \nu/\epsilon$ is much smaller than the bounce frequency of these particles, $\omega_b \sim \epsilon^{1/2} v_T/qR$,

$$\nu_* = \frac{\nu_{\text{eff}}}{\omega_b} \ll 1, \tag{11.1}$$

and we must face up to the full drift kinetic equation (8.16),

$$v_{\parallel} \nabla_{\parallel} f_{a1} + \mathbf{v}_d \cdot \nabla f_{a0} - \frac{e_a v_{\parallel} E_{\parallel}^{(A)}}{T_a} f_{a0} = C_a(f_{a1}),$$

$$\mathbf{v}_d \cdot \nabla f_{a0} = I v_{\parallel} \nabla_{\parallel} \left(\frac{v_{\parallel}}{\Omega_a} \right) \frac{\partial f_{a0}}{\partial \psi}. \tag{11.2}$$

We recall that the expansion $f_a = f_{a0} + f_{a1} + \cdots$ refers to the smallness of the Larmor radius, and that gradients are taken at fixed energy $\mathscr{E} = m_a v^2/2 + e_a \Phi$ and magnetic moment $\mu = m_a v_{\perp}^2/2B$, so that the mirror force is accounted for by the parallel gradient. For each species f_{a0} is a Maxwellian, and the drift kinetic equation (11.2) is to be solved for f_{a1}. It is clear that there are two driving terms. The first one is the cross-field drift, $\mathbf{v}_d \cdot \nabla f_{a0}$, which is proportional to the radial gradients in $\partial f_{a0}/\partial \psi$, and the second one is associated with the induced electric field $E_{\parallel}^{(A)}$. The latter can be conveniently accounted for by using the Spitzer function f_{as} defined by

$$C_a(f_{as}) = -\frac{e_a v_{\parallel} E_{\parallel}^{(A)}}{T_a} f_{a0}. \tag{11.3}$$

In Chapter 4 we solved this equation using a model collision operator. The Spitzer problem has been solved more accurately in the literature,

187

where tables and analytic approximations of the Spitzer function can be found (Spitzer and Härm, 1953; Hirshman and Sigmar, 1976; Hirshman, 1980). Thus, we can regard it as a known function and write the drift kinetic equation as

$$v_\parallel \nabla_\parallel (f_{a1} - F_a) = C_a(f_{a1} - f_{as}), \tag{11.4}$$

where

$$F_a \equiv -\frac{I v_\parallel}{\Omega_a} \frac{\partial f_{a0}}{\partial \psi} = -\frac{I v_\parallel}{\Omega_a} \left[\frac{d \ln n_a}{d\psi} + \frac{e_a}{T_a} \frac{d\Phi}{d\psi} + \left(\frac{m_a v^2}{2 T_a} - \frac{3}{2} \right) \frac{d \ln T_a}{d\psi} \right] f_{a0}, \tag{11.5}$$

and where the parallel electric field no longer appears explicitly.

To proceed analytically from here we invoke the banana regime assumption (11.1) of low collisionality (Rutherford, 1970; Rosenbluth, Hazeltine and Hinton, 1972). Accordingly, we make a *subsidiary* expansion of f_{a1} in the smallness of the collision frequency, $f_{a1} = f_{a1}^{(0)} + f_{a1}^{(1)} + \cdots$, and endeavour to solve

$$v_\parallel \nabla_\parallel (f_{a1}^{(0)} - F_a) = 0,$$
$$v_\parallel \nabla_\parallel f_{a1}^{(1)} = C_a(f_{a1}^{(0)} - f_{as}). \tag{11.6}$$

The first of these equations is evidently solved by

$$f_{a1}^{(0)} = g_a + F_a,$$

where g_a is an integration constant independent of θ, i.e., $\nabla_\parallel g_a = 0$ and $g_a = g_a(\psi, \mathscr{E}, \mu, \sigma)$. For passing particles, the function g_a can be determined from the solubility constraint obtained by multiplying the next-order equation (11.6) by $B/v_\parallel$ and taking the flux-surface average,

$$\left\langle \frac{B}{v_\parallel} C_a(g_a + F_a - f_{as}) \right\rangle = 0. \qquad \text{(passing)} \tag{11.7}$$

Here we have used (7.6) to annihilate the left-hand side of (11.6) and also noted that $f_{a1}^{(1)}$ must be periodic in θ at fixed $(\psi, \mathscr{E}, \mu, \sigma)$.

While g_a is determined by (11.7) in the circulating domain, it vanishes in trapped particle space,

$$g_a = 0. \qquad \text{(trapped)}$$

To demonstrate this statement, we first observe that in the trapped domain g_a must be an even function of $\sigma = v_\parallel / |v_\parallel|$,

$$g_a(\psi, \mathscr{E}, \mu, \sigma) = g_a(\psi, \mathscr{E}, \mu, -\sigma), \qquad \text{(trapped)}$$

since this must be true at the bounce points, $\theta = \pm\theta_b$, where $v_\parallel = 0$, and therefore holds everywhere as $\partial g_a/\partial\theta = 0$. Next, we rewrite (11.6) as

$$\frac{\partial f_{a1}^{(1)}}{\partial\theta} = \frac{B}{\sigma|v_\parallel||\mathbf{B}\cdot\nabla\theta}C_a(g_a + F_a - f_{as}).$$

and integrate between the bounce points at fixed $(\psi, \mathscr{E}, \mu, \sigma)$,

$$\int_{-\theta_b}^{\theta_b}\frac{d\theta}{\mathbf{B}\cdot\nabla\theta}\frac{B}{\sigma|v_\parallel|}C_a(g_a + F_a - f_{as}) = f_{a1}^{(1)}(\theta_b) - f_{a1}^{(1)}(-\theta_b).$$

The difference between the $\sigma = \pm 1$ versions of this relation is

$$\int_{-\theta_b}^{\theta_b}\frac{d\theta}{\mathbf{B}\cdot\nabla\theta}\frac{B}{|v_\parallel|}C_a\left[g_a(\sigma = 1) + g_a(\sigma = -1)\right] = 0, \qquad (11.8)$$

where the terms containing F_a and f_{as} have dropped out since these functions are odd in σ. This equation, which holds for each set of coordinates (v, λ, ψ), shows that g_a must also be odd in σ. Since we already know that g_a is even, it follows that it vanishes identically in the trapped domain.

Thus, the trapped particle response to the radial gradients is entirely described by $f_{a1} = F_a = -(Iv_\parallel/\Omega_a)\partial f_{a0}/\partial\psi$, which has a simple physical interpretation. Since the toroidal canonical momentum

$$p_\varphi = m_a Rb_\varphi v_\parallel - e_a\psi = e_a\left(\frac{Iv_\parallel}{\Omega_a} - \psi\right) \qquad (11.9)$$

is constant along a particle orbit, see Eq. (7.19), the excursion from the flux surface defined by the banana tips, where $v_\parallel = 0$, is $\Delta\psi = Iv_\parallel/\Omega_a$, and (neglecting f_{as})

$$f_{a0} + f_{a1} = f_{a0} - \Delta\psi\frac{\partial f_{a0}}{\partial\psi} \simeq f_{a0}(\psi - \Delta\psi). \qquad \text{(trapped)}$$

In other words, the distribution function of trapped particles is equal to a Maxwellian evaluated at the banana tips.

The distribution of passing particles is determined by the collisional equilibrium they establish with the trapped population. Mathematically, our remaining task is to solve (11.7) for g_a in the passing region, which is done separately for ions and electrons. For electrons, the source term associated with the electric field is generally of the same order as that associated with radial gradients, i.e.,

$$f_{es} \sim F_e.$$

Since the latter is derived from the cross-field drift, it is proportional to the Larmor radius, and is therefore much larger for the ions. It will therefore dominate over f_{is}, which does not scale with gyroradius,

$$f_{is} \ll F_i.$$

In the ion equation we may therefore neglect the parallel electric field. While this is common practice in neoclassical theory and is usually well justified for calculating radial transport, recent work has shown that the induced electric field can, in fact, drive a substantial poloidal ion flow under experimentally relevant conditions (Catto *et al.* 2000).

Regarding the electric field, we may note at this point that the details of its poloidal variation over the flux surface are to some extent unimportant since the the distribution function is determined by Eq. (11.7), where only the average $\langle E_\parallel B \rangle$ appears. In the definition (11.3) of the Spitzer function, we may thus replace $E_\parallel$ by the modified parallel electric field

$$E'_\parallel = \frac{\langle E_\parallel B \rangle B}{\langle B^2 \rangle}.$$

This is generally true in all neoclassical theory, regardless of the collisionality.

The parallel particle flux associated with F_a,

$$\int v_\parallel F_a d^3v = \frac{I n_a \omega_a}{B}, \tag{11.10}$$

where we recall the definition (8.22) of ω_a,

$$\omega_a(\psi) = -\frac{d\Phi}{d\psi} - \frac{1}{n_a e_a}\frac{dp_a}{d\psi},$$

has the character of a diamagnetic flow in the following sense. If we consider the banana orbits that pass through a particular point in the outer midplane, we realize from Eq. (11.9) that those with $v_\parallel > 0$ have their bounce points closer to the centre of the plasma and those with $v_\parallel < 0$ have their bounce points further out, see Fig. 11.1. If the density is centrally peaked, the former are more numerous, and hence a parallel 'diamagnetic' flow is produced. Unlike the usual diamagnetic flow, which has to do with Larmor orbits, this flow is associated with banana orbits. From general arguments we have earlier established that the parallel flow velocity of each species is given by (8.23) and the total flow by (8.24), regardless of the collisionality. It is then clear from (11.10) that, in the banana regime, poloidal rotation is entirely contained in g_a and f_{as},

$$K_a = \frac{1}{B}\int (g_a + f_{as})v_\parallel d^3v. \tag{11.11}$$

For ions, only the first term matters in the ordering we have adopted, and since g_i vanishes in the trapped region, we conclude that only passing ions contribute to the poloidal rotation of the plasma. This is natural since trapped particles are locked in the magnetic well.

The principal mathematical problem of the banana regime is to solve the constraint equation (11.7). Before attempting to do that, we close this section by showing another, perhaps more intuitive, way of deriving this equation, following Catto, Bernstein and Tessarotto (1987). The drift kinetic equation can be written as

$$\frac{df_a}{dt} = C_a(f_a - f_{as} - f_{a0}), \tag{11.12}$$

where the left-hand side denotes the time derivative along the collisionless orbit. On the right-hand side we have linearized the collision operator around a Maxwellian $f_{a0}(\psi)$ which is constant on flux surfaces. This is permissible since we know that the distribution function must be close to $f_{a0}(\psi)$, see Eq. (8.11). Let us define a function f_{a*} as the Maxwellian taken at the location $\psi_* = -p_\varphi/e_a = \psi - Iv_\parallel/\Omega_a$. For trapped particles this is the position of the bounce points. Thus we define

$$f_{a*} \equiv f_{a0}(\psi = \psi_*).$$

Since $\psi \simeq \psi_*$ and $f_a \simeq f_{a0}$ it follows that f_a is close to f_{a*}, i.e.,

$$f_a = f_{a*} + g_a,$$

where $g_a \ll f_a$. The drift kinetic equation (11.12) can now written as

$$\frac{dg_a}{dt} = C_a(f_{a*} - f_{a0} + g_a - f_{as}). \tag{11.13}$$

In the banana regime the collision frequency is small, which implies that the left side vanishes, $dg_a/dt = 0$, to lowest order in ν. This means that g_a is a function only of constants of motion, $g_a = g_a(\psi_*, \mathscr{E}, \mu, \sigma)$. Since (v, λ, ψ) are approximate constants of motion we may write $g_a = g_a(v, \lambda, \psi, \sigma)$ to the requisite accuracy. Taking the integral of Eq. (11.13) one turn around the orbit annihilates the left side, leaving

$$\oint C_a(g_a + F_a - f_{as})dt = 0,$$

since $f_{a*} - f_{a0} \simeq (\psi_* - \psi)\partial f_{a0}/\partial\psi = F_a$. This equation is equivalent to Eq. (11.7) for passing particles since $dt = d\theta/v_\parallel \nabla_\parallel \theta$.

11.2 Ion transport

We now calculate the transport of ions by solving the ion version of (11.7), where ion–electron collisions are negligible, $C_i \simeq C_{ii}$, and we may also neglect the parallel electric field by disregarding f_{is}. Before going into the details of this kinetic problem, we make one important observation.

If there is no temperature gradient, $dT_i/d\psi = 0$, the solution to (11.7) is simply $g_i = 0$. This follows from the fact that $C_{ii}(F_i) = 0$ when $dT_i/d\psi = 0$ since $C_{ii}(cv_\parallel f_{i0}) = 0$ for any constant c, according to (3.62) and (3.63). As the equation (11.7) is linear, we thus expect its solution to be proportional to the temperature gradient, $g_i \propto dT_i/d\psi$.

Proceeding to find this solution approximately, we employ the model operator (3.69) for ion–ion collisions,

$$C_{ii}(f_{i1}) = v_D^{ii}(v) \left(\mathcal{L}(f_{i1}) + \frac{m_i v_\parallel u_i}{T_i} f_{i0} \right). \qquad (11.14)$$

Here we have anticipated that the momentum-restoring vector $\mathbf{u}$ is parallel to the magnetic field. The Lorentz operator is

$$\mathcal{L} = \frac{2hv_\parallel}{v^2} \frac{\partial}{\partial \lambda} \lambda v_\parallel \frac{\partial}{\partial \lambda},$$
$$h \equiv B_0/B, \qquad (11.15)$$

when written in terms of the variable

$$\lambda = \frac{\mu B_0}{\mathcal{E} - Ze\Phi} = \frac{v_\perp^2 B_0}{v^2 B}$$

we introduced in (7.20). The normalizing magnetic field is in principle arbitrary, but we chose it to be $B_0 \equiv \left\langle B^2 \right\rangle^{1/2}$. Note that $\mathcal{L}(v_\parallel) = -v_\parallel$ and hence $\mathcal{L}(F_i) = -F_i$. Our kinetic problem (11.7),

$$\left\langle \frac{B}{v_\parallel} \left[\mathcal{L}(g_i + F_i) + \frac{m_i v_\parallel u_i}{T_i} f_{i0} \right] \right\rangle = 0,$$

can now be written as

$$\frac{\partial}{\partial \lambda} \lambda \left\langle v_\parallel \right\rangle \frac{\partial g_i}{\partial \lambda} = -\frac{v^2}{2} s_i(v, \psi) f_{i0}, \qquad (11.16)$$

$$s_i(v, \psi) \equiv \frac{I}{\hat{\Omega}_i} \frac{\partial \ln f_{i0}}{\partial \psi} + \left\langle \frac{u_i}{h} \right\rangle \frac{m_i}{T_i}, \qquad (11.17)$$

with $\hat{\Omega}_i \equiv h\Omega_i = e_i B_0/m_i$.

Note that we have now succeeded in reducing the original kinetic equation (2.1) from a seven-dimensional partial differential equation to an ordinary differential equation! The basic key to success was the assumption of small Larmor radius (8.1) and the transport ordering (8.3), which removed gyromotion and time variation from the problem. We were thus able to consider the time-independent drift kinetic equation rather than the full Fokker–Planck equation. Since transport along the magnetic field is faster than that across the field, the transport problem could be considered at one flux surface at a time, which eliminates the radial coordinate. Furthermore, since the plasma is axisymmetric the toroidal coordinate φ is ignorable. Finally, the adoption of a model collision operator enabled us to ignore one of the velocity dimensions. Of course, the model operator is strictly accurate only in the Lorentz limit, $Z \to \infty$, where pitch-angle scattering dominates in the collision dynamics. In general, the full Coulomb collision must be used. However, as we shall see, the model operator turns out to give accurate results also in the limit of large aspect ratio, regardless of the ion charge.

The ordinary differential equation (11.16) is solved by integrating twice over λ. Since we already know that g_i vanishes in the trapped domain, we need only consider the passing domain, $0 \leq \lambda \leq B_0/B_{max}$, where $B_{max}(\psi)$ is the maximum magnetic field strength on the flux surface ψ. Furthermore, since g_i is continuous, it vanishes at the trapped–passing boundary $\lambda_c \equiv B_0/B_{max}$. Hence

$$g_i = H(\lambda_c - \lambda)V_{\parallel}s_if_{i0}, \tag{11.18}$$

where

$$V_{\parallel}(\lambda, v, \psi) \equiv \frac{v^2}{2} \int_{\lambda}^{\lambda_c} \frac{d\lambda'}{\langle v_{\parallel}(\lambda')\rangle} = \frac{\sigma v}{2} \int_{\lambda}^{\lambda_c} \frac{d\lambda'}{\langle\sqrt{1 - \lambda'/h(\theta)}\rangle}, \tag{11.19}$$

and H is the Heaviside step function, so that $H(\lambda_c - \lambda) = 1$ in the passing region and $H(\lambda_c - \lambda) = 0$ in the trapped region. In a large-aspect-ratio torus the magnetic field strength is almost constant, $h = B_0/B = 1 + O(\epsilon)$, and the trapped–passing boundary is located at $\lambda_c = 1 - \epsilon$. The quantity we have denoted by $V_{\parallel}$ is therefore approximately equal to the parallel velocity, $V_{\parallel} = v_{\parallel}[1 + O(\epsilon^{1/2})]$ in most of velocity-space, $1 - \lambda \gg \epsilon$. In fact, if $\epsilon \to 0$ then

$$V_{\parallel} \to \frac{\sigma v}{2} \int_{\lambda}^{1} \frac{d\lambda'}{\sqrt{1 - \lambda'}} = \sigma v\sqrt{1 - \lambda} = v_{\parallel}.$$

The full ion distribution function $f_{i1} = g_i + F_i$ is obtained by piecing together Eqs. (11.5), (11.17) and (11.18),

$$f_{i1} = -\frac{I}{\hat{\Omega}_i}(hv_{\parallel} - HV_{\parallel})\frac{\partial f_{i0}}{\partial \psi} + \frac{m_iHV_{\parallel}}{T_i}\left\langle\frac{u_i}{h}\right\rangle f_{i0}, \tag{11.20}$$

and the coefficient u_i is determined by (3.70), reflecting momentum conservation in ion–ion collisions. Recalling the velocity-space average (4.26), we can write the equation for u_i as

$$\{v_D^{ii}\}\left\langle\frac{u_i}{h}\right\rangle = \left\langle\frac{1}{hn_i}\int v_D^{ii}v_{\parallel}f_{i1}d^3v\right\rangle. \tag{11.21}$$

To evaluate the integral on the right of this equation using the solution (11.20), it is useful to note that for any function $A(v)$,

$$\left\langle \int A(v) \frac{m_a v_\parallel H V_\parallel}{h T_a} f_{a0} d^3 v \right\rangle = \left\langle h^{-2} \right\rangle \sum_\sigma \int_0^\infty A(v) f_{a0} \frac{\pi m_a v^3 dv}{T_a} \int_0^{\lambda_c} H V_\parallel d\lambda$$

$$= \pi \int_0^\infty \frac{m_a v^4}{T_a} f_{a0} dv \int_0^{\lambda_c} d\lambda \int_\lambda^{\lambda_c} \frac{d\lambda'}{\left\langle \sqrt{1 - \lambda'/h} \right\rangle}$$

$$= (1 - f_t) n_a \{A\}, \tag{11.22}$$

so that

$$\left\langle \int A(v) \frac{m_a v_\parallel (h v_\parallel - H V_\parallel)}{h T_a} f_{a0} d^3 v \right\rangle = f_t n_a \{A\}, \tag{11.23}$$

where we have used the velocity-space volume element (7.21) and noted that $\left\langle h^{-2} \right\rangle = 1$, which follows from $h = B_0/B$ and $B_0^2 \equiv \left\langle B^2 \right\rangle$. We have also introduced the 'effective fraction of trapped particles'

$$\boxed{f_t = 1 - \frac{3}{4} \int_0^{\lambda_c} \frac{\lambda d\lambda}{\left\langle \sqrt{1 - \lambda/h} \right\rangle} \simeq 1.46 \sqrt{\epsilon},} \tag{11.24}$$

which is a quantity that appears repeatedly in banana-regime transport theory. The last, approximate relation in (11.24) holds in a circular torus with large aspect ratio, as shown in an exercise at the end of this chapter. In another exercise, an efficient way of calculating f_t is derived for plasmas of arbitrary cross section. The number $f_c = 1 - f_t$ is called the effective fraction of circulating particles for obvious reasons. It is useful to note that Eq. (11.22) also holds if we replace h by $1/h$, so that

$$\left\langle h \int A(v) \frac{m_a v_\parallel H V_\parallel}{T_a} f_{a0} d^3 v \right\rangle = f_c n_a \{A\}. \tag{11.25}$$

Returning to the equation (11.21) for u_i and inserting the solution now gives

$$\left\langle \frac{u_i}{h} \right\rangle = -\frac{I T_i}{m_i \hat{\Omega}_i} \frac{\{v_D^{ii} \, \partial \ln f_{i0}/\partial \psi\}}{\{v_D^{ii}\}}$$

$$\simeq -\frac{I T_i}{m_i \hat{\Omega}_i} \left(\frac{d \ln p_i}{d\psi} + \frac{Z e}{T_i} \frac{d\Phi}{d\psi} - 1.173 \frac{d \ln T_i}{d\psi} \right), \tag{11.26}$$

where $-1.173 = \{v_D^{ii}(x^2 - 5/2)\}/\{v_D^{ii}\}$ according to velocity-space averages given in the Appendix. We can now finally write our solution (11.20) to the ion drift kinetic equation as

$$f_{i1} = -\frac{Iv_{\parallel}}{\Omega_i}\frac{\partial f_{i0}}{\partial \psi} + \frac{IHV_{\parallel}}{\hat{\Omega}_i}\left(\frac{m_i v^2}{2T_i} - 1.33\right)\frac{\partial \ln T_i}{\partial \psi}f_{i0}. \tag{11.27}$$

Note that $g_i = f_{i1} - F_i$ is proportional to the ion temperature gradient, as anticipated, and that the distribution function has a discontinuous derivative $\partial f_{i1}/\partial \lambda$ at the trapped–passing boundary since the Heaviside function $H(\lambda_c - \lambda)$ has a step there. The distribution function itself is still continuous since $V_{\parallel}(\lambda_c) = 0$, but $\mathcal{L}(f_{i1})$ has a delta function at $\lambda = \lambda_c$. Mathematically, this is what gives rise to the large neoclassical enhancement of the parallel friction and the cross-field flux (8.32).

Knowing the ion distribution function, we are now in a position to calculate quantities of physical interest by taking moments of f_{i1}. For instance, the flow velocity (8.24) of the ions within the flux surface can be calculated using (11.11), which implies

$$K_i\left\langle B^2 \right\rangle = \left\langle \frac{B_0}{h}\int g_i v_{\parallel} d^3 v \right\rangle.$$

The quantity K_i determining the poloidal flow becomes

$$K_i = 1.17 f_c \frac{n_i I}{m_i \hat{\Omega}_i B_0}\frac{dT_i}{d\psi}. \tag{11.28}$$

The poloidal flow $n_i \mathbf{V}_{i\theta} = K_i \mathbf{B}_\theta$ is thus simply proportional to the temperature gradient. The gradients of the pressure and electrostatic potential determine the toroidal flow velocity according to (8.24), and the parallel flow becomes

$$V_{i\parallel} = -\frac{IT_i}{m_i \hat{\Omega}_i}\left[h\left(\frac{d\ln p_i}{d\psi} + \frac{Ze}{T_i}\frac{d\Phi}{d\psi}\right) - 1.17 h^{-1} f_c \frac{d\ln T_i}{d\psi}\right]. \tag{11.29}$$

A word of caution is appropriate here. Like most of the results derived in this chapter, these relations are strictly accurate only in the limits of very large and very tight aspect ratio, i.e., $\epsilon \ll 1$ or $\epsilon = 1$. For instance, the expression (11.28) for the poloidal flow is exact in the limits $f_t = 0$ and $f_t = 1$, and joins these end points in a qualitatively correct way, but because of the approximate nature of the model operator we have employed the coefficient in Eq. (11.28) is not reliable at intermediate aspect ratios. A more accurate expression for the poloidal ion flow velocity will be derived later in this chapter.

We now turn to transport across flux surfaces. The particle flux vanishes to leading order in $(m_e/m_i)^{1/2}$ since ion–electron collisions are neglected in this order and (8.27) therefore vanishes. (There is no force $F_{i\varphi}$ among the ions themselves because of momentum conservation in ion–ion collisions, and the electric field is negligible for the ions.) This is only true in a pure plasma, however. As usual, the situation changes if impurities are introduced. Ion–ion collisions give rise to a heat flux, which can be calculated by evaluating (8.32). Simple, accurate results are available in the limits of large aspect ratio and unit aspect ratio, which we now consider separately.

Large aspect ratio

So far we have kept the flux-surface geometry general, but we now specialize to the case of a large-aspect-ratio plasma with circular flux surfaces. Only at large aspect ratio (or high ion charge) is it possible to rigorously justify the use of the model operator (11.14).

If $\epsilon \ll 1$ so that $h = 1 + O(\epsilon)$, and if the flux surfaces are circular so that $d\psi = RB_\theta dr$, the parallel mean ion velocity (11.29) becomes

$$V_{i\parallel} = -\frac{I\,T_i}{m_i\Omega_{i\theta}}\left(\frac{d\ln p_i}{dr} + \frac{Ze}{T_i}\frac{d\Phi}{dr} - 1.17\frac{d\ln T_i}{dr}\right), \qquad (11.30)$$

where $\Omega_{i\theta} = ZeB_\theta/m_i$ is the poloidal ion gyrofrequency. The poloidal rotation of the plasma is given by

$$V_{i\theta} = \frac{K_i B_\theta}{n_i B} = \frac{1.17}{m_i\Omega_i}\frac{dT_i}{dr},$$

and is thus of order $V_{i\theta} \sim v_{Ti}\rho_i/L$, which is typically smaller than the parallel velocity given in (11.30), which is $V_{i\parallel} \sim v_{Ti}\rho_{i\theta}/L$.

To obtain the radial heat flux, we use the model collision operator (11.14) in the expression (8.32) to find

$$\langle \mathbf{q}_i \cdot \nabla\psi \rangle = -\frac{I}{ZeB_0}\left\langle h\int \frac{m_i v^2}{2}m_i v_\parallel v_D^{ii}\left(-f_{i1} + \frac{m_i v_\parallel u_i}{T_i}f_{i0}\right)d^3v \right\rangle,$$

where we have made use of the self-adjointness of the Lorentz operator and recalled that $\mathcal{L}(v_\parallel) = -v_\parallel$. Note that the convective heat flux is negligible, being of the same order $(m_e/m_i)^{1/2}$ as the particle flux. By inserting the distribution function (11.27) here and using (11.23) with $h \to 1$ and (11.26) we obtain, to the requisite accuracy, the cross-field heat flux

$$\langle \mathbf{q}_i \cdot \nabla\psi \rangle = -f_t\frac{n_i T_i I^2}{m_i\Omega_i^2}\frac{dT_i}{d\psi}\left\{v_D^{ii}x^2(x^2-1.33)\right\} = -0.92f_t\frac{n_i I^2 T_i}{m_i\Omega_i^2\tau_i}\frac{dT_i}{d\psi}.$$

$$(11.31)$$

Here $x^2 \equiv m_i v^2 / 2T_i$, we have used velocity-space averages of v_D^{ii} given in the Appendix, and we have recalled the definition of the Braginskii collision time $\tau_i = \tau_{ii}\sqrt{2}$. In a circular tokamak, where $f_t \simeq 1.46\epsilon^{1/2}$ and $\nabla\psi = RB_\theta \nabla r$, the radial heat flux is

$$q_{ir} = \frac{\langle \mathbf{q}_i \cdot \nabla\psi \rangle}{|\nabla\psi|} = -1.35\epsilon^{1/2}\frac{n_i T_i}{m_i \Omega_{i\theta}^2 \tau_i}\frac{dT_i}{dr}, \qquad (\epsilon \ll 1). \qquad (11.32)$$

Note that the banana heat diffusivity scales as anticipated in (1.10), i.e.,

$$\chi_i \sim f_t \frac{\rho_{i\theta}^2}{\tau_i}, \qquad (11.33)$$

where $\rho_{i\theta} = m_i v_{Ti}/ZeB_\theta$ is the poloidal ion gyroradius. While both results (11.31) and (11.32) assume large aspect ratio, the former is more general in that the shape of the flux surface is arbitrary.

Let us now pause to discuss the accuracy of the results just found and to seek justification for employing the simple model operator (11.14). In the case of transport in a large-aspect-ratio torus this operator is actually sufficiently accurate to give results identical to those found with the full Fokker–Planck operator. The reason for this is that the structure of the distribution function (11.27) is localized to the trapped and barely passing region, $1 - \lambda = O(\epsilon)$ in the sense that f_{i1} varies with the pitch-angle variable λ much more rapidly there than elsewhere. While the pitch-angle derivative is modest, $\partial f_{i1}/\partial\lambda = O(1)$, in most of velocity space, $1 - \lambda \ll \epsilon$, it is much larger, $\partial f_{i1}/\partial\lambda = O(\epsilon^{-1})$, close to and inside the trapped region, $1 - \lambda = O(\epsilon)$. There, the collision operator is dominated by its pitch-angle scattering part, $C_{ii} \simeq v_D^{ii}\mathcal{L}$, and it follows from the form of the Lorentz operator (11.15) that

$$C_{ii}(f_{i1}) = O(v_D^{ii}/\epsilon), \qquad 1 - \lambda = O(\epsilon)$$

in the trapped and barely passing region. On the other hand, in the well passing region the distribution function $f_{i0} + f_{i1}$ differs from a shifted Maxwellian only by $O(\epsilon^{1/2})$ since $V_\parallel = v_\parallel[1 + O(\epsilon^{1/2})]$. Therefore

$$C_{ii}(f_{i1}) = O(v_D^{ii}\epsilon^{1/2}), \qquad 1 - \lambda \ll \epsilon$$

in the well passing region. This has two important consequences. First, it is indeed satisfactory to use the model operator in the trapped (and nearly trapped) region of velocity space. Second, the integral for the radial, neoclassical heat flux (8.32) is dominated by the contribution from this region, although it occupies only a small fraction $O(\epsilon^{1/2})$ of phase space. Accordingly, the heat flux given by Eqs. (11.31) and (11.32) is accurate at large aspect ratio.

Unit aspect ratio

In the edge of a spherical tokamak, the magnetic field is much stronger on the inboard side of the torus than on the outboard side. All particles are therefore trapped, and there is practically no passing population. It follows immediately from the considerations in Section 11.1 that $g_i = 0$ and that such a plasma cannot rotate poloidally, see Eq. (11.11). As we shall see in the next chapter, this has to do with the fact that the neoclassical parallel viscosity is infinite at unit aspect ratio. The ion distribution is given exactly by $f_{i1} = F_i = (11.5)$, and we can calculate the neoclassical heat flux (8.32) using the exact, linearised collision operator (Hazeltine, Hinton and Rosenbluth, 1973). (The model operator is no longer accurate enough.) From (8.32) and (11.5), the neoclassical heat flux becomes

$$\langle \mathbf{q}_i \cdot \nabla \psi \rangle^{\text{neo}} = \left\langle \frac{I^2}{m_i \Omega_i^2} \right\rangle \int \left(\frac{m_i v^2}{2} - \frac{5T_i}{2} \right) m_i v_\parallel C_{ii} \left(v_\parallel \frac{\partial f_{i0}}{\partial \psi} \right) d^3 v$$

$$= \frac{m_i I^2 p_i \left\langle B^{-2} \right\rangle}{Z^2 e^2 \tau_{ii}} \frac{dT_i}{d\psi} (M_{ii}^{11} + N_{ii}^{11})$$

where we have recalled the matrix elements (4.30) of the collision operator. Note that only the temperature gradient term in $\partial f_{i0}/\partial \psi$ contributes to the heat flux because $M_{ii}^{0k} + N_{ii}^{0k} = M_{ii}^{k0} + N_{ii}^{k0} = 0$ for all k, see the Appendix. Since $\tau_{ii} = \tau_i/\sqrt{2}$ and $M_{ii}^{11} + N_{ii}^{11} = -\sqrt{2}$, the heat flux is

$$\boxed{\langle \mathbf{q}_i \cdot \nabla \psi \rangle^{\text{neo}} = -\frac{2n_i m_i T_i I^2}{Z^2 e^2 \tau_i} \left\langle B^{-2} \right\rangle \frac{dT_i}{d\psi}. \qquad (\epsilon = 1)} \qquad (11.34)$$

The heat diffusivity is thus of the order $\chi_i \sim \rho_{i\theta}^2/\tau_i$. At the end of this chapter, we shall return to the ion conductivity and show how the gap between the limits $\epsilon \ll 1$ and $\epsilon = 1$, where there are exact analytical results, can be bridged by an approximate calculation.

11.3 Electron transport

In the electron transport problem there are two new ingredients: collisions between unlike species (electron–ion collisions) and the presence of an inductive electric field. Both these elements give rise to cross-field particle transport according to Eq. (8.29). Unlike the ion problem just considered, we shall therefore calculate both particle and heat fluxes. We also compute the parallel current, where very important new effects emerge: the bootstrap current and a reduction of conductivity due to particle trapping. From a practical point of view, these predictions are possibly the most important results of neoclassical theory.

To solve (11.7) for electrons we again use the model operator (3.69) for electron–electron collisions, but must now also include electron–ion collisions (3.28), as well as the parallel electric field appearing in form of the Spitzer function f_{es}. As we have defined it in Eq. (11.3), this function satisfies

$$C_{ee}(f_{es}) + v_D^{ei}\left(\mathscr{L}(f_{es}) + \frac{m_e v_\parallel}{T_e}V_{i\parallel}f_{e0}\right) = \frac{ev_\parallel E_\parallel^{(A)}}{T_e}f_{e0}.$$

It is convenient to remove the ion flow velocity $V_{i\parallel}$ from f_{es} by writing

$$f_{es} = \frac{m_e v_\parallel}{T_e}V_{i\parallel}f_{e0} + f_s,$$

where f_s is the Spitzer function in the ion rest frame and satisfies

$$C_{ee}(f_s) + v_D^{ei}\mathscr{L}(f_s) = C_e(f_{es}) = \frac{ev_\parallel E_\parallel^{(A)}}{T_e}f_{e0}. \tag{11.35}$$

In the kinetic equation (11.7) for electrons, the collision operator is thus

$$C_e(f_{e1} - f_{es}) = (v_D^{ee} + v_D^{ei})\mathscr{L}(f_{e1} - f_s) + \frac{m_e v_\parallel}{T_e}(v_D^{ee}u_e + v_D^{ei}V_{i\parallel})f_{e0}, \tag{11.36}$$

which is more complicated than in the ion problem. However, the kinetic equation is still very similar to the one already solved for the ions. Instead of (11.16) one obtains the equation

$$\frac{\partial}{\partial\lambda}\lambda\langle v_\parallel\rangle\frac{\partial g_e}{\partial\lambda} = \frac{v^2}{2}s_e(v,\psi)f_{e0},$$

with

$$s_e = \frac{I}{\hat{\Omega}_e}\frac{\partial\ln f_{e0}}{\partial\psi} + \frac{m_e}{T_e}\left\langle\frac{v_D^{ee}u_e + v_D^{ei}V_{i\parallel}}{h(v_D^{ee} + v_D^{ei})}\right\rangle + \left\langle\frac{\hat{f}_s}{h}\right\rangle,$$

where we have written $f_s = v_\parallel\hat{f}_s(v)f_{e0}$, noting that the Spitzer function must be proportional to $v_\parallel$. In analogy with (11.18) and (11.20) we thus obtain the electron distribution function

$$f_{e1} = -\frac{I}{\hat{\Omega}_e}(hv_\parallel - HV_\parallel)\frac{\partial f_{e0}}{\partial\psi} + \left\langle\frac{m_e}{hT_e}\frac{v_D^{ee}u_e + v_D^{ei}V_{i\parallel}}{v_D^{ee} + v_D^{ei}} + \frac{\hat{f}_s}{h}\right\rangle HV_\parallel f_{e0}. \tag{11.37}$$

The effective ion charge, Z, determines the relative importance between electron–ion and electron–electron collisions. In the following, we give results for two important limits: $Z = 1$ (pure plasma) and $Z \to \infty$ (Lorentz limit). In the former limit, electron–electron collisions are important and we need to calculate the constant u_e from momentum conservation in these collisions. This is particularly easy to do in the case of large aspect ratio, $\epsilon \ll 1$. As already mentioned, in this limit the quantities denoted by $V_\parallel$ and $v_\parallel$ are nearly equal in most of velocity-space: away from the trapped region $V_\parallel = v_\parallel[1 + O(\epsilon^{1/2})]$. In view of (11.37), this implies that the mean parallel electron velocity is approximately equal to

$$\frac{v_D^{ee} u_e + v_D^{ei} V_{i\parallel}}{v_D^{ee} + v_D^{ei}}, \tag{11.38}$$

if the contribution from f_{es} is excluded. The error is of order $\epsilon^{1/2}$. On the other hand, away from the trapped region in velocity-space, the constraint equation (11.7) reduces to $C_e(f_{e1} - f_{es}) \simeq 0$ since $v_\parallel$ is nearly constant for most circulating electrons. This means that the mean ion velocity must be close to the mean velocity associated with the distribution function $f_{e1} - f_{es}$. As this velocity is given by Eq. (11.38), it follows that

$$u_e = V_{i\parallel}[1 + O(\epsilon^{1/2})].$$

Thus, within errors of order $\epsilon^{1/2}$ the distribution function (11.37) is equal to

$$f_{e1} = -\frac{I}{\hat{\Omega}_e}(hv_\parallel - HV_\parallel)\frac{\partial f_{e0}}{\partial \psi} + \left\langle \frac{m_e V_{i\parallel}}{T_e} + \frac{\hat{f}_s}{h} \right\rangle HV_\parallel f_{e0}. \tag{11.39}$$

This result is sufficiently accurate to enable us to calculate radial particle and heat fluxes in the limit $\epsilon \ll 1$. The three terms in (11.39) represent the electron response to radial gradients, ion flow and parallel electric field, respectively. The friction force and the electric force both try to drag the electron population along the magnetic field, but in a torus this is only partly successful. For trapped electrons, these forces are balanced by the mirror force, which keeps these particles stationary in the magnetic well. This is why these terms in (11.39) are not proportional to $v_\parallel$, but to $H(\lambda_c - \lambda)V_\parallel$, which vanishes in the trapped region.

Particle flux

The radial electron particle flux is related to the parallel friction by (8.29). It is convenient to use the relation

$$\int m_e v_\parallel C_e(f_{es}) d^3v = n_e e E_\parallel^{(A)},$$

which follows from the definition (11.3) of the Spitzer function f_{es}, to write the flux as

$$\langle \boldsymbol{\Gamma}_e \cdot \nabla \psi \rangle = \left\langle \frac{I}{eB} \int m_e v_\parallel C_e (f_{e1} - f_{es}) d^3 v \right\rangle \tag{11.40}$$

$$= \left\langle \frac{I}{eB} \int m_e v_\parallel \left[-(v_D^{ee} + v_D^{ei})(f_{e1} - f_s) \right. \right.$$

$$\left. \left. + \frac{m_e v_\parallel}{T_e}(v_D^{ee} u_e + v_D^{ei} V_{i\parallel}) f_{e0} \right] d^3 v \right\rangle,$$

where we have used the collision operator (11.36). Again, $u_e \simeq V_{i\parallel}$ when $\epsilon \ll 1$, which may be used to make the simplification

$$\langle \boldsymbol{\Gamma}_e \cdot \nabla \psi \rangle = -\left\langle \frac{I}{eB} \int m_e v_\parallel (v_D^{ee} + v_D^{ei}) \left(f_{e1} - f_s - \frac{m_e v_\parallel}{T_e} V_{i\parallel} f_{e0} \right) d^3 v \right\rangle$$

$$= -f_t \frac{n_e I}{m_e \Omega_e} \left\{ (v_D^{ee} + v_D^{ei}) \left(\frac{I T_e}{\Omega_e} \frac{\partial \ln f_{e0}}{\partial \psi} + T_e \hat{f}_s + m_e V_{t\parallel} \right) \right\},$$

where we have used the neoclassical average (11.23) and the distribution function (11.37) in the second line. Finally, substituting the parallel ion velocity from (11.30) gives the radial particle flux

$$\langle \boldsymbol{\Gamma}_e \cdot \nabla \psi \rangle = -f_t \frac{n_e I^2 T_e}{m_e \Omega_e^2} \left\{ (v_D^{ee} + v_D^{ei}) \left[\left(1 + \frac{T_i}{Z T_e} \right) \frac{d \ln n_e}{d\psi} \right. \right.$$

$$\left. \left. + (x^2 - \frac{3}{2}) \frac{d \ln T_e}{d\psi} - \frac{0.173}{Z T_e} \frac{d T_i}{d\psi} \right] \right\}$$

$$- f_t \frac{n_e T_e I}{m_e \Omega_e} \left\{ (v_D^{ee} + v_D^{ei}) \hat{f}_s \right\}, \tag{11.41}$$

where $x^2 = m_e v^2 / 2 T_e$.

This flux clearly depends on the effective ion charge Z. The Lorentz limit, $Z \to \infty$, is particularly simple since electron–electron collisions do not matter,

$$v_D^{ee} \ll v_D^{ei} = \frac{3\pi^{1/2}}{4 \tau_{ei} x^3}.$$

The velocity-space averages needed to evaluate (11.41) are simple to work out,

$$\{ v_D^{ei} x^{2n} \} = \frac{2}{\tau_{ei}} \int_0^\infty x^{2n+1} e^{-x^2} dx = \frac{n!}{\tau_{ei}}. \tag{11.42}$$

The Spitzer function is simply $f_s = -ev_\parallel E_\parallel f_{e0}/v_D^{ei} T_e$, and we obtain

$$
\langle \mathbf{\Gamma}_e \cdot \nabla \psi \rangle = -f_t \frac{n_e I^2 T_e}{m_e \Omega_e^2 \tau_e} \left[\left(1 + \frac{T_i}{Z T_e} \right) \frac{d \ln n_e}{d\psi} \right.
$$
$$
\left. - \frac{1}{2} \frac{d \ln T_e}{d\psi} - \frac{0.173}{Z T_e} \frac{d T_i}{d\psi} \right] - f_t \frac{n_e I E_\parallel^{(A)}}{B_0}. \quad (\epsilon \ll 1, Z \to \infty)
$$

In the case of a pure hydrogen plasma, $Z = 1$, electron–electron collisions occur with the frequency v_D^{ee}, and we need the velocity-space averages of v_D^{ee} which are given in the Appendix, and

$$
\{(v_D^{ee} + v_D^{ei})\hat{f}_{es}\} = -1.66 \frac{e E_\parallel}{T_e},
$$

(calculated numerically from the exact Spitzer function) to obtain the cross-field flux

$$
\langle \mathbf{\Gamma}_e \cdot \nabla \psi \rangle = -f_t \frac{n_e I^2 T_e}{m_e \Omega_e^2 \tau_e} \left[1.53 \left(1 + \frac{T_i}{T_e} \right) \frac{d \ln n_e}{d\psi} \right.
$$
$$
\left. - 0.59 \frac{d \ln T_e}{d\psi} - \frac{0.26}{T_e} \frac{d T_i}{d\psi} \right] - 1.66 f_t \frac{n_e I E_\parallel^{(A)}}{B_0}. \quad (\epsilon \ll 1, Z = 1)
$$
$$
\tag{11.43}
$$

Thus, the electron particle flux, which equals the ion flux, is driven by four terms: the density gradient, the electron and ion temperature gradients, and the parallel electric field. The term proportional to the density gradient has the character of ordinary diffusion, with a diffusion coefficient of order

$$
D_e \sim f_t \frac{\rho_{e\theta}^2}{\tau_{ei}},
$$

similar to the ion heat diffusivity (11.33), but with the poloidal gyroradius and collision time for electrons rather than ions. In a pure plasma, therefore, neoclassical particle transport is slower than heat transport, $D_e \sim (m_e/m_i)^{1/2} \chi_i$. The temperature gradient terms in (11.43) tend to produce an inward flux, and so does the last term, which involves the inductive electric field and is normally called the *Ware pinch*. In a torus with circular cross section and large aspect ratio, where $f_t = 1.46\epsilon^{1/2}$ and $d\psi = R B_\theta dr$, the Ware pinch becomes

$$
\Gamma_r = -c_W n_e E_\parallel^{(A)}/B_\theta,
$$

with $c_W = 2.44$ for $Z = 1$, and $c_W = 1.46$ for $Z \to \infty$.

A physical explanation of the Ware pinch

The Ware pinch is qualitatively similar to the radial $\mathbf{E} \times \mathbf{B}$ drift produced by a toroidal electric field,

$$\frac{\mathbf{E}_\varphi \times \mathbf{B}}{B^2} \cdot \nabla\psi = -\left(\frac{B_\theta}{B}\right)^2 RE_\varphi,$$

but is much larger. The Ware pinch exceeds the $\mathbf{E} \times \mathbf{B}$ by a factor of about $f_t(B/B_\theta)^2 \sim q^2\epsilon^{-3/2}$.

Physically, the Ware pinch has to do with the circumstance that in the presence of an induced electric field, $E_\varphi^{(A)} = -\partial A_\varphi/\partial t$, the poloidal magnetic flux $\psi = -RA_\varphi$ at any given point varies in time,

$$\frac{\partial\psi}{\partial t} = RE_\varphi^{(A)}. \tag{11.44}$$

Accordingly, the flux surface labelled by ψ moves slowly with a velocity $\mathbf{V}_\psi$ satisfying

$$\frac{\partial\psi}{\partial t} + \mathbf{V}_\psi \cdot \nabla\psi = 0.$$

In a torus with circular cross section and large aspect ratio, where $d\psi = RB_\theta$ and $E_\parallel^{(A)} \simeq E_\varphi$, the radial velocity of the flux surface becomes

$$V_\psi \simeq -E_\parallel^{(A)}/B_\theta. \tag{11.45}$$

Now, since toroidal canonical momentum, $p_\varphi = mRb_\varphi v_\parallel - Ze\psi$, is a conserved quantity, the bounce points, $v_\parallel = 0$, of a trapped particle remain on the flux surface labelled by the same $\psi = -p_\varphi/Ze$. The bounce points move with the flux surface, and the particle therefore experiences an average inward radial drift (11.45), which is independent of its mass and charge.

Passing particles react rather differently to the electric field since the latter tends to increase the parallel velocity, $m\dot{v}_\parallel = ZeE_\parallel$, which does not vanish on averaging over a passing orbit. The constancy of $p_\varphi = mRb_\varphi v_\parallel - Ze\psi$ then implies that $\dot{\psi} = mRb_\varphi\dot{v}_\parallel/Ze = RE_\varphi^{(A)}$. According to (11.44), this is also how ψ changes at any given, stationary point, and we conclude that passing particles do not experience any radial drift.

Thus, trapped particles drift inwards while passing ones do not. The former constitute a fraction $O(\epsilon^{1/2})$ of the total population, and the net radial flow is of the order

$$\Gamma_r^{\text{Ware}} \sim -\epsilon^{1/2}E_\parallel/B_\theta,$$

with a coefficient of order unity determined by the collisional equilibrium between the trapped and passing populations, as derived from (11.37).

Heat flux

The heat flux is obtained in a very similar way to the particle flux, by evaluating the 'heat friction', see (8.32). In analogy with (11.41), we can immediately write

$$\langle \mathbf{q}_e \cdot \nabla \psi \rangle = -f_t \frac{n_e I^2 T_e^2}{m_e \Omega_e^2} \left\{ (v_D^{ee} + v_D^{ei}) \left(x^2 - \frac{5}{2} \right) \left[\left(1 + \frac{T_i}{Z T_e} \right) \frac{d \ln n_e}{d \psi} \right. \right.$$

$$\left. \left. + \left(x^2 - \frac{3}{2} \right) \frac{d \ln T_e}{d \psi} - \frac{0.173}{Z T_e} \frac{d T_i}{d \psi} \right] \right\}$$

$$- f_t \frac{n_e T_e^2 I}{m_e \Omega_e} \left\{ (v_D^{ee} + v_D^{ei}) \left(x^2 - \frac{5}{2} \right) \hat{f}_{es} \right\},$$

which differs from the expression (11.41) for the particle flux only by a factor $(x^2 - 5/2)T_e$. Again, we evaluate the velocity-space averages in the limits $Z \to \infty$ and $Z = 1$. The former limit is the simpler one since we may neglect electron–electron collisions. Using velocity-space averages of v_D^{ee} from the Appendix and (11.42), we find the electron heat flux

$$\langle \mathbf{q}_e \cdot \nabla \psi \rangle = -f_t \frac{n_e I^2 T_e^2}{m_e \Omega_e^2 \tau_{ei}} \left[-\frac{3}{2} \left(1 + \frac{T_i}{Z T_e} \right) \frac{d \ln n_e}{d \psi} \right.$$

$$\left. + \frac{7}{4} \frac{d \ln T_e}{d \psi} + \frac{0.26}{Z T_e} \frac{d T_i}{d \psi} \right]. \qquad (\epsilon \ll 1, Z \to \infty)$$

In the case $Z = 1$, we need the additional velocity-space average

$$\left\{ (v_D^{ee} + v_D^{ei}) x^2 \hat{f}_s \right\} = -2.96 e E_{\parallel}^{(A)} / T_e$$

(again obtained numerically) to establish

$$\langle \mathbf{q}_e \cdot \nabla \psi \rangle = -f_t \frac{n_e I^2 T_e^2}{m_e \Omega_e^2 \tau_{ei}} \left[-2.12 \left(1 + \frac{T_i}{T_e} \right) \frac{d \ln n_e}{d \psi} + 2.51 \frac{d \ln T_e}{d \psi} \right.$$

$$\left. + \frac{0.37}{T_e} \frac{d T_i}{d \psi} \right] + 1.19 f_t \frac{n_e T_e I E_{\parallel}^{(A)}}{B}. \qquad (\epsilon \ll 1, Z = 1)$$

$$(11.46)$$

We note that the parallel electric field drives a conductive heat flux when $Z = 1$, but not when $Z \to \infty$. This heat flux is in the opposite direction to the Ware pinch.

11.4 Bootstrap current

We have already noted that there is no current across the flux surfaces since the transport is ambipolar. Within the flux surface there is a diamagnetic current perpendicular to the magnetic field and a parallel current (8.18). In the collisional regime, the latter consists of the Pfirsch–Schlüter current and the Ohmic current, but in a collisionless plasma a much larger non-inductive current, known as the bootstrap current, arises.

Unit aspect ratio

In the banana regime, the current is particularly simple to calculate in the limit of unit aspect ratio, $\epsilon = 1$. In this limit, there are no passing particles, $g_a = 0$, and the first-order distribution function is

$$f_{a1} = -\frac{I v_\parallel}{\Omega_a}\frac{\partial f_{a0}}{\partial \psi} = (11.5), \tag{11.47}$$

as we have already seen. The parallel current thus becomes

$$j_\parallel = e \int (Z f_{i1} - f_{e1}) v_\parallel d^3 v = -\frac{I}{B}\frac{dp}{d\psi},$$

where Z is the ion charge number and $p = p_e + p_i$ the total pressure. If we compare this expression to the general relations (8.17) and (8.18), it is clear that the total current is entirely in the toroidal direction,

$$\boxed{\mathbf{j} = -\frac{dp}{d\psi} R\hat{\varphi}. \qquad (\epsilon = 1)}$$

Note that there is no current in response to the toroidal electric field; the resistivity is infinite since there are no passing particles and the trapped ones cannot move in the toroidal direction. Instead, the plasma spontaneously sets up a current proportional to the pressure gradient. This current is comparable to the usual Pfirsch–Schlüter current, and its origin is easy to understand: it is the diamagnetic current associated with the banana orbits. The deviation of a banana orbit from its mean radial position is $\Delta\psi = I v_\parallel / \Omega$. If we consider a specific flux surface, and take the difference in the current carried by the co-moving ($v_\parallel > 0$) and the counter-moving ($v_\parallel < 0$) particles, we obtain exactly the same result as just found,

$$j_\parallel = \sum_a e_a \int [f_{a0}(\psi + \Delta\psi) - f_{a0}(\psi)] v_\parallel d^3 v$$

$$\simeq \sum_a e_a \int v_\parallel \Delta\psi \frac{\partial f_{a0}}{\partial \psi} d^3 v = -\frac{I}{B}\frac{dp}{d\psi}.$$

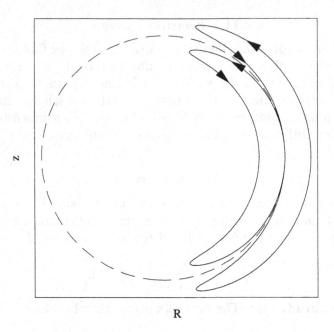

Fig. 11.1. There are more particles on the inner banana orbit than on the outer one if $dn/dr < 0$. There is therefore a net flow of trapped particles on the dashed flux surface, giving rise to a parallel current of trapped particles. In addition, if $\epsilon < 1$, circulating particles on this surface experience friction in the direction of this flow. Their motion amplifies the trapped-particle current.

In other words, the current arises because the co-moving guiding centres are more numerous (if $dn/d\psi < 0$) and hotter (if $dT/d\psi < 0$) than the counter-moving ones, see Fig. 11.1.

Large aspect ratio

The case of large aspect ratio, $\epsilon \to 0$, is more subtle since there are both trapped and passing particles. The behaviour of the former is dominated by the diamagnetic effect, but the kinetics of the latter is also influenced by the parallel electric field. To calculate the current, it is instructive to split off the part of the current that would arise in a straight magnetic field,

$$j_\parallel^{\text{Spitzer}} = n_i Z e V_{i\parallel} - e \int f_{es} v_\parallel d^3 v.$$

The remainder then becomes

$$j_\parallel - j_\parallel^{\text{Spitzer}} = n_i Z e V_{i\parallel} - e \int f_{e1} v_\parallel d^3 v - j_\parallel^{\text{Spitzer}} = -e \int (f_{e1} - f_{es}) v_\parallel d^3 v.$$

To calculate this integral directly would require accurate knowledge of f_{e1}, which can be obtained only by solving the electron drift kinetic equation using the full Fokker–Planck collision operator. Our model-operator solution is not sufficiently accurate for this purpose. This difficulty can, however, be circumvented by using the definition of the Spitzer function (11.35) and the self-adjointness of the collision operator to write

$$j_\parallel - j_\parallel^{\text{Spitzer}} = -\frac{T_e}{E_\parallel^{(A)}} \int C_e(f_s) \frac{f_{e1} - f_{es}}{f_{e0}} d^3v = -\frac{T_e}{E_\parallel^{(A)}} \int \frac{f_s}{f_{e0}} C_e(f_{e1} - f_{es}) d^3v.$$

To evaluate this expression, we need to know $C_e(f_{e1} - f_{es})$, rather than f_{e1}. This is a relief, since at large aspect ratio the $C_e(f_{e1} - f_{es})$ is localized in the nearly trapped region $1 - \lambda = O(\epsilon)$, and the model operator (11.36) may be used, as explained in the previous section. Hence, in analogy with (11.40) and (11.41) we have

$$\left\langle j_\parallel - j_\parallel^{\text{Spitzer}} \right\rangle = -\frac{T_e}{E_\parallel^{(A)}} \left\langle \int v_\parallel \hat{f}_s C_e(f_{e1} - f_{es}) d^3v \right\rangle$$

$$= -f_t \frac{n_e I T_e^2}{m_e \Omega_e E_\parallel^{(A)}} \left\{ (\nu_D^{ee} + \nu_D^{ei}) \hat{f}_s \left[\left(1 + \frac{T_i}{Z T_e} \right) \frac{d \ln n_e}{d\psi} \right. \right.$$

$$\left. \left. + \left(x^2 - \frac{3}{2} \right) \frac{d \ln T_e}{d\psi} - \frac{0.173}{Z T_e} \frac{dT_i}{d\psi} \right] \right\}$$

$$- f_t \frac{n_e T_e^2}{m_e E_\parallel^{(A)}} \left\{ (\nu_D^{ee} + \nu_D^{ei}) \hat{f}_{es}^2 \right\}.$$

In the Lorentz limit, where $\hat{f}_s = eE_\parallel^{(A)}/\nu_D^{ei} T_e$, this current becomes

$$\left\langle j_\parallel - j_\parallel^{\text{Spitzer}} \right\rangle = -f_t n_e T_e R \left[\left(1 + \frac{T_i}{Z T_e} \right) \frac{d \ln n_e}{d\psi} + \frac{d \ln T_e}{d\psi} - \frac{0.173}{Z T_e} \frac{dT_i}{d\psi} \right]$$

$$- f_t \sigma E_\parallel^{(A)}, \qquad (\epsilon \ll 1, Z \to \infty)$$

where $\sigma = 32 n_e e^2 \tau_e / 3\pi m_e$ is the Spitzer conductivity (for $Z \gg 1$). In the opposite limit, $Z = 1$, we need to evaluate the velocity-space averages involving Spitzer functions numerically, and obtain

$$\left\langle j_\parallel - j_\parallel^{\text{Spitzer}} \right\rangle = -f_t n_e T_e R \left[1.66 \left(1 + \frac{T_i}{Z T_e} \right) \frac{d \ln n_e}{d\psi} \right.$$

$$\left. + 0.47 \frac{d \ln T_e}{d\psi} - \frac{0.29}{Z T_e} \frac{dT_i}{d\psi} \right] - 1.31 f_t \sigma E_\parallel^{(A)},$$

$$(\epsilon \ll 1, Z = 1)$$

where the conductivity is now $\sigma = 1.96 n_e e^2 \tau_e / m_e$.

These expressions for the non-Spitzer current consist of terms involving radial gradients, and a term proportional to $E_\parallel^{(A)}$. The latter is negative and proportional to f_t. Physically, it represents the reduction in electrical conductivity due to trapping because trapped electrons do not carry any current. Since they make up a fraction $O(\epsilon^{1/2})$ of the total population, it is natural to expect that the conductivity should be reduced by an amount of this magnitude. This is an experimentally important prediction of neoclassical theory. The Ohmic heating of a tokamak plasma in the banana regime with given toroidal current is larger than one would expect from classical transport theory.

The terms involving radial gradients are even more interesting. They constitute the so-called 'bootstrap current', which is perhaps the most important result of neoclassical theory. This current arises spontaneously in response to radial gradients, and is in the same direction as the Ohmic current. Thus, it helps to confine the plasma by the additional poloidal field it generates. The improved confinement may cause the pressure gradient to steepen further, leading to an even larger bootstrap current. If this process continues, the bootstrap current could constitute most (but by itself probably not all) of the total plasma current.

As we have already seen, in the limit of unit aspect ratio the bootstrap current is entirely diamagnetic in nature, and is equal to $j_\parallel = -(I/B)dp/d\psi$. In the case of large aspect ratio the physical interpretation is more subtle. Because of the diamagnetic effect, the co-moving, trapped population is larger than the counter-moving one by an amount

$$-\epsilon^{1/2}\frac{dn}{d\psi}\Delta\psi,$$

where $\Delta\psi \sim \epsilon^{1/2}Iv_T/\Omega_a$ is the banana width and $\epsilon^{1/2}$ the approximate fraction of trapped particles. Note that the parallel velocity of a trapped particle is of order $\epsilon^{1/2}v_T$. The trapped particles are in collisional equilibrium with the passing ones, whose co-passing population therefore exceeds the counter-passing one by

$$-\frac{dn}{d\psi}\Delta\psi.$$

The resulting current is of the order

$$j_{BS} \sim -v_T e_a \frac{dn}{d\psi}\Delta\psi \sim -\epsilon^{1/2} R T_a \frac{dn_a}{d\psi}.$$

Thus, at large aspect ratio the bootstrap current is mostly carried by the *passing* particles, although it is ultimately caused by the diamagnetic effect of the *trapped* ones.

The bootstrap current is of order

$$j_{BS} \sim \frac{\epsilon^{1/2}p}{rB_P},$$

and compares in the following way with the Ohmic current (8.20),

$$j_{BS}/j_{OH} \sim \epsilon^{1/2}\beta_p,$$

where $\beta_p = 2\mu_0 p/B_\theta^2$ is the poloidal beta. In the standard tokamak ordering $\beta \sim \epsilon^2 \Rightarrow \beta_p \sim 1$, the bootstrap current is thus formally smaller than the Ohmic current by a factor $\epsilon^{1/2}$. In practice, β and $\epsilon^{1/2}$ are sometimes not very small, and the bootstrap current is often comparable to, or even larger than, the Ohmic current.

Need for a seed current

Nevertheless, it is usually not possible to replace all current by bootstrap current. Of course, this would be desirable from a practical point of view since it would lead to an inherently steady-state tokamak. The problem occurs near the magnetic axis, where the trapped-particle fraction $f_t \simeq 1.46\epsilon^{1/2}$ vanishes. If there were no current other than bootstrap current, the safety factor $q(r)$ would become infinite on the magnetic axis. While this may not be completely unacceptable, it then becomes difficult to maintain a transport equilibrium near the magnetic axis. To see this, we consider the Grad–Shafranov equation (7.10), which near the axis reduces to the simple Ampère's law

$$\nabla^2\psi = \frac{1}{r}\frac{d}{dr}r\frac{d\psi}{dr} = \mu_0 j_\parallel, \tag{11.48}$$

and assume that the current is given exclusively by the bootstrap current,

$$j_\parallel \sim -f_t\frac{dp}{d\psi} \propto -r^{1/2}\frac{dr}{d\psi}\frac{dp}{dr}. \tag{11.49}$$

Here we have taken the current to be proportional to the pressure gradient, rather than having individual contributions from the density and temperature gradients. This is not crucial but simplifies the argument. By definition, the poloidal flux vanishes at the magnetic axis. Suppose therefore that $\psi \propto r^\mu$ near the axis for some number $\mu > 0$, and Eqs. (11.48)–(11.49) then imply that

$$\frac{dp}{dr} \propto r^{2\mu-7/2} \tag{11.50}$$

for any equilibrium where the entire near-axis current is supplied by the bootstrap current. On the other hand, the pressure gradient must also satisfy the energy transport equation

$$\frac{3}{2r}\frac{d}{dr}r\chi\frac{dp}{dr} = -S,$$

where S is an energy source describing plasma heating, and we have neglected the convective heat flux. The neoclassical heat flux is given by (11.32) and (11.46), so that the heat conductivity scales as

$$\chi \propto \frac{\epsilon^{1/2}}{B_\theta^2}.$$

Realistically, the heating term S must be taken to be constant over a small region around the magnetic axis, which implies that the pressure gradient there is proportional to

$$\frac{dp}{dr} = -\frac{Sr}{3\chi} \propto r^{1/2}B_\theta^2 \propto r^{2\mu-3/2}$$

since $RB_\theta = d\psi/dr \propto r^{\mu-1}$. As this contradicts the scaling (11.50) there are no simultaneous solutions of the Grad–Shafranov equation and the neoclassical energy transport equation. The point is that the Grad–Shafranov equation fixes the pressure gradient $dp/d\psi$, which will then not satisfy the transport equation in general, since the heat conductivity (whether neoclassical or anomalous) is governed by different physics.

It has been suggested that other internal processes in the plasma may fill in the 'hole' left by the bootstrap current. Of course, the plasma current will not strictly vanish at $r = 0$, but will be determined by some physics we have not considered. For instance, very near the magnetic axis conventional neoclassical theory is not valid as trapped electrons follow potato orbits rather than banana orbits. As we have seen in Chapter 7, this leads to a finite fraction of trapped particles at the magnetic axis and should, similarly, lead to a finite bootstrap current. Unfortunately, the region covered by electron potato orbits tends to be very small and can easily fall in the plateau regime, even in very hot plasmas. In this regime, the bootstrap current is proportional to $\epsilon^{1/2}v_*^{-1}dp/d\psi$, see (12.52) and therefore vanishes on the magnetic axis. It has also been proposed that fusion-produced alpha particles, which have much wider potato orbits, could provide the necessary current, but these particles are relatively few. The 'potato current' thus tends to be quite small. If it is too small, some form of external current drive (by RF waves or neutral-beam injection) is necessary near the magnetic axis. This 'seed current' is then amplified by the bootstrap effect farther away from the axis.

11.5 Variational principle

Instead of directly solving the kinetic equation, as we have done, it is also possible to calculate the neoclassical transport by using a variational principle. Not only does this bring a deeper insight to the problem, but it can also be used to rigorously prove that the model operator we have employed gives exact results in the limit of large aspect ratio (Rosenbluth, Hazeltine and Hinton, 1972). To briefly illustrate the technique, we focus on the electron transport and consider the functional

$$S_e = -\left\langle \int \frac{f_{e1} - f_{es}}{f_{e0}} C_e(f_{e1} - f_{es}) d^3v \right\rangle.$$

This is reminiscent of the entropy production functional (8.33) in Section 8.6, but with the Spitzer contribution subtracted from the distribution function.

Now let us vary $f_{e1} = g_e + F_e$ slightly as in the calculus of variations, $\delta f_{e1} = \delta g_e$. The term F_e is not varied as it represents the given driving term in the kinetic equation (11.4) that we are trying to solve. Since the linearized collision operator $C_e = C_{ee} + C_{ei}$ is self-adjoint, the variation of S_e then becomes

$$\delta S_e = -2\left\langle \int \frac{\delta g_e}{f_{e0}} C_e(f_{e1} - f_{es}) d^3v \right\rangle.$$

We now split this integral into two parts, taken over the untrapped and trapped regions, respectively. Recalling the velocity-space volume element (7.21) and the definition of the flux-surface average (7.5), we have

$$\delta S_e = -2\sum_\sigma \int_0^\infty \pi v^3 dv \left(\int_0^{1-\epsilon} d\lambda \int_0^{2\pi} \frac{AB}{|v_\parallel| B_0} \frac{d\theta}{\mathbf{B} \cdot \nabla\theta} \right.$$

$$\left. + \int_{1-\epsilon}^{1+\epsilon} d\lambda \int_{-\theta_b}^{\theta_b} \frac{\Lambda B}{|v_\parallel| B_0} \frac{d\theta}{\mathbf{B} \cdot \nabla\theta} \right) \bigg/ \oint \frac{d\theta}{\mathbf{B} \cdot \nabla\theta},$$

where $A = (\delta g_e / f_{e0}) C_e(f_{e1} - f_{es})$.

Let us now constrain the variation by requiring that δg should only depend on constants of motion, $\partial \delta g/\partial \theta = 0$, so that $\delta g(v, \lambda, \psi, \sigma)$ can be pulled outside the θ-integral. We then obtain

$$\delta S_e = -2\sum_\sigma \int_0^\infty \frac{\pi v^3 dv}{B_0 f_{e0}} \left[\int_0^{1-\epsilon} \delta g_e \left\langle \frac{B}{|v_\parallel|} C_e(f_{e1} - f_{es}) \right\rangle d\lambda \right.$$

$$\left. + \int_{1-\epsilon}^{1+\epsilon} \delta g_e d\lambda \int_{-\theta_b}^{\theta_b} \frac{B}{|v_\parallel|} \frac{C_e(f_{e1} - f_{es}) d\theta}{\mathbf{B} \cdot \nabla\theta} \right] \bigg/ \oint \frac{d\theta}{\mathbf{B} \cdot \nabla\theta}.$$

It follows from this expression that δS_e vanishes if the banana-regime constraint equations (11.7) and (11.8) are satisfied. Vice versa, if $\delta S_e = 0$ for all $\delta g_e(v, \lambda, \psi, \sigma)$, then *both* these constraints must hold. The banana-regime transport problem is thus completely equivalent to this variational principle.

By an argument analogous to the one presented in Section 8.6 it follows that S_e is equal to

$$S_e = -\sum_{j=1}^{3} A_j^e \Gamma_j^a,$$

where the 'fluxes' Γ_j are defined by

$$\Gamma_1^e = \langle \mathbf{\Gamma}_e \cdot \nabla \psi \rangle,$$

$$\Gamma_2^e = \frac{\langle \mathbf{q}_e \cdot \nabla \psi \rangle}{T_e},$$

$$\Gamma_3^e = \left\langle j_{e\parallel} - j_{es} \right\rangle,$$

and the thermodynamic forces are

$$A_1^e = \frac{d \ln p_e}{d\psi}$$

$$A_2^e = \frac{d \ln T_e}{d\psi}$$

$$A_3^e = \frac{E_\parallel^{(A)}}{T_e}.$$

For simplicity we are only considering the case of large aspect ratio, so that the electron current $j_{e\parallel}$ and the induced electric field $E_\parallel^{(A)}$ are constant over a flux surface. We also ignore the ion temperature gradient drive to keep the discussion as simple as possible. The general case is slightly more complicated and is discussed by Sugama and Horton (1996).

We have seen in this chapter that the transport fluxes Γ_j^e are linear functions of the forces A_k^e,

$$\Gamma_j^a = \sum_{b,k} L_{jk}^{ab} A_k^b,$$

where the transport coefficients L_{jk}^{ab} depend on the plasma parameters and the geometry of the magnetic field. The entropy production can thus be written as a quadratic form

$$\dot{S}_e = -\sum_{jk} L_{jk}^{ee} A_j^e A_k^e,$$

which summarizes all the neoclassical transport properties of the electrons. This form is necessarily positive definite, by the second law of thermodynamics. It is also symmetric in the off-diagonal components – another manifestation of Onsager symmetry, which we first encountered in Chapter 4. For instance, the Ware pinch and the bootstrap current are Onsager symmetric processes in the same sense that the thermal force $\mathbf{R}_T$ and the friction-driven heat flux $\mathbf{q}_u$ are in Braginskii's equations.

Improved treatment of ion transport

The variational principle just described can of course also be used for ion transport, by maximizing ion entropy production

$$S_i = -\left\langle \int \frac{g_i + F_i}{f_{i0}} C_{ii}(g_i + F_i)d^3v \right\rangle \tag{11.51}$$

with respect to g_i. The proof of this is entirely analogous to that given for the electrons. We conclude this chapter by showing how to use this variational principle, together with an improved model collision operator, to calculate the ion thermal conductivity more accurately at intermediate aspect ratio, $0 < \epsilon < 1$, than we achieved earlier.

In the model operator (11.14) used in Section 11.2, the momentum restoring coefficient u_i was taken to be constant. The most obvious way to improve the operator is to allow this term to depend on velocity, thus writing

$$C_{ii}^{\text{model}}(g_i) = v_D^{ii}(\mathscr{L}(g_i) + M(v, \psi, \theta)v_\| f_{i0}). \tag{11.52}$$

We can use the additional freedom in the new momentum restoring coefficient M to make the operator not only to produce the right ion friction, i.e., conserve momentum, but also to describe the 'heat friction' and higher-order moments correctly. That is, we choose $M(v, \psi, \theta)$ to satisfy

$$\int m_i v_\| L_j^{(3/2)}(x^2) C_{ii}^{\text{model}}(g_i)d^3v = \int m_i v_\| L_j^{(3/2)}(x^2) C_{ii}^{\text{exact}}(g_i)d^3v \tag{11.53}$$

for all j, where $L_j^{(3/2)}$ are Sonine polynomials, see Section 4.5. Clearly, this improves the accuracy of the model operator and, in fact, makes it exact when operating on functions of the form $g_i = v_\| f(v)$.

The actual distribution function, however, has a more complicated pitch-angle dependence, so the model collision operator (11.52) is still an approximation. In fact, extending the model operator in this way does not affect the form of the solution, which can still be written in the form (11.18),

$$g_i = H(\lambda_c - \lambda)V_\| s_i(v)f_{i0},$$

but the function s_i will now be more complicated. Since we thus know the approximate pitch-angle structure of the distribution function, it appears worthwhile to use this g_i as a *trial function* in the variational principle, taking for s_i a Sonine polynomial expansion,

$$s_i = \sum_j a_j L_j^{(3/2)}(x^2) = a_0 + a_1 \left(\frac{5}{2} - x^2\right) + \cdots,$$

which we shall truncate after two terms. Inserting this expansion in (11.53) gives

$$\frac{f_c}{\tau_{ii}} \sum_k a_k (M_{ii}^{jk} + N_{ii}^{jk}) = -f_c \sum_k a_k v_{jk} + \left\{ \left\langle \frac{M}{h} \right\rangle v_D^{ii} L_j^{(3/2)} \right\}, \qquad (11.54)$$

where we have recalled the matrix elements (4.30), the neoclassical average (11.25), and introduced

$$v_{jk} = \left\{ v_D^{ii} L_j^{(3/2)} L_k^{(3/2)} \right\}.$$

Equation (11.54) determines the velocity dependence of M since it specifies its velocity-space averages when multiplied by any Sonine polynomial.

Having thus defined an improved collision operator, we now use this operator in the variational principle, minimizing the entropy production with respect to the variational parameters a_0 and a_1. The entropy production functional (11.51) depends quadratically on $f_{i1} = g_i + F_i$. The part that depends only on F_i is

$$S_i(F_i, F_i) = -\left\langle \int \frac{F_i}{f_{i0}} C_{ii}(F_i) d^3v \right\rangle$$

$$= -\left(\frac{I T_i'}{\hat{\Omega}_i T_i}\right)^2 \left\langle h^2 \int v_\parallel L_1^{(3/2)}(x^2) C_{ii} \left[v_\parallel L_1^{(3/2)}(x^2) f_{i0} \right] d^3v \right\rangle$$

$$= -\frac{p_i X^2}{m_i \tau_{ii}} \left\langle h^2 \right\rangle (M_{ii}^{11} + N_{ii}^{11}) = \frac{2p_i X^2}{m_i \tau_i} \left\langle h^2 \right\rangle, \qquad (11.55)$$

where $X = -(I/\hat{\Omega}_i) d \ln T_i / d\psi$ and $M_{ii}^{11} + N_{ii}^{11} = -\sqrt{2}$ in analogy with (11.34), and the cross term is equal to

$$2S_i(g_i, F_i) = 2X \left\langle h \int H V_\parallel s_i C_{ii} \left[v_\parallel L_1^{(3/2)}(x^2) f_{i0} \right] d^3v \right\rangle$$

$$= f_c \frac{2p_i X}{m_i \tau_{ii}} \sum_j a_j (M_{ii}^{j1} + N_{ii}^{j1}) = -f_c \frac{4p_i X}{m_i \tau_i} a_1, \qquad (11.56)$$

where we have neglected terms higher than a_1 in the Sonine polynomial expansion. To evaluate the term quadratic in g_i,

$$S_i(g_i, g_i) = -\sum_\sigma \left\langle \int \frac{g_i}{f_{i0}} C_{ii}(g_i) \frac{\pi v^3 dv d\lambda}{h|v_\parallel|} \right\rangle = -\left\langle h \int \frac{v_\parallel g_i}{f_{i0}} \left\langle \frac{C_{ii}(g_i)}{hv_\parallel} \right\rangle d^3 v \right\rangle,$$

we use the model operator (11.52) to establish

$$\left\langle \frac{C_{ii}(g_i)}{hv_\parallel} \right\rangle = \left\langle \frac{C_{ii}(H V_\parallel s_i f_{i0})}{hv_\parallel} \right\rangle = v_D^{ii} \left\langle \frac{M}{h} - s_i \right\rangle f_{i0},$$

where we have noted that $\langle \mathcal{L}(V_\parallel)/hv_\parallel \rangle = -1$. Hence

$$S_i(g_i, g_i) = \left\langle h \int s_i H V_\parallel v_\parallel v_D^{ii} \left\langle s_i - \frac{M}{h} \right\rangle f_{i0} d^3 v \right\rangle$$

$$= \frac{f_c p_i}{m_i} \sum_j a_j \left\{ v_D^{ii} L_j^{(3/2)} \left(\sum_k a_k L_k^{(3/2)} - \left\langle \frac{M}{h} \right\rangle \right) \right\}$$

$$= \frac{f_c p_i}{m_i} \sum_{j,k} a_j a_k \left(f_t v_{jk} - \frac{f_c \left(M_{ii}^{jk} + N_{ii}^{jk} \right)}{\tau_{ii}} \right), \qquad (11.57)$$

where we have used (11.54) in the last step.

Summing up the contributions from (11.55), (11.56) and (11.57), gives the total entropy production rate

$$S_i = \frac{2p_i}{m_i \tau_i} \left[X^2 \langle h^2 \rangle - 2f_c X a_1 + f_c^2 a_1^2 + \frac{f_c f_t \tau_i}{2} (a_0^2 v_{00} + 2a_0 a_1 v_{01} + a_1^2 v_{11}) \right].$$

This rate is minimized by solving the equations

$$\frac{\partial S_i}{\partial a_0} = \frac{\partial S_i}{\partial a_1} = 0,$$

giving $a_0 v_{00} + a_1 v_{01} = 0$ and

$$a_1 = \frac{X}{f_c + f_t(v_{11} - v_{01}^2/v_{00})\tau_i/2} = \frac{X}{f_c + 0.462 f_t}.$$

When thus minimized, the entropy production rate becomes

$$S_i = -\frac{\langle \mathbf{q}_i \cdot \nabla T_i \rangle}{T_i^2} = \frac{2p_i X^2}{m_i \tau_i} \left(\langle h^2 \rangle - \frac{f_c}{f_c + 0.462 f_t} \right),$$

and the heat flux is thus

$$\boxed{ \langle \mathbf{q}_i \cdot \nabla \psi \rangle = -\frac{2p_i I^2}{m_i \hat{\Omega}_i^2 \tau_i} \left(\left\langle \frac{B_0^2}{B^2} \right\rangle - \frac{f_c}{f_c + 0.462 f_t} \right) \frac{dT_i}{d\psi}. }$$

At large aspect ratio, $f_t \to 0$, and unit aspect ratio, $f_c \to 0$, this expression reduces to our earlier results (11.31) and (11.34), and it provides a fairly reliable interpolation formula between these limits. In fact, this expression, which was derived by Taguchi (1988), is the most accurate calculation of the ion banana heat flux in the literature.

Finally, we calculate the poloidal rotation of the plasma,

$$V_{i\theta} = \frac{K_i B_\theta}{n_i},$$

where K_i is given by (11.28) and now becomes

$$K_i = \frac{1}{B_0} \left\langle \frac{1}{h} \int g_i v_\parallel d^3 v \right\rangle = \frac{f_c p_i a_0}{m_i B_0} = \frac{1.17 f_c}{f_c + 0.462 f_t} \frac{n_i I}{Z e \langle B^2 \rangle} \frac{dT_i}{d\psi}. \quad (11.58)$$

Again, this agrees with our earlier results for $\epsilon \to 0$ and $\epsilon \to 1$, but is more accurate at intermediate aspect ratios than (11.29). In the next chapter, we shall see how to derive this result with considerably less effort by using the moment approach.

Further reading

Galeev and Sagdeev (1968) published the first theory of transport in the banana regime, building on ideas by Tamm and Sakharov (1961). In a truly remarkable work, Rosenbluth, Hazeltine and Hinton (1972) were able to calculate all the transport coefficients exactly in the limit of large aspect ratio, and later (Hazeltine, Hinton and Rosenbluth, 1973) in the opposite limit. An early paper by Rutherford (1970) is also well worth reading. The subject is of course given plenty of room in the texts by Hinton and Hazeltine (1976), Hirshman and Sigmar (1981), and Balescu (1988). A very useful discussion emphasizing physics, rather than mathematical details, is given by Hazeltine and Meiss (1992). A paper by Taguchi (1988) contains the solution of the ion drift kinetic equation in the banana regime and its application to the cross-flux surface ion heat flux in a tokamak with arbitrary aspect ratio. His calculation is mathematically equivalent to that presented in Section 11.5. This kinetic theory derivation agrees with the interpolation formula of Chang and Hinton (1986) for arbitrary aspect ratio.

Exercises

1. Calculate the effective fraction of trapped particles (11.24) in a large-aspect-ratio tokamak with circular cross section.

(a) First, evaluate the flux-surface average $\langle \xi \rangle = \sigma \langle \sqrt{1 - \lambda(1 - \epsilon \cos \theta)} \rangle$ for circulating particles appearing in the denominator of (11.24). Express the result in terms of the elliptic integral

$$E(k) = \int_0^{\pi/2} dx \sqrt{1 - k^2 \sin^2 x}.$$

Solution: Using the trapping parameter (7.25) and taking $\epsilon \ll 1$, we can write

$$\frac{1}{2\pi} \int_0^{2\pi} d\theta \sqrt{1 - \lambda(1 - \epsilon \cos \theta)}$$

$$= \sqrt{1 - \lambda(1 - \epsilon)} \frac{1}{2\pi} \int_0^{2\pi} d\theta \sqrt{1 - k^{-2} \sin^2 \theta/2} \simeq \sqrt{\frac{2\epsilon}{k^{-2} + 2\epsilon}} \frac{2E(k^{-1})}{\pi}.$$

(b) Perform the remaining integral (11.24) by changing the integration variable to k^{-2} and using the numerical result

$$I = \int_0^1 \left[1 - \frac{\pi}{2E(k^{-1})} \right] \frac{dk^{-2}}{k^{-3}} = -0.621.$$

Solution: After noting that

$$\lambda \simeq \frac{k^{-2}}{k^{-2} + 2\epsilon} \quad \Rightarrow \quad d\lambda \simeq \frac{2\epsilon dk^{-2}}{(k^{-2} + 2\epsilon)^2},$$

we obtain

$$f_t = 1 - \frac{3\sqrt{2\epsilon}}{4} \int_0^1 \frac{\pi}{2E(k^{-1})} \frac{k^{-2} dk^{-2}}{(k^{-2} + 2\epsilon)^{5/2}}$$

$$= 1 - \frac{3\sqrt{2\epsilon}}{4} \left(\int_0^1 \frac{k^{-2} dk^{-2}}{(k^{-2} + 2\epsilon)^{5/2}} - I \right) + O(\epsilon) = 1.462\sqrt{\epsilon} + O(\epsilon).$$

2. Since the definition of f_t involves two nested integrals, this quantity is difficult to calculate for realistic equilibria. An approximate formula is therefore useful. An excellent approximation to f_t is given by

$$f_t \simeq 0.25 f_{tl} + 0.75 f_{tu},$$

where

$$f_{tl} = 1 - \frac{3}{4} \int_0^{\lambda_c} \left\langle \frac{1}{\sqrt{1 - \lambda/h}} \right\rangle \lambda d\lambda,$$

$$f_{tu} = 1 - \frac{3}{4} \int_0^{\lambda_c} \frac{\lambda d\lambda}{\sqrt{1 - \lambda \langle h^{-1} \rangle}},$$

are rigorous upper and lower bounds on f_t (Lin-Liu and Miller, 1995).

(a) Start by proving that for any function $A(\theta)$

$$\langle A \rangle \le \langle A^2 \rangle^{1/2},$$

$$\frac{1}{\langle A \rangle} \le \left\langle \frac{1}{A} \right\rangle.$$

Solution: Apply Schwarz's inequality,

$$\left(\oint fg \, d\theta \right)^2 \le \oint f^2 \, d\theta \oint g^2 \, d\theta,$$

to the functions $f = A$, $g = 1$, and $f = A^{1/2}$, $g = A^{-1/2}$, respectively.

(b) Show that $f_{tl} \le f_t \le f_{tu}$ for any magnetic equilibrium.
    Solution: Apply these inequalities to $A = \sqrt{1 - \lambda/h}$.

(c) Show that

$$f_{tl} = 1 - \langle y^2 \rangle \left\langle y^{-2} \left[1 - \left(1 + \frac{y}{2} \right) \sqrt{1 - y} \right] \right\rangle,$$

where $y = B/B_{\max} = \lambda_c/h$, by interchanging the flux-surface average and the integral in the definition of f_{tl}.

(d) Apply these formulas to the standard, large-aspect-ratio equilibrium, $h = B_0/B = 1 + \epsilon \cos \theta$, and calculate f_{tl} and f_{tu} to lowest order in $\epsilon^{1/2}$.

Solution:

$$f_{tl} = \frac{3\sqrt{2\epsilon}}{\pi} \simeq 1.35\epsilon^{1/2} + O(\epsilon),$$

$$f_{tu} = \frac{3\epsilon^{1/2}}{2} + O(\epsilon),$$

$$f_t \simeq 0.25 f_{tl} + 0.75 f_{tu} = 1.463\epsilon^{1/2} + O(\epsilon).$$

3. Verify Onsager symmetry of the electron transport coefficients in the banana regime.

   Solution: Onsager symmetry becomes manifest when the transport fluxes are expressed in terms of pressure and temperature gradients. From the expressions given in this chapter for the particle flux, electron heat flux, and current it follows that the electron transport matrix is proportional to

$$L_{jk}^{ee} \propto \begin{pmatrix} 1.53 & -2.12 & 1.67 \\ -2.12 & 4.63 & -1.19 \\ 1.67 & -1.19 & 1.31 \end{pmatrix},$$

which is obviously symmetric.

12

The moment approach to neoclassical theory

The theory of plasma transport in a torus developed in the last two chapters relies mostly on kinetic theory; the drift kinetic equation was solved rather than moment equations. A kinetic treatment was necessary since the mean-free path is long in the banana–plateau regime. Accordingly, Braginskii's collisional closure of the moment equations, described in Chapter 4, is not applicable. With kinetic, long-mean-free-path results in hand it is, however, now instructive to inspect their implications in terms of moment equations. Not only does this exercise shed light on the physics behind neoclassical transport, it is also highly useful in generalizing the theory to include the case of several ion species. The essential advantage of the moment approach to neoclassical transport in a multicomponent plasma is that it largely decouples the kinetics of different particle species from each other, which simplifies the calculations considerably. In this chapter we give a broad outline of the most important elements of the theory, which has been reviewed in full detail by Hirshman and Sigmar (1981).

12.1 The parallel viscous force

We begin by recalling that the cross-field particle flux can be obtained by taking the toroidal projection of the momentum equation, as in Section 7.4. After splitting off the classical flux (due to perpendicular friction) and the $\mathbf{E} \times \mathbf{B}$ flux, the neoclassical flux (8.29) remains. We now decompose the latter into the *Pfirsch–Schlüter flux* and the *banana–plateau* flux,

$$\langle \Gamma_a \cdot \nabla \psi \rangle^{\text{neo}} = \langle \Gamma_a \cdot \nabla \psi \rangle^{PS} + \langle \Gamma_a \cdot \nabla \psi \rangle^{BP},$$

219

with

$$\langle \Gamma_a \cdot \nabla \psi \rangle^{PS} \equiv -I \left\langle \frac{R_{a\parallel} + n_a e_a E_\parallel^{(A)}}{e_a B} \left(1 - \frac{B^2}{\langle B^2 \rangle} \right) \right\rangle,$$

$$\langle \Gamma_a \cdot \nabla \psi \rangle^{BP} \equiv -I \frac{\left\langle B(R_{a\parallel} + n_a e_a E_\parallel^{(A)}) \right\rangle}{e_a \langle B^2 \rangle}. \tag{12.1}$$

Some details of these decompositions vary in the literature. The neoclassical heat flux (8.32), (9.14) is decomposed similarly, into

$$\langle \mathbf{q}_a \cdot \nabla \psi \rangle^{\text{neo}} = \langle \mathbf{q}_a \cdot \nabla \psi \rangle^{PS} + \langle \mathbf{q}_a \cdot \nabla \psi \rangle^{BP},$$

with

$$\langle \mathbf{q}_a \cdot \nabla \psi \rangle^{PS} \equiv -I T_a \left\langle \frac{H_{a\parallel}}{e_a B} \left(1 - \frac{B^2}{\langle B^2 \rangle} \right) \right\rangle,$$

$$\langle \mathbf{q}_a \cdot \nabla \psi \rangle^{BP} \equiv -I T_a \frac{\langle B H_{a\parallel} \rangle}{e_a \langle B^2 \rangle}. \tag{12.2}$$

In this chapter we focus on the banana–plateau fluxes, which turn out to be dominant at low collisionality, $v \ll v_T/qR$, and large aspect ratio, $\epsilon \ll 1$.

The banana–plateau particle flux is driven by the *parallel viscous force*. To see this, we consider the scalar product of the momentum equation in the form (2.10) with **B**, and take the flux-surface average. Since the time derivative is small in the transport ordering (8.3), we have

$$\left\langle B(R_{a\parallel} + n_a e_a E_\parallel) \right\rangle = \langle \mathbf{B} \cdot \nabla \cdot \mathbf{\Pi}_a \rangle, \tag{12.3}$$

where we recall the vector notation $\mathbf{B} \cdot \nabla \cdot \mathbf{\Pi} = \sum_{j,k} B_j \partial_k \Pi_{jk}$, and the definition

$$\Pi_{jk} \equiv \int m \left(v_j v_k - \frac{v^2}{3} \delta_{jk} \right) f d^3 v.$$

In Chapter 2, we defined $\mathbf{\Pi}$ without the term $(v^2/3)\delta_{jk}$; this difference is unimportant here since this term does not survive the flux-surface average (7.6). We also note from (2.13) that

$$\Pi_{jk} = \pi_{jk} + mn V_j V_k.$$

Since we have assumed small flow velocities, $V \sim \delta v_T$, as is appropriate for most naturally occurring flows in a tokamak, $\mathbf{\Pi}$ is approximately equal to the usual viscosity π; the difference is only $O(\delta^2 p)$, and we shall make no distinction between them. It is thus clear that the banana–plateau particle flux (12.1) is proportional to the parallel viscous force (12.3), averaged over the flux surface,

$$\langle \mathbf{\Gamma}_a \cdot \nabla \psi \rangle^{BP} \equiv -I \frac{\langle \mathbf{B} \cdot \nabla \cdot \mathbf{\Pi}_a \rangle}{e_a \langle B^2 \rangle}. \tag{12.4}$$

In a completely analogous way, the heat flux across the flux surface, which is related to the heat friction by (12.2), as in (9.14), can be expressed in terms of the 'heat flux tensor'

$$\Theta_{jk} \equiv \int m \left(v_j v_k - \frac{v^2}{3} \delta_{jk} \right) \left(\frac{mv^2}{2T} - \frac{5}{2} \right) f d^3 v,$$

by

$$\langle \mathbf{q} \cdot \nabla \psi \rangle^{BP} = -I T_a \frac{\langle B H_{a\parallel} \rangle}{e_a \langle B^2 \rangle} = -I T_a \frac{\langle \mathbf{B} \cdot \nabla \cdot \mathbf{\Theta}_a \rangle}{e_a \langle B^2 \rangle}. \tag{12.5}$$

The parallel viscous force can be related to a pressure anisotropy in the following way. From the end of Chapter 6 we know that the distribution function f depends only weakly on the gyroangle ϑ,

$$\frac{\partial f}{\partial \vartheta} \sim \delta f,$$

if the Larmor radius is small. Therefore, in a coordinate system aligned with the magnetic field, off-diagonal elements of $\mathbf{\Pi}$ are small. For instance, if the coordinates (x, y, z) are orthogonal, with z in the direction of $\mathbf{B}$, a typical off-diagonal element is

$$\Pi_{xy} = \int m v_x v_y f \, d^3 v = \int m v_\perp^2 \sin \vartheta \cos \vartheta \, f \, d^3 v = O(\delta).$$

Hence

$$\int m \mathbf{v} \mathbf{v} f d^3 v = \begin{pmatrix} p_\perp & 0 & 0 \\ 0 & p_\perp & 0 \\ 0 & 0 & p_\parallel \end{pmatrix} = p_\perp (\mathbf{I} - \mathbf{b} \mathbf{b}) + p_\parallel \mathbf{b} \mathbf{b}.$$

Here, $\mathbf{I}$ is the unit tensor, $I_{jk} = \delta_{jk}$, $\mathbf{b} = \mathbf{B}/B$ is the unit vector in the direction of the magnetic field, and the parallel and perpendicular pressures are defined by

$$\begin{pmatrix} p_\parallel \\ p_\perp \end{pmatrix} = \int m \begin{pmatrix} v_\parallel^2 \\ v_\perp^2/2 \end{pmatrix} f \, d^3 v.$$

Note that the usual pressure is the mean of the pressures in the three perpendicular directions, $p = (2p_\perp^2 + p_\parallel)/3$. It now follows from the definition of $\mathbf{\Pi}$ that we can express this quantity in terms of the parallel and perpendicular pressures,

$$\mathbf{\Pi} = p_\perp(\mathsf{I} - \mathbf{bb}) + p_\parallel\mathbf{bb} - p\mathsf{I} = (p_\parallel - p_\perp)\left(\mathbf{bb} - \frac{1}{3}\mathsf{I}\right). \qquad (12.6)$$

Finally, to evaluate $\langle\mathbf{B}\cdot\nabla\cdot\mathbf{\Pi}\rangle$, we note that for any scalar a, we have $\mathbf{B}\cdot\nabla\cdot(a\mathsf{I}) = \mathbf{B}\cdot\nabla a$ and

$$\mathbf{B}\cdot\nabla\cdot(a\mathbf{bb}) = B_j\partial_k(ab_jb_k) = B_jb_j\partial_k(ab_k) + B_jb_ka\partial_kb_j$$
$$= B\nabla\cdot\left(\frac{a\mathbf{B}}{B}\right) + a\mathbf{B}\cdot[(\mathbf{b}\cdot\nabla)\mathbf{b}] = B^2\nabla_\parallel\left(\frac{a}{B}\right) = B\nabla_\parallel a - a\nabla_\parallel B,$$

where we have noted that the curvature $\boldsymbol{\kappa} = (\mathbf{b}\cdot\nabla)\mathbf{b}$ is perpendicular to the magnetic field, see (6.17). The parallel viscous force thus becomes

$$\mathbf{B}\cdot\nabla\cdot\mathbf{\Pi} = (p_\perp - p_\parallel)\nabla_\parallel B + \frac{2}{3}B\nabla_\parallel(p_\parallel - p_\perp),$$

where the second term does not contribute to the flux-surface average (see (7.6))

$$\langle\mathbf{B}\cdot\nabla\cdot\mathbf{\Pi}\rangle = \left\langle(p_\perp - p_\parallel)\nabla_\parallel B\right\rangle. \qquad (12.7)$$

We conclude that the banana–plateau flux is driven by parallel variation of magnetic field strength, $\nabla_\parallel B \neq 0$, in combination with a difference between the parallel and perpendicular pressures. This difference can be written in terms of the Legendre polynomial $P_2(\xi) = (3\xi^2 - 1)/2$,

$$p_\parallel - p_\perp = \int mv^2 P_2(\xi)f\,d^3v. \qquad (12.8)$$

12.2 Plasma flows

Parallel viscosity coefficients

As we have seen, the banana–plateau cross-field transport fluxes can be expressed in terms of parallel viscous forces. Our next step is to relate these forces to flows within the flux surface. We learned in Chapter 8 that the requirements that (i) perpendicular fluxes are diamagnetic and (ii) total fluxes within the flux surface are divergence free forces the parallel fluxes of particles and heat to have the form (9.11), (9.12). Let us write these relations as

$$V_{a\parallel} = V_{1a} + u_{a\theta}(\psi)B, \qquad (12.9)$$

$$q_{a\parallel} = \frac{5p_a}{2}V_{2a} + q_{a\theta}(\psi)B, \qquad (12.10)$$

where

$$V_{1a} \equiv -\frac{I T_a}{m_a \Omega_a} \left(\frac{d \ln p_a}{d\psi} + \frac{e_a}{T_a} \frac{d\Phi}{d\psi} \right),$$

$$V_{2a} \equiv -\frac{I}{m_a \Omega_a} \frac{d T_a}{d\psi},$$

and

$$u_{a\theta}(\psi) = \frac{\mathbf{V}_a \cdot \nabla\theta}{\mathbf{B} \cdot \nabla\theta}, \tag{12.11}$$

$$q_{a\theta}(\psi) = \frac{\mathbf{q}_a \cdot \nabla\theta}{\mathbf{B} \cdot \nabla\theta} \tag{12.12}$$

are contravariant components of the flow velocity and the heat flux. They are related to the notation in (8.24) and (8.26) by $u_{a\theta} = K_a/n_a$, $q_{a\theta} = L_a$. Note that BV_{1a} and BV_{2a} are flux functions, so that the two terms in the parallel fluxes (12.9) and (12.10) vary over the flux surface in different ways. The first term is inversely proportional to B and the second term is directly proportional to B.

In general, the quantities $u_{a\theta}$ and $q_{a\theta}$ must be determined from kinetic theory. They are essentially the fluxes associated with the part g_a of the distribution function f_{a1} when the latter is decomposed as

$$f_{a1} = g_a + F_a,$$

with F_a defined by (11.5). Indeed, if g_a is expanded in Sonine polynomials,

$$g_a = f_{a0} \frac{m_a v_\parallel}{T_a} \sum_{j=0}^{\infty} u_{aj} L_j^{(3/2)}(x^2), \tag{12.13}$$

then

$$u_{a0} = u_{a\theta}(\psi)B,$$

$$u_{a1} = -\frac{2}{5 p_a} q_{a\theta}(\psi)B. \tag{12.14}$$

It is clear that if the coefficients u_{aj} of all species were known, we would have complete knowledge of the distribution functions, and we could thus calculate the friction force $R_{a\parallel}$, or equivalently, the parallel viscous force by (12.3). In a similar way, the heat friction $H_{a\parallel}$ and the heat flux tensor Θ could also be calculated. In practice, sufficient accuracy is often obtained by including only the first two terms. This truncation is known as the 13-moment approximation in the literature on kinetic gas theory. Hence, the basic problem of neoclassical theory is to calculate the coefficients

μ_{aj} in

$$\langle \mathbf{B} \cdot \nabla \cdot \mathbf{\Pi}_a \rangle = 3 \left\langle (\nabla_\parallel B)^2 \right\rangle \left(\mu_{a1} u_{a\theta} + \mu_{a2} \frac{2q_{a\theta}}{5p_a} \right), \tag{12.15}$$

$$\langle \mathbf{B} \cdot \nabla \cdot \mathbf{\Theta}_a \rangle = 3 \left\langle (\nabla_\parallel B)^2 \right\rangle \left(\mu_{a2} u_{a\theta} + \mu_{a3} \frac{2q_{a\theta}}{5p_a} \right), \tag{12.16}$$

where the overall multiplier $3\langle(\nabla_\parallel B)^2\rangle$ has been chosen in order to match to Braginskii's terminology in the appropriate limit, as we shall see in Section 12.3. The unknown coefficients μ_{aj} are often referred to as *neo-classical parallel viscosity coefficients* and summarize most of the kinetic information necessary to evaluate the neoclassical transport. It is sometimes practical to write them as

$$\mu_{a1} = K_{11}^a,$$

$$\mu_{a2} = K_{12}^a - \frac{5}{2}K_{11}^a, \tag{12.17}$$

$$\mu_{a3} = K_{22}^a - 5K_{12}^a + \frac{25}{4}K_{11}^a.$$

The new coefficients K_{jk}^a thus defined turn out to be positive definite and are easier to interpolate between different collisionality regimes.

Parallel flow

We now use neoclassical parallel viscosity coefficients to construct the parallel flow velocity in a torus with large aspect ratio, $\epsilon \ll 1$. In this limit the parallel flow velocities of all species are approximately equal and parallel heat fluxes are small (compared with $n_a T_a V_{1a}$ or $n_a T_a V_{2a}$). This is intuitively clear from the circumstance that in a cylinder (the limit $\epsilon \to 0$) all species must share a common parallel flow velocity. There is then no toroidicity to drive a particle flow or a heat flux in either direction along the field. In a torus, the relative flow velocity between different species is finite but small if $\epsilon \ll 1$. As found in the previous chapter it is of order $O(\epsilon^{1/2})V_1$, and we can thus write

$$\left\langle V_{a\parallel} B \right\rangle \simeq \langle VB \rangle, \tag{12.18}$$

$$\left\langle q_{a\parallel} B \right\rangle \simeq 0, \tag{12.19}$$

where V is the same for all species if ϵ is small. This common flow V can be determined from momentum balance,

$$\sum_a \langle \mathbf{B} \cdot \nabla \cdot \mathbf{\Pi}_a \rangle = \sum_a \left\langle B(R_{a\parallel} + n_a e_a E_\parallel) \right\rangle = 0,$$

where we substitute (12.9), (12.10) and (12.15), and solve for V,

$$V = \frac{\sum(\mu_{a1}V_{1a} + \mu_{a2}V_{2a})}{\sum \mu_{a1}}.$$

As the viscosity coefficients turn out to be smaller for electrons than for ions by the ratio $(m_e/m_i)^{1/2}$, only the latter need to be included in the sums over species index a. In a pure plasma with only one ion species, we have simply

$$V = V_{1i} + \frac{\mu_{i2}}{\mu_{i1}}V_{2i}.$$

Hence and from (12.9) it follows that, in such a plasma, the poloidal flow velocity is simply proportional to the temperature gradient,

$$V_{i\theta} = u_{i\theta}(\psi)B = V_{i\parallel} - V_{1i} = \frac{\mu_{i2}}{\mu_{i1}}V_{2i}. \tag{12.20}$$

We also note that the parallel and toroidal flow velocities within the flux surface depend on density and temperature gradients as well as on the radial electric field (through $d\Phi/d\psi$ in V_{1a}). The latter does not affect cross-field fluxes (as long as it is small enough to comply with our orderings, which preclude sonic flows), to which we now turn our attention.

Cross-field transport

When the aspect ratio is large and the collisionality low the largest contribution to cross-field transport comes from the banana–plateau fluxes (12.4), (12.5). These fluxes can be obtained directly from the fundamental relations (12.15), (12.16) once the poloidal flows are known. In the approximation (12.18), (12.19), which holds for $\epsilon \to 0$, the latter are given by

$$u_{a\theta}(\psi) = \frac{\langle(V - V_{1a})B\rangle}{\langle B^2\rangle},$$

$$\frac{2q_{a\theta}(\psi)}{5p_a} = -\frac{V_{2a}B}{\langle B^2\rangle}.$$

Note that although the radial electric field enters through $d\Phi/d\psi$ in V_{1a} it disappears from the poloidal flow $u_{a\theta}$ and from the banana–plateau cross-field fluxes of particles and heat, Eqs. (12.4) and (12.5), which become

$$\langle\mathbf{\Gamma}_a \cdot \nabla\psi\rangle^{BP} = -3\left\langle\left(\nabla_\parallel B\right)^2\right\rangle\frac{IB}{e_aB_0^4}$$

$$\times\left[\frac{\mu_{a1}}{\mu_1}\sum_b(\mu_{b1}V_{1b} + \mu_{b2}V_{2b}) - \mu_{a1}V_{1a} - \mu_{a2}V_{2a}\right],$$

$$\langle \mathbf{q}_a \cdot \nabla \psi \rangle^{BP} = -3 \left\langle (\nabla_\parallel B)^2 \right\rangle \frac{I T_a B}{e_a B_0^4}$$

$$\times \left[\frac{\mu_{a2}}{\mu_1} \sum_b (\mu_{b1} V_{1b} + \mu_{b2} V_{2b}) - \mu_{a2} V_{1a} - \mu_{a3} V_{2a} \right],$$

with $B_0^2 = \langle B^2 \rangle$ and

$$\mu_1 \equiv \sum_a \mu_{a1}.$$

Finally, these transport laws may be summarized in a compact form by introducing the thermodynamic forces

$$A_1^a \equiv \frac{d \ln p_a}{d\psi} + \frac{e_a}{T_a} \frac{d\Phi}{d\psi},$$

$$A_2^a \equiv \frac{d \ln T_a}{d\psi},$$

so that $V_{ja} = -(I T_a / e_a B) A_1^a$, and the cross-field fluxes

$$I_1^a \equiv \langle \mathbf{\Gamma}_a \cdot \nabla \psi \rangle^{BP},$$

$$I_2^a \equiv \langle \mathbf{q}_a \cdot \nabla \psi \rangle^{BP} / T_a.$$

Their relation to one another is then given by

$$I_j^a = \sum_{b,k} L_{jk}^{ab} A_k^b,$$

$$L_{jk}^{ab} = 3 \left\langle (\nabla_\parallel B)^2 \right\rangle \frac{I^2 T_b}{e_a e_b B_0^4} \left(\frac{\mu_{aj} \mu_{bk}}{\mu_1} - \mu_{a,j+k-1} \delta_{ab} \right). \qquad (12.21)$$

Note that the transport coefficients $L_{jk}^{ab} = L_{kj}^{ab}$ are Onsager symmetric. These laws essentially summarize the neoclassical transport of a plasma with an arbitrary number of ion species. All kinetic information necessary to evaluate the transport at large aspect ratio is contained in the viscosity coefficients μ_{aj}. The advantage of this formulation is that the viscosity coefficients can be determined relatively easily for one species at a time since they depend on the other species only through collision frequencies. The fluxes are more complex quantities, depending in a complicated way on the parallel flows of all species.

In the next three sections, we determine the viscosity coefficients in the Pfirsch–Schlüter, plateau and banana regimes from kinetic theory. Of course, much of the work has already been done in the two previous chapters, where radial fluxes were calculated directly. Here, by focusing on viscosity coefficients rather than the fluxes themselves, we are able to simplify the calculations and to treat plasmas with an arbitrary number of ion species.

12.3 Collisional regime

Since Braginskii has solved the kinetic equation and calculated the viscosity tensor for a collisional plasma, it is natural to think that the neoclassical viscosity coefficients in the Pfirsch–Schlüter regime could be extracted from his results. This is, however, not completely true. The reason is that the ordering of the flow velocity is different in Braginskii's equations and in standard neoclassical theory. Braginskii takes the flow velocity to be of the order of the ion thermal speed, $V \sim v_{Ti}$ (see (4.6)), whereas we have assumed a smaller flow velocity, $V \sim \delta v_{Ti}$ (see (8.4)), in the neoclassical transport theory. There is no guarantee that the results should agree, and as we shall see there is indeed some discrepancy.

First, let us inspect Braginskii's viscosity tensor (4.50)–(4.55). In the limit of small gyroradius ($\Omega \to \infty$), the non-vanishing viscosity coefficients are $\eta_0^{(i)}$ and $\eta_0^{(e)}$ (given by (4.57) and (4.60)) and the viscosity tensor becomes

$$
\Pi_a = -\eta_0^{(a)} \begin{pmatrix} (W_{xx} + W_{yy})/2 & 0 & 0 \\ 0 & (W_{xx} + W_{yy})/2 & 0 \\ 0 & 0 & W_{zz} \end{pmatrix}
$$

$$
= -\eta_0^{(a)} W_{zz} \begin{pmatrix} -1/2 & 0 & 0 \\ 0 & -1/2 & 0 \\ 0 & 0 & 1 \end{pmatrix} = -\frac{3}{2}\eta_0^{(a)} W_{zz} \left(\mathbf{bb} - \frac{1}{3}\mathbf{I} \right),
$$

where the coordinate system (x, y, z) has the z-axis parallel to the magnetic field, and we have noted that the rate-of-strain tensor is traceless, $W_{xx} + W_{yy} + W_{zz} = 0$. A comparison with (12.6) shows that its components are proportional to the pressure anisotropy,

$$
p_\perp - p_\parallel = \frac{3}{2}\eta_0 W_{zz},
$$

and from (12.7) we see that the parallel viscous force is

$$
\langle \mathbf{B} \cdot \nabla \cdot \Pi \rangle = \frac{3}{2}\eta_0 \left\langle W_{zz} \nabla_\parallel B \right\rangle. \tag{12.22}
$$

We need therefore to determine W_{zz} when the velocity is given by

$$
\mathbf{V}_a = \omega_a(\psi) R \hat{\varphi} + u_{a\theta}(\psi)\mathbf{B}, \tag{12.23}
$$

see (8.24) and (12.11). As explained in Chapter 8, this is the most general form for an incompressible velocity field that lies entirely within flux surfaces. We recall the expression (4.34) for the rate-of-strain tensor, and isolate the component of interest,

$$W_{zz} = b_j b_k \left[\partial_j V_k + \partial_k V_j - \frac{2}{3}(\nabla \cdot \mathbf{V})\delta_{jk} \right],$$

where as usual b_k denotes the kth component of $\mathbf{b} = \mathbf{B}/B$, and summation over repeated indices is understood. In the present situation, the flow (12.22) is incompressible, $\nabla \cdot \mathbf{V} = 0$, so that

$$W_{zz} = 2b_j b_k \partial_j V_k = 2b_j \left[\partial_j(b_k V_k) - V_k \partial_j b_k \right] = 2(\nabla_\parallel V_\parallel - \mathbf{V} \cdot \boldsymbol{\kappa}),$$

where $\boldsymbol{\kappa} = \nabla_\parallel \mathbf{b}$ is the magnetic curvature vector (6.17). Inserting the flow velocity (12.23) gives

$$W_{zz} = 2 \left[\left(u_{a\theta} - \frac{\omega_a I}{B^2} \right) \nabla_\parallel B - \omega_a R \hat{\boldsymbol{\varphi}} \cdot \boldsymbol{\kappa} \right],$$

since $V_\parallel = \omega_a I/B + u_{a\theta} B$ and $\mathbf{B} \cdot \boldsymbol{\kappa} = 0$. This expression can be simplified by noting that it follows from (6.22) that the toroidal component of the curvature vector is

$$\hat{\boldsymbol{\varphi}} \cdot \boldsymbol{\kappa} = \frac{\hat{\boldsymbol{\varphi}} \cdot \nabla_\perp B}{B} = -\frac{\hat{\boldsymbol{\varphi}} \cdot \mathbf{b}}{B} \nabla_\parallel B = -\frac{I}{RB^2} \nabla_\parallel B,$$

so that the rate of strain simplifies to

$$W_{zz} = 2u_{a\theta} \nabla_\parallel B.$$

Thus, the first term on the right-hand side of (12.23) does not contribute to the rate of strain or the parallel viscosity. This is not surprising since this term represents rigid toroidal rotation of the flux surface as a whole. The parallel viscosity (12.22) finally becomes

$$\langle \mathbf{B} \cdot \nabla \cdot \mathbf{\Pi}_a \rangle = 3\eta_0 \left\langle (\nabla_\parallel B)^2 \right\rangle u_{a\theta},$$

and we conclude by comparing with (12.15) that in Braginskii's ordering the first neoclassical viscosity coefficients are

$$\mu_{a1}^{\text{Brag}} = \eta_0^{(a)},$$

$$\mu_{a2}^{\text{Brag}} = 0.$$

The choice of the mysterious factor $3\langle (\nabla_\parallel B)^2 \rangle$ in Eq. (12.15) is now clear. This factor was chosen so that the neoclassical parallel viscosity coefficient μ_{a1} reduces to Braginskii's η_0 in the appropriate limit. Note that in Braginskii's theory the parallel viscosity is determined by the particle flux $u_{a\theta}$ alone. There is no contribution from the heat flux $q_{a\theta}$, as we expected in (12.15). The reason for this lies in Braginskii's large-flow ordering ($V \sim v_T$), which is different from the neoclassical ordering $V \sim \delta v_T$. The heat-flux term in (12.15) is simply overwhelmed by the term $\mu_{a1} u_{a\theta}$ if the flow velocity is ordered large. When the flow velocity is small, $V \sim \delta v_T$, the two terms are comparable and a more careful calculation is necessary.

To perform this calculation, we recall the kinetic analysis of Pfirsch–Schlüter transport in Section 9.2, where we expanded the drift kinetic equation,

$$v_{\parallel} \nabla f_{a1} + \mathbf{v}_d \cdot \nabla f_{a0} - \frac{e_a v_{\parallel} E_{\parallel}^{(A)}}{T_a} f_{a0} = C_a(f_{a1}),$$

in the smallness of the mean-free path. Thus writing $f_{a1} = f_{a1}^{(-1)} + f_{a1}^{(0)} + \cdots$, we then had

$$C_a(f_{a1}^{(-1)}) = 0,$$

$$C_a(f_{a1}^{(0)}) = v_{\parallel} \nabla_{\parallel} f_{a1}^{(-1)}.$$

The first of these equations implies that $f_{a1}^{(-1)}$ is a Maxwellian, and the second equation, which has the form of a Spitzer problem, shows that the pitch-angle dependence of $f_{a1}^{(0)}$ is that of the Legendre polynomial $P_1(\xi) = v_{\parallel}/v$. Neither of these pieces of the distribution function carries any parallel viscosity, which is contained in the component of f_{a1} that is proportional to $P_2(\xi)$, as shown in Eqs. (12.7) and (12.8). The point is that Legendre polynomials are orthogonal to each other in the sense

$$\int_{-1}^{1} P_j(\xi) P_k(\xi) d\xi = \frac{2\delta_{jk}}{2k+1},$$

so that only the piece of f_{a1} that is proportional to $P_2(\xi)$ contributes to the viscosity (12.8).

We must therefore go to next order and consider the equation

$$C_a(f_{a1}^{(1)}) = v_{\parallel} \nabla_{\parallel} f_{a1}^{(0)} + \mathbf{v}_d \cdot \nabla f_{a0} - \frac{e_a v_{\parallel} E_{\parallel}^{(A)}}{T_a} f_{a0} = v_{\parallel} \nabla_{\parallel} (f_{a1}^{(0)} - F_a).$$

In the second equality we have used the expression (8.14) for the cross-field drift and the definition (11.5) of $F_a = -Iv_{\parallel} \partial f_{a0}/\partial \psi$. We have also ignored the driving term proportional to the parallel electric field as this term is proportional to $P_1(\xi)$ and hence does not produce any viscosity. It is important to remember that the gradient $\nabla_{\parallel}$ is taken at constant magnetic moment rather than constant $v_{\parallel}$, so that, e.g.,

$$\nabla_{\parallel} v_{\parallel} = \frac{\nabla_{\parallel} v_{\parallel}^2}{2v_{\parallel}} = \frac{\nabla_{\parallel}(\mathscr{E} - \mu B)}{m v_{\parallel}} = -\frac{v_{\perp}^2}{2v_{\parallel}} \nabla_{\parallel} \ln B.$$

Recalling the expansion (12.13), (12.14) of $g_a = f_{a1}^{(0)} - F_a$ in Sonine polynomials, we now obtain

$$C_a(f_{a1}^{(1)}) = v_\| \nabla g_a = f_{a0} \frac{m_a v_\|}{T_a} \nabla_\|(v_\| B) \left(u_{a\theta} L_0^{(3/2)} - \frac{2q_{a\theta}}{5p_a} L_1^{(3/2)} + \cdots \right)$$

$$= f_{a0} \frac{m_a v^2}{T_a} P_2(\xi) \left(u_{a\theta} L_0^{(3/2)} - \frac{2q_{a\theta}}{5p_a} L_1^{(3/2)} + \cdots \right) \nabla_\| B, \quad (12.24)$$

where we have used

$$v_\| \nabla_\|(v_\| B) = \left(v_\|^2 - \frac{v_\perp^2}{2} \right) \nabla_\| B = v^2 P_2(\xi) \nabla_\| B. \quad (12.25)$$

Hence it follows that $f_{a1}^{(1)}$ has a pitch-angle dependence that is proportional to $P_2(\xi)$, as desired. This makes the kinetic equation mathematically different from the various Spitzer-type problems we have solved before, where the driving terms always were proportional to $P_1(\xi)$. For instance, if we were to solve (12.24) by expanding $f_{a1}^{(1)}$ in Sonine polynomials, we would need matrix elements involving $P_2(\xi)$ rather than the ones we defined in Section 4.5. The result of this straightforward calculation can be found in the review paper by Hirshman and Sigmar (1981). Hazeltine (1974) did a similar calculation using the variational principle. Here we shall be content with a simpler, but non-rigorous procedure, by approximating the collision operator by a Krook operator

$$C_a(f_{a1}^{(1)}) = -\nu_T^a(v) f_{a1}^{(1)}, \quad (12.26)$$

with a suitably chosen collision frequency $\nu_T^a(v)$. Although this cannot be justified rigorously, it tends to produce quite accurate results in practice, while drastically reducing the burden of the calculation. In Exercise 2 at the end of Chapter 3, it was shown that the pressure anisotropy associated with $P_2(\xi)$ is relaxed by collisions with a Maxwellian background at the rate

$$\nu_T^a = \sum_b (2\nu_s^{ab} - \nu_\|^{ab} + \nu_D^{ab}),$$

where the collision frequencies on the right were defined in Eqs. (3.45)–(3.47). It appears natural to define the frequency entering in the Krook operator in this way. Trivially solving (12.24) for $f_{a1}^{(1)}$ and taking the viscosity moment (12.8) then gives

$$\langle \mathbf{B} \cdot \nabla \cdot \mathbf{\Pi}_a \rangle = -\left\langle (\nabla_\| B) \int f_{a1}^{(1)} m_a v^2 P_2(\xi) d^3 v \right\rangle$$

$$= \frac{6p_a}{5} \left\langle (\nabla_\| B)^2 \right\rangle \left\{ \frac{x^2}{\nu_T^a} \left(u_{a\theta} L_0^{(3/2)} - \frac{2q_{a\theta}}{5p_a} L_1^{(3/2)} + \cdots \right) \right\},$$

where we have used the velocity-space integral $\{\cdots\}$ introduced in Eq. (4.26). Similarly,

$$\langle \mathbf{B} \cdot \nabla \cdot \mathbf{\Theta}_a \rangle = \frac{6p_a}{5} \left\langle (\nabla_{\parallel} B)^2 \right\rangle \left\{ \frac{x^2}{v_T^a} \left(x^2 - \frac{5}{2} \right) \left(u_{a\theta} L_0^{(3/2)} - \frac{2q_{a\theta}}{5p_a} L_1^{(3/2)} + \cdots \right) \right\}.$$

Thus the parallel viscosity coefficients defined in (12.15) and (12.16) become

$$\mu_{ak} = \frac{2p_a}{5} \left\{ \frac{x^2}{v_T^a(x)} \left(x^2 - \frac{5}{2} \right)^{k-1} \right\},$$

and the modified viscosity coefficients K_{jk}^a defined in (12.17) are equal to

$$\boxed{K_{jk}^a \simeq \frac{2p_a}{5} \left\{ x^{2(j+k-1)}/v_T^a(x) \right\}.} \qquad (12.27)$$

For hydrogen ions in a pure plasma these approximate Pfirsch–Schlüter viscosity coefficients resulting from the use of the Krook operator (12.26) are remarkably close to the exact ones (Hirshman and Sigmar, 1981),

$$\begin{pmatrix} K_{11}^i \\ K_{12}^i \\ K_{22}^i \end{pmatrix}_{\text{Eq (12.27)}} = \frac{p_i \tau_i}{\sqrt{2}} \begin{pmatrix} 1.26 \\ 5.99 \\ 34.9 \end{pmatrix}, \qquad \begin{pmatrix} K_{11}^i \\ K_{12}^i \\ K_{22}^i \end{pmatrix}_{\text{Exact}} = \frac{p_i \tau_i}{\sqrt{2}} \begin{pmatrix} 1.36 \\ 5.72 \\ 28.9 \end{pmatrix}.$$

We close with a comment on the parallel momentum balance equation $\langle BR_{\parallel a} + \mathbf{B} \cdot \nabla \cdot \mathbf{\Pi}_a \rangle = 0$ in the Pfirsch–Schlüter regime. The parallel viscosity is inversely proportional to the collision frequency ν_a while the friction force is directly proportional to ν_a. Therefore, in the literature the viscous force is often neglected, so that the average parallel friction force vanishes approximately, $\langle BR_{\parallel a} \rangle \simeq 0$. While this relation is sometimes useful, the parallel viscosity is always needed for calculating poloidal plasma rotation since only the relative velocity between different species, and not the common flow V, enters in the friction force.

12.4 Plateau regime

In the plateau regime, it is convenient to write the distribution function as

$$f_{a1} = -\frac{I v_{\parallel}}{\Omega_a} \frac{\partial f_{a0}}{\partial \psi} + \frac{v_{\parallel} B}{v B_0} s_a(v, \psi) f_{a0} + h_a,$$

where we require the unknown function h_a to be localized around $\xi = v_{\parallel}/v = 0$ in the sense discussed in Chapter 9. This Ansatz (Shaing and Callen, 1983; Coronado and Wobig, 1986) is motivated by the structure of the parallel flows (12.9) and (12.10), where the first terms on the right correspond to $-(Iv_\parallel/\Omega_a)\partial f_{a0}/\partial\psi$, and the second terms to $(v_\parallel B/vB_0)s_a f_{a0}$. In fact, if we expand s_a in Sonine polynomials,

$$s_a(v,\psi) = \frac{m_a v}{T_a}\sum_j u_{aj}(\psi)L_j^{3/2}(x^2),$$

as in Eq. (12.13), then

$$u_{a0} = u_{a\theta}(\psi)B_0,$$

$$u_{a1} = -\frac{2q_{a\theta}}{5p_a}B_0,$$

in analogy with (12.11) and (12.12).

The drift kinetic equation (8.16) becomes

$$v_\parallel\nabla_\parallel\left(\frac{v_\parallel B}{vB_0}s_a(v,\psi)f_{a0} + h_a\right) = C_a(f_{a1}),$$

where we have neglected the inductive electric field and recall that the parallel gradient is taken at constant $(\mathscr{E},\mu)$. If we instead regard (r,θ,v,ξ) as independent variables, the mirror force is introduced as the second term on the right of

$$\nabla_\parallel = \frac{1}{qR}\frac{\partial}{\partial\theta} - (1-\xi^2)\frac{\nabla_\parallel\ln B}{2}\frac{\partial}{\partial\xi},$$

see (10.1) and (10.2). In a torus with circular cross-section $\nabla_\parallel\ln B = \epsilon\sin\theta/qR$, and the kinetic equation becomes

$$\frac{v}{qR}\left(\xi\frac{h_a}{\partial\theta} - \epsilon\sin\theta\frac{1-\xi^2}{2}\frac{\partial h_a}{\partial\xi} - \epsilon\sin\theta\frac{1-3\xi^2}{2}s_a f_{a0}\right) = C_a(f_{a1}),$$

where we have used (12.25) in the last term on the left. By the same reasoning as in Section 10.2, we may neglect the action of the mirror force on h_a represented by the second term in this equation. Also, because pitch-angle scattering of h_a dominates in the collision operator, we make the approximation $v_a C_a(f_{a1}) \simeq (v_a/2)\partial^2 h_a/\partial\xi^2$. Finally, since $\xi \ll 1$ in the resonance region, the kinetic equation is simplified to

$$\xi\frac{\partial h_a}{\partial\theta} - \frac{v_a qR}{2v}\frac{\partial^2 h_a}{\partial\xi^2} = \frac{\epsilon\sin\theta}{2}s_a f_{a0}.$$

This equation is similar to (10.8), which was solved in Exercise 1, Chapter 10, and has the solution

$$h_a^{\text{even}} = \frac{\pi\epsilon}{2}\sin\theta\,\delta(\xi)s_a f_{a0}$$

in the limit $v_a \to 0$. Accordingly, the mean parallel viscous force (12.7) becomes

$$\langle \mathbf{B} \cdot \nabla \cdot \mathbf{\Pi}_a \rangle = \left\langle (\nabla_\parallel B) \int m_a \left(\frac{v_\perp^2}{2} - v_\parallel^2 \right) h_a d^3 v \right\rangle = \frac{\pi^2 \epsilon^2 m_a B}{4qR} \int_0^\infty s_a f_{a0} v^4 dv.$$

The corresponding expression for the parallel heat stress is similar, but has an additional factor $(x^2 - 5/2)$ in the integral. Employing the Sonine polynomial expansion of s_a and evaluating the integrals gives

$$\langle \mathbf{B} \cdot \nabla \cdot \mathbf{\Pi}_a \rangle = \frac{\pi^{1/2} \epsilon^2 p_a B}{v_{Ta} qR} \left(u_{a0} + \frac{1}{2} u_{a1} \right),$$

$$\langle \mathbf{B} \cdot \nabla \cdot \mathbf{\Theta}_a \rangle = \frac{\pi^{1/2} \epsilon^2 p_a B}{v_{Ta} qR} \left(\frac{1}{2} u_{a0} + \frac{13}{4} u_{a1} \right),$$

where as usual $v_{Ta} = (2T_a/m_a)^{1/2}$. Finally, since $\langle (\nabla_\parallel B)^2 \rangle = (\epsilon B/qR)^2/2$ the plateau viscosity coefficients become

$$\begin{pmatrix} \mu_{a1} \\ \mu_{a2} \\ \mu_{a3} \end{pmatrix} = \frac{2\pi^{1/2} p_a qR}{3v_{Ta}} \begin{pmatrix} 1 \\ 1/2 \\ 13/4 \end{pmatrix}. \tag{12.28}$$

This can be particularly compactly expressed in terms of the modified viscosity coefficients (12.17)

$$\boxed{K_{jk}^a = \frac{\pi^{1/2} p_a qR}{3v_{Ta}} (j+k)!} \tag{12.29}$$

12.5 Banana regime

We now calculate viscosity coefficients in the banana regime for a multi-species plasma. The simplest way to do so is to determine the parallel friction force and the heat friction, and to relate them to viscosity using (12.3) and (12.5). Thus we need to calculate

$$W_{aj} \equiv \left\langle B \int m_a v_\parallel L_j^{(3/2)}(x^2) C_a(f_{a1} - f_{as}) d^3 v \right\rangle \tag{12.30}$$

for $j = 0, 1$, where f_{as} is the Spitzer function (11.3). We remind ourselves that $f_{a1} = g_a + F_a$, with g_a unknown and F_a given by (11.5). As a suitable approximation for the collision operator we use

$$C_a(f_{a1} - f_{as}) = v_D^a \left[\mathcal{L}(g_a) + v_\parallel M_a(v, \psi, \theta) f_{a0} \right], \tag{12.31}$$

where

$$v_D^a = \sum_b v_D^{ab},$$

and where $M_a(v, \psi, \theta)$ is a coefficient chosen so as to make this collision operator resemble the exact one as closely as possible, in a way we leave unspecified. This resembles the model collision operator (3.69) for self-collisions, where M_a corresponds to the momentum-restoring term, but generalizes it substantially by allowing $M_a(v, \psi, \theta)$ to be an arbitrary function of v in the same way as we did in (11.52). Note that since the functions F_a and f_{as} are both proportional to $v_\parallel$, the contribution $C_a(F_a - f_{as})$ can be included conveniently in the term $v_\parallel M f_{a0}$ on the right-hand side of (12.31).

The approximation (12.31) of the collision operator can be justified in the following way. Suppose we expand the distribution function in Legendre polynomials $P_k(\xi)$,

$$f_{a1} = \sum_{k=1}^{\infty} f^{(k)}(v, \psi, \theta) P_k(\xi).$$

Then, clearly, F_a contributes only to the $k = 1$ term, and the terms with high k describe any rapid variation with pitch angle present in g_a. The action of the collision operator on the latter terms can be expected to be well approximated by pitch-angle scattering. If we take this to be true for all $k \geq 2$, a collision operator of the form (12.31) results. In other words, this operator corresponds to approximating all $k \geq 2$ terms in a Legendre polynomial expansion by pitch-angle scattering. Of course, the function $M_a(v, \psi, \theta)$ will be very complicated in general, but it turns out that the viscous forces W_{aj} can be calculated without actually knowing $M_a(v, \psi, \theta)$! The moment approach thus allows us to use a more accurate collision operator than we did in the previous chapter, at no computational expense.

Mathematically, the task of kinetic theory in the banana regime is to solve the equation (11.7), which determines the distribution function $f_{a1} = g_a + F_a$. With the approximation (12.31) for the collision operator, this equation becomes

$$\left\langle \frac{B}{v_\parallel} v_D^a (\mathscr{L}(g_a) + v_\parallel M_a f_{a0}) \right\rangle = 0,$$

where the Lorentz operator is given in (11.15). Thus, following exactly the same algebra as in the beginning of Section 11.2, we obtain the equation

$$\frac{\partial}{\partial \lambda} \lambda \langle v_\parallel \rangle \frac{\partial g_a}{\partial \lambda} = -\frac{v^2}{2} s_a(v, \psi) f_{a0},$$

where

$$s_a = \frac{\langle M_a B \rangle}{B_0}.$$

This equation is similar to (11.16) and therefore has the solution

$$g_a = H(\lambda_c - \lambda)V_\| s_a(v, \psi) f_{a0}, \tag{12.32}$$

in analogy with (11.18). While we still do not know M_a, we nevertheless proceed to calculate the parallel friction (12.30). Again employing the model operator (12.31) we find

$$W_{aj} = \left\langle B \int m_a v_\| L_j^{(3/2)}(x^2) v_D^a (v_\| M_a f_{a0} - g_a) d^3 v \right\rangle$$

$$= B_0 \left\langle \int v_D^a s_a L_j^{(3/2)}(x^2) m_a v_\| (v_\| - HV_\|/h) f_{a0} d^3 v \right\rangle,$$

where we have taken advantage of the self-adjointness of the Lorentz operator. The last expression is of the same form as the neoclassical average (11.23), and we obtain the result

$$W_{aj} = f_t B_0 p_a \left\{ v_D^a s_a L_j^{(3/2)}(x^2) \right\} = f_t B_0 p_a \left\{ \frac{v_D^a g_a}{HV_\| f_{a0}} L_j^{(3/2)}(x^2) \right\}. \tag{12.33}$$

Finally, we expand g_a in Sonine polynomials and then obtain from Eq. (12.32)

$$g_a = \frac{m_a HV_\| B_0}{T_a} \frac{f_{a0}}{f_c} \left[u_{a\theta} L_0^{(3/2)}(x^2) - \frac{2q_{a\theta}}{5p_a} L_1^{(3/2)}(x^2) + \cdots \right],$$

where the coefficients follow from a comparison with (12.13) and (11.23), and $f_c = 1 - f_t$ is the effective fraction of circulating particles. Hence

$$W_{aj} = m_a n_a B_0^2 \frac{f_t}{f_c} \left\{ v_D^a \left(u_{a\theta} L_0^{(3/2)}(x^2) - \frac{2q_{a\theta}}{5p_a} L_1^{(3/2)}(x^2) + \cdots \right) \right\}.$$

It now becomes evident from (12.3), (12.5), (12.15) and (12.16) that the neoclassical viscosity coefficients in the banana regime are equal to

$$\mu_{ak} = \frac{m_a n_a \langle B^2 \rangle}{3 \langle (\nabla_\| B)^2 \rangle} \frac{f_t}{f_c} \left\{ v_D^a \left(x^2 - \frac{5}{2} \right)^{k-1} \right\}. \tag{12.34}$$

The coefficients K_{jk}^a defined in Eq. (12.17) thus become

$$\boxed{K_{jk}^a = \frac{m_a n_a \langle B^2 \rangle}{3 \langle (\nabla_\| B)^2 \rangle} \frac{f_t}{f_c} \left\{ v_D^a x^{2(j+k-2)} \right\}.} \tag{12.35}$$

12.6 Interpolation between different regimes

An exact analytical treatment of neoclassical transport is only possible when the plasma is in one of the three asymptotic regimes we have considered. In practice, this is rarely the case. The core of a hot tokamak is usually in the banana regime (except very close to the magnetic axis, where the plateau regime may be more appropriate), while the edge is collisional. It may also be argued that at any one point in the plasma, particles with high and low energies can be in different collisionality regimes. Thus there is a practical need to consider transport in intermediate collisionality regimes, and for this purpose interpolation formulas have been developed which describe transport reasonably accurately across all three regimes. Such expressions must necessarily be non-rigorous, and depending on the required precision different interpolation formulas of various complexity have been constructed.

As can be seen from Eqs. (12.27), (12.29) and (12.35), the neoclassical viscosity coefficients first increase with increasing collisionality in the banana regime, then reach a plateau, and finally fall off in the Pfirsch–Schlüter regime. Indeed, since at large aspect ratio $f_t \sim \epsilon^{1/2}$ and

$$\left\langle (\nabla_\parallel B)^2 \right\rangle \simeq \frac{\epsilon^2 B^2}{2q^2 R^2},$$

the viscosity coefficients scale as

$$(K^a)^B = c_1 \frac{p_a}{\omega_{ta}} v_*,$$

$$(K^a)^{Pl} = c_2 \frac{p_a}{\omega_{ta}},$$

$$(K^a)^{PS} = c_3 \frac{p_a}{\omega_{ta} v_* \epsilon^{3/2}},$$

in the regimes. Here c_j are three numbers of order unity which can be obtained by direct comparison with the expressions (12.27), (12.29) and (12.35), $\omega_{ta} = v_{Ta}/qR$ is the transit frequency of well-circulating particles of species a, and $v_* = 1/\omega_{ta}\tau_{aa}\epsilon^{3/2}$ is the collisionality.

A simple interpolation formula which describes the asymptotic behaviour in the limits of small and large collisionality is

$$K_{jk}^a = \frac{(K_{jk}^a)^B (K_{jk}^a)^{PS}}{(K_{jk}^a)^B + (K_{jk}^a)^{PS}}, \tag{12.36}$$

where $(K_{jk}^a)^{PS}$ and $(K_{jk}^a)^B$ refer to the results (12.27) and (12.35), respectively. This expression completely ignores the plateau region, but as we have seen earlier, this region tends to be quite small in practice (and is non-existent at finite aspect ratio). It follows from the scalings above that this interpolation can be written as

$$K_{jk}^a = \frac{pa}{\omega_{ta}} \frac{c_1 c_3 v_*}{c_3 + c_1 v_*^2 \epsilon^{3/2}},$$

which reduces to the banana limit when $v_* \ll 1$ and to the Pfirsch–Schlüter result when $v_* \gg \epsilon^{-3/2}$. It provides a reasonably accurate estimate in intermediate regimes.

This example illustrates the general philosophy of interpolation formulas for the viscosity. If higher precision is warranted, one may use the following more carefully constructed formula by Shaing *et al.* (1996), which takes the plateau regime more carefully into account and which also allows for arbitrary cross section in this regime. Their interpolation formula is more accurate than Eq. (12.36) and has been benchmarked against a numerical solution of the linearized drift kinetic equation, but it should nevertheless be used with some caution. Like any interpolation formula, it is strictly accurate only in the various limits that are analytically tractable, and in the Pfirsch–Schlüter limit it should be remembered that the approximate viscosity coefficients (12.27) to which it is fitted are only accurate within 20%.

Chang–Hinton formula

It is sometimes useful to have simple, reasonably accurate expressions for transport coefficients (rather than viscosity coefficients) across all collisionality regimes. The most widely used such formula is the Chang–Hinton expression for ion thermal conductivity, which is the neoclassical radial transport quantity of most practical interest (Chang and Hinton, 1986). This formula interpolates between the $\epsilon \ll 1$ and $\epsilon = 1$ results (11.31) and (11.34), and the heat fluxes in the plateau and Pfirsch–Schlüter regimes, Eqs. (9.10) and (10.11). In addition, it accounts for the effect of any heavy, highly charged impurities that may be present in the plasma. Plasma shaping is also accounted for in an approximate way.

For a plasma with arbitrary cross section, a radial coordinate $r(\psi)$ may be defined by

$$B_{t0} \pi r^2(\psi) = \int B_\varphi dS,$$

where the integral on the right-hand side is taken over the poloidal cross section inside the flux surface ψ, and $B_{t0} = I/R_0$ with $I = RB_\varphi$ fixed and R_0 the radius of the magnetic axis. Thus, $r(\psi)$ is the radius of an equivalent circular flux surface enclosing the same toroidal flux, if the toroidal magnetic field were constant and equal to B_{t0}. A typical poloidal gyroradius is

$$\rho_{i\theta}^2 = \frac{2m_i T_i}{e^2 B_{\theta 0}^2},$$

with

$$B_{\theta 0} = \frac{1}{R_0} \frac{\partial \psi}{\partial r}.$$

The Chang–Hinton formula for ion heat flux is

$$\langle \mathbf{q}_i \cdot \nabla r \rangle = -\epsilon^{1/2} n_i \frac{\rho_{i\theta}^2}{\tau_i} \left(\frac{K_1}{n_i} \frac{dp_i}{dr} + K_2 \frac{dT_i}{dr} \right),$$

where $\epsilon = r/R$,

$$K_1 = -\alpha \frac{0.83 + 0.42\alpha}{0.58 + \alpha} \frac{0.74\mu_*\epsilon^{3/2}}{1 + 0.74\mu_*\epsilon^{3/2}} F(r),$$

$$K_2 = \frac{0.66(1 + 1.54\alpha) + (1.88\epsilon^{1/2} - 1.54\epsilon)(1 + 3.75\alpha)}{1 + 1.03\mu_*^{1/2} + 0.31\mu_*} \left\langle \frac{B_{t0}^2}{B^2} \right\rangle$$

$$+ 1.6 \left(1 + 1.33\alpha \frac{1 + 0.6\alpha}{1 + 1.79\alpha} \right) \frac{0.74\mu_*\epsilon^{3/2}}{1 + 0.74\mu_*\epsilon^{3/2}} F(r),$$

$$F(r) = \frac{1}{2\epsilon^{1/2}} \left\langle \frac{B_{t0}^2}{B^2} - 1 \right\rangle,$$

$\alpha = n_Z Z^2/n_i = Z_{\text{eff}} - 1$, and $\mu_* = (1 + 1.54\alpha)qR/v_{Ti}\tau_i\epsilon^{3/2}$. The coefficient K_1 (which does not appear in Chang and Hinton's paper, but which should be added for constistency) and the second term in K_2 describe the Pfirsch–Schlüter contributions to the heat flux and are dominant when collisionality is high, $\mu_* \gg 1$. The first term in K_2 represents banana–plateau transport and is constructed so as to match exact analytical results in the banana limit ($\mu_* \to 0$), the plateau limit ($1 \ll \mu_* \ll \epsilon^{-3/2}$), and intermediate boundary-layer calculations at large aspect ratio (Hinton and Hazeltine, 1976).

12.7 Ion rotation and bootstrap current at finite aspect ratio

Several times in this book we have seen that neoclassical transport calculations are most easily performed in the limit of large aspect ratio. In fact, in the plateau and banana regimes, exact analytical results only exist in this limit and (for the banana regime) in the opposite limit of unit aspect ratio. One of the merits of the moment approach is that it allows approximate results to be derived for intermediate aspect ratios in a systematic way. The results are approximate because the expansion in orthogonal polynomials is truncated, but in practice the errors incurred are frequently very small, and the accuracy can always be improved by including more terms in the expansions.

As we have seen in Section 12.2, in the limit of large aspect ratio all plasma species share a common parallel flow velocity within the flux surface. The individual flow speeds differ from the common flow by a small factor of order $O(\epsilon^{1/2})$. If we wish to calculate this small difference, e.g., to obtain the bootstrap current, or if indeed $\epsilon = O(1)$, the parallel flow velocity of each species must instead be determined individually from the system of parallel momentum equations

$$\langle \mathbf{B} \cdot \nabla \cdot \mathbf{\Pi}_a \rangle = \langle B(R_{a\parallel} + n_a e_a E_\parallel) \rangle, \tag{12.37}$$

$$\langle \mathbf{B} \cdot \nabla \cdot \mathbf{\Theta}_a \rangle = \langle B H_{a\parallel} \rangle. \tag{12.38}$$

The friction force $R_{a\parallel}$ and heat friction $H_{a\parallel}$ appearing here can be calculated from the distribution functions by the usual formulas,

$$R_{a\parallel} = \sum_b R_{ab\parallel} = \sum_b \int m_a v_\parallel [C_{ab}(f_{a1}, f_{b0}) + C_{ab}(f_{a0}, f_{b1})] d^3 v. \tag{12.39}$$

$$H_{a\parallel} = \sum_b H_{ab\parallel} = \sum_b \int m_a v_\parallel \left(x^2 - \frac{5}{2} \right) [C_{ab}(f_{a1}, f_{b0}) + C_{ab}(f_{a0}, f_{b1})] d^3 v. \tag{12.40}$$

By expanding f_{a1} and f_{b1} in Sonine polynomials as in (4.29),

$$f_{a1} = f_{a0} \frac{m_a v_\parallel}{T_a} \left[V_{a\parallel} L_0^{(3/2)}(x^2) - \frac{2 q_{a\parallel}}{5 p_a} L_1^{(3/2)}(x^2) + \cdots \right],$$

these forces can be expressed in terms of parallel flows,

$$R_{a\parallel} = \sum_b \left(l_{11}^{ab} V_{b\parallel} - l_{12}^{ab} \frac{2 q_{b\parallel}}{5 p_b} + \cdots \right), \tag{12.41}$$

$$H_{a\parallel} = \sum_b \left(-l_{21}^{ab} V_{b\parallel} + l_{22}^{ab} \frac{2 q_{b\parallel}}{5 p_b} + \cdots \right). \tag{12.42}$$

The coefficients l_{jk}^{ab} can be determined by comparing with the matrix elements (4.30) and isolating terms in the friction force involving one fixed species b,

$$l_{jk}^{ab} = m_a n_a \left(\delta_{ab} \sum_c \frac{M_{ac}^{j-1,k-1}}{\tau_{ac}} + \frac{N_{ab}^{j-1,k-1}}{\tau_{ab}} \right). \tag{12.43}$$

Thus, these coefficients can be regarded as known. Together with the neo-classical viscosity coefficients μ_{aj}, they summarize all kinetic information needed to evaluate the transport. The poloidal flow velocities are determined from the system of equations that is obtained from (12.9), (12.10), (12.37) and (12.38),

$$3 \left\langle (\nabla_\| B)^2 \right\rangle \begin{pmatrix} \mu_{a1} & \mu_{a2} \\ \mu_{a2} & \mu_{a3} \end{pmatrix} \begin{pmatrix} u_{a\theta} \\ \frac{2q_{a\theta}}{5p_a} \end{pmatrix}$$

$$= \sum_b \begin{pmatrix} l_{11}^{ab} & -l_{12}^{ab} \\ -l_{12}^{ab} & l_{22}^{ab} \end{pmatrix} \begin{pmatrix} V_{1b}B + u_{b\theta}\left\langle B^2 \right\rangle \\ V_{2b}B + \frac{2q_{b\theta}}{5p_b}\left\langle B^2 \right\rangle \end{pmatrix} + \begin{pmatrix} n_a e_a \left\langle E_\|^{(A)} B \right\rangle \\ 0 \end{pmatrix}.$$

$$(12.44)$$

Thus, a finite aspect ratio introduces some algebraic complexity in the calculation of parallel flows, but no fundamental obstacles are encountered.

Ions: poloidal plasma rotation

We first apply these equations to ions in a pure plasma (i.e., having only one species). Ignoring ion–electron collisions, the lowest non-vanishing friction coefficient (12.43) in the expansions (12.41) and (12.42) is $l_{22}^{ii} = -2^{1/2}m_i n_i/\tau_{ii}$, so that the system (12.44) simply reduces to

$$\hat{\mu}_{i1}u_{i\theta} + \hat{\mu}_{i2}\frac{2q_{i\theta}}{5p_i} = 0, \tag{12.45}$$

$$\hat{\mu}_{i2}u_{i\theta} + \hat{\mu}_{i3}\frac{2q_{i\theta}}{5p_i} = l_{22}^{ii}\left(\frac{V_{2i}B}{\left\langle B^2 \right\rangle} + \frac{2q_{i\theta}}{5p_i}\right),$$

where we have introduced the normalized viscosity coefficients

$$\hat{\mu}_{aj} = \frac{3\left\langle (\nabla_\| B)^2 \right\rangle}{\left\langle B^2 \right\rangle}\mu_{aj}.$$

This gives the following expression for the poloidal plasma rotation at arbitrary aspect ratio and collisionality,

$$u_{i\theta} = \alpha_i \frac{V_{2i}B}{\left\langle B^2 \right\rangle}, \tag{12.46}$$

$$\alpha_i = \frac{-l_{22}^{ii}}{\hat{\mu}_{i1}(\hat{\mu}_{i3} - l_{22}^{ii})/\hat{\mu}_{i2} - \hat{\mu}_{i2}}. \tag{12.47}$$

For instance, for a pure plasma in the banana regime the viscosity coefficients (12.34) are

$$\begin{pmatrix} \hat{\mu}_{i1} \\ \hat{\mu}_{i2} \\ \hat{\mu}_{i3} \end{pmatrix} = \frac{f_t}{f_c} \frac{m_i n_i}{\tau_{ii}} \begin{pmatrix} 0.533 \\ -0.625 \\ 1.387 \end{pmatrix} \tag{12.48}$$

so that the poloidal rotation coefficient becomes $\alpha_i = -1.173/(1+0.462y)$, with $y = f_t/f_c$ in agreement with our earlier result (11.58).

Note that the ion poloidal rotation (12.46) in a pure plasma is always proportional to the ion temperature gradient but is independent of the density gradient. For instance, it vanishes in an isothermal plasma. The fundamental reason for this is that

$$f_{i1} = F_i = -\frac{I v_\parallel}{\Omega_i} \left(\frac{d \ln p_i}{d\psi} + \frac{Ze}{T_i} \frac{d\Phi}{d\psi} \right) f_{i0}$$

is an *exact* solution to the ion drift kinetic equation (8.13) if the ion temperature gradient vanishes and if the parallel electric field and collisions with electrons are ignored. However, the poloidal velocity of *impurity ions*, which is more easily and commonly measured experimentally, depends on both density and temperature gradients.

If the aspect ratio is very large, $f_t \to 0$, the neoclassical viscosity coefficients are small in the sense that $\hat{\mu}_{jk}^{ii} \ll l_{22}^{ii}$, so that $\alpha_i \to \mu_{i2}/\mu_{i1}$ and Eq. (12.47) reduces to our earlier result (12.20). In the opposite limit of very tight aspect ratio, $f_t \to 1$, the viscosity coefficients become very large, $\hat{\mu}_{jk}^{ii} \gg l_{22}^{ii}$, so that Eq. (12.47) implies that the poloidal rotation stops, $u_{i\theta} \to 0$. In the banana regime, this follows from the fact that the viscosity coefficients (12.34) are proportional to the ratio f_t/f_c, which approaches infinity as $\epsilon \to 1$. It also agrees with our conclusion in the beginning of Section 11.4, where the solution (11.47) to the drift kinetic equation shows that there is no poloidal rotation. In the Pfirsch–Schlüter regime the viscosity coefficients approach infinity since they contain the factor $\langle (\nabla_\parallel B)^2 \rangle$, which can be estimated as

$$\langle (\nabla_\parallel B)^2 \rangle = \oint \left(\frac{\mathbf{B} \cdot \nabla \theta}{B} \frac{\partial B}{\partial \theta} \right)^2 \frac{d\theta}{\mathbf{B} \cdot \nabla \theta} \bigg/ \oint \frac{d\theta}{\mathbf{B} \cdot \nabla \theta}$$

$$> \frac{C^2}{2\pi} \oint \left(\frac{\partial \ln B}{\partial \theta} \right)^2 d\theta \to \infty \tag{12.49}$$

if the poloidal angle is chosen so that the contravariant component of the magnetic field is bounded from below, $C < \mathbf{B} \cdot \nabla \theta$. Thus, in the limit where the aspect ratio is made so small that the magnetic field variation B_{max}/B_{min} goes to infinity, parallel viscosity becomes large, again preventing poloidal rotation.

Electrons: electrical conductivity and bootstrap current

We now turn our attention to the electrons and calculate the conductivity and bootstrap current in a plasma with arbitrary aspect ratio and effective ion charge, Z, following Hirshman (1988). This is done by considering the force balance (12.44) for electrons. Evaluating the friction coefficients (12.43) for electrons gives

$$
\begin{pmatrix} l_{11}^{ee} & l_{12}^{ee} \\ l_{21}^{ee} & l_{22}^{ee} \end{pmatrix} = -\frac{m_e n_e}{\tau_{ei}} \begin{pmatrix} 1 & 3/2 \\ 3/2 & 13/4 + \sqrt{2}/Z \end{pmatrix},
$$

$$
\begin{pmatrix} l_{11}^{ei} & l_{12}^{ei} \\ l_{21}^{ei} & l_{22}^{ei} \end{pmatrix} = \frac{m_e n_e}{\tau_{ei}} \begin{pmatrix} 1 & 0 \\ 3/2 & 0 \end{pmatrix},
$$

where zero entries denote coefficients of order $(m_e/m_i)^{1/2}$. We can thus write the required flux-surface average of the friction force (12.39) and the heat friction on the electrons as

$$
\left\langle \begin{matrix} BR_{e\parallel} \\ BH_{e\parallel} \end{matrix} \right\rangle = \begin{pmatrix} l_{11} & l_{12} \\ l_{21} & l_{22} \end{pmatrix} \left\langle \begin{matrix} (V_{i\parallel} - V_{e\parallel})B \\ -2q_{e\parallel}B/5p_e \end{matrix} \right\rangle
$$

with $l_{11} = m_e n_e/\tau_{ee}$, $l_{12} = l_{21} = -3l_{11}/2$, and $l_{22} = (13/4 + \sqrt{2}/Z)l_{11}$. Recalling Eq. (12.9), we can then express the force balance (12.44) as

$$
\begin{pmatrix} \hat{\mu}_{e1} + l_{11} & \hat{\mu}_{e2} + l_{12} \\ \hat{\mu}_{e2} + l_{21} & \hat{\mu}_{e3} + l_{22} \end{pmatrix} \left\langle \begin{matrix} (V_{i\parallel} - V_{e\parallel})B \\ -2q_{e\parallel}B/5p_e \end{matrix} \right\rangle
$$

$$
= -\begin{pmatrix} \hat{\mu}_{e1} & \hat{\mu}_{e2} \\ \hat{\mu}_{e2} & \hat{\mu}_{e3} \end{pmatrix} \begin{pmatrix} (V_{1e} - V_{1i})B - u_{i\theta}\left\langle B^2 \right\rangle \\ V_{2e}B \end{pmatrix} + n_e e \left\langle \begin{matrix} E_{\parallel}B \\ 0 \end{matrix} \right\rangle.
$$

Solving this system of equations for the current $\langle j_{\parallel}B \rangle = \langle e(Z n_i V_{i\parallel} - n_e V_{\parallel})B \rangle = n_e e \langle (V_{i\parallel} - V_{e\parallel})B \rangle$ gives

$$
\left\langle j_{\parallel}B \right\rangle = -\frac{n_e e}{D} \left[d_1(V_{1e}B - V_{1i}B - u_{i\theta}\left\langle B^2 \right\rangle) + d_2 V_{2e}B - d_3 \frac{e\tau_{ee}}{m_e} \left\langle E_{\parallel}B \right\rangle \right],
$$

(12.50)

where we have introduced the constants

$$
D = \left(\frac{\tau_{ee}}{m_e n_e} \right)^2 \left[(\hat{\mu}_{e1} + l_{11})(\hat{\mu}_{e3} + l_{22}) - (\hat{\mu}_{e2} + l_{12})^2 \right],
$$

$$
d_1 = \left(\frac{\tau_{ee}}{m_e n_e} \right)^2 \left[\hat{\mu}_{e1}(\hat{\mu}_{e3} + l_{22}) - \hat{\mu}_{e2}(\hat{\mu}_{e2} + l_{12}) \right],
$$

$$
d_2 = \left(\frac{\tau_{ee}}{m_e n_e} \right)^2 (\hat{\mu}_{e2}l_{22} - \hat{\mu}_{e3}l_{12}),
$$

$$
d_3 = \frac{\tau_{ee}}{m_e n_e}(\hat{\mu}_{e3} + l_{22}).
$$

Finally, combining Eqs. (12.47) and (12.50) gives

$$\langle j_\parallel B \rangle = -\frac{I p_e}{D} \left[d_1 \frac{d \ln p_e}{d\psi} + d_2 \frac{d \ln T_e}{d\psi} + d_1 \frac{T_i}{Z T_e} \left(\frac{d \ln p_i}{d\psi} + \alpha_i \frac{d \ln T_i}{d\psi} \right) \right]$$

$$+ \frac{d_3}{D} \frac{n_e e^2 \tau_{ee}}{m_e} \langle E_\parallel B \rangle, \qquad (12.51)$$

where we have recalled the definitions (12.9) and (12.10). Equation (12.51) is an approximate expression for the parallel current that is valid for arbitrary aspect ratio, ion charge and collisionality. To obtain a fully explicit result we need only insert the appropriate viscosity and friction coefficients. In the banana regime this gives

$$D = 1.414Z + Z^2 + y(0.754 + 2.657Z + 2Z^2) + y^2(0.348 + 1.243Z + Z^2),$$

$$d_1 = y[0.754 + 2.21Z + Z^2 + y(0.348 + 1.243Z + Z^2)],$$

$$d_2 = -y(0.884 + 2.074Z),$$

$$d_3 = 1.414 + 3.25Z + y(1.387 + 3.25Z),$$

where again $y = f_t/f_c$. A comparison with the exact results given in Section 11.4 shows that the coefficients are correct within a few per cent (!) when the current is expressed in terms of pressure (rather than density) gradients and temperature gradients, as has been done here.

It is also interesting to note from Eq. (12.51) that in the plateau regime, where the viscosity coefficients (12.28) are of order

$$\hat{\mu}_{ej} = \frac{3 \langle (\nabla_\parallel B)^2 \rangle}{\langle B^2 \rangle} \mu_{ej} \sim \frac{\epsilon^2 p_e}{v_{te} q R} \sim \frac{\epsilon^{1/2} m_e n_e}{v_* \tau_{ee}},$$

with $v_* = qR/v_{Te}\tau_{ei}\epsilon^{3/2}$ as usual, the plateau bootstrap current is of order

$$\langle j_\parallel B \rangle \sim -\frac{\epsilon^{1/2} I}{v_*} \frac{dp_e}{d\psi}, \qquad (12.52)$$

which is much smaller than in the banana regime since $v_* > 1$. Because the region close to the magnetic axis tends to be in the plateau regime (or in the Pfirsch–Schlüter regime if collisionality is very high), this means that the bootstrap current is proportional to r^2 near the centre of a tokamak.

Further reading

The moment approach to neoclassical transport theory was initiated by Hirshman (1978) and presented in depth by Hirshman and Sigmar (1981), which remains the most complete review of impurity transport. Balescu (1988) also discusses the moment approach in considerable depth. A synoptic formulation for all transport coefficients using the moment approach can also be found in a paper by Sugama and Horton (1996), which focuses on entropy production and Onsager symmetry as underlying unifying principles. The calculation of plasma flows within magnetic flux surfaces has been the subject of several papers, including Kim, Diamond and Groebner (1991), and Zhu, Horton and Sugama (1999). The most reliable transport interpolation formulas in the literature have been published by Shaing *et al.* (1996), and Angioni and Sauter (2000).

Exercises

1. Derive the banana-regime ion heat flux (11.31) in a pure plasma with large aspect ratio, by using the moment approach.

   Solution: According to Eq. (12.21), the neoclassical ion heat flux is

   $$\langle \mathbf{q} \cdot \nabla\psi \rangle = L_{22}^{ii} \frac{dT_i}{d\psi},$$

   $$L_{22}^{ii} = 3 \left\langle (\nabla_\parallel B)^2 \right\rangle \frac{I^2 T_i}{e^2 B^4} \left(\frac{\mu_{i2}^2}{\mu_{i1}} - \mu_{i3} \right).$$

   Using (12.34) for the viscosity coefficients in the banana regime, we obtain

   $$L_{22}^{ii} = \frac{f_t}{f_c} \frac{m_i n_i I^2 T_i}{e^2 B^2} \left(\frac{\{v_D^{ii}(x^2 - 5/2)\}^2}{\{v_D^{ii}\}} - \{v_D^{ii}(x^2 - 5/2)^2\} \right)$$

   $$= -0.92 \frac{f_t}{f_c} \frac{m_i n_i I^2 T_i}{\tau_i e^2 B^2}.$$

2. In the same way, calculate the electron particle and heat fluxes at low collisionality and large aspect ratio.

   Solution: Both electron–electron and electron–ion collisions matter. In (12.34) we thus need to take $v_D^e = v_D^{ee} + v_D^{ei}$, giving

   $$3 \left\langle (\nabla_\parallel B)^2 \right\rangle \begin{pmatrix} \mu_{e1} \\ \mu_{e2} \\ \mu_{e3} \end{pmatrix} = m_e n_e B^2 \frac{f_t}{f_c} \begin{pmatrix} 0.533\tau_{ee}^{-1} + \tau_{ei}^{-1} \\ -0.625\tau_{ee}^{-1} - 3\tau_{ei}^{-1}/2 \\ 1.387\tau_{ee}^{-1} + 13\tau_{ei}^{-1}/4 \end{pmatrix},$$

   where $\tau_{ee} = \tau_{ei}$ in a pure plasma. Thus from (12.21) we obtain

   $$L_{jk}^{ee} = -\frac{f_t}{f_c} \frac{m_e n_e}{\tau_{ei}} \frac{I^2 T_e}{e^2 B^2} \begin{pmatrix} 1.53 & -2.12 \\ -2.12 & 4.64 \end{pmatrix}$$

   $$L_{jk}^{ei} = -1.53 \frac{f_t}{f_c} \frac{m_e n_e}{\tau_{ei}} \frac{I^2 T_i}{e^2 B^2} \begin{pmatrix} 1 & 1.17 \\ 1 & -1.17 \end{pmatrix},$$

   and both (11.43) and (11.46) follow.

3. Calculate the neoclassical transport of highly charged impurities in the banana regime. Express the flux in terms of the impurity strength parameter $\alpha = Z_{\mathrm{eff}} - 1 = n_Z Z^2 / n_i$.

Solution: In analogy with the electron–ion problem, the ion collision frequency is $v_D^i = v_D^{ii} + v_D^{iZ}$ and

$$3 \left\langle (\nabla_\parallel B)^2 \right\rangle \begin{pmatrix} \mu_{i1} \\ \mu_{i2} \\ \mu_{i3} \end{pmatrix} = m_i n_i B^2 \frac{f_t}{f_c} \begin{pmatrix} 0.533 \tau_{ii}^{-1} + \tau_{iZ}^{-1} \\ -0.625 \tau_{ii}^{-1} - 3\tau_{iZ}^{-1}/2 \\ 1.387 \tau_{ii}^{-1} + 13\tau_{iZ}^{-1}/4 \end{pmatrix},$$

where $\tau_{ii}/\tau_{iZ} = \alpha$. Since $v_D^{Zi} \gg v_D^{ii}$, the impurity viscosity coefficients μ_{Zk} are much larger than the ion ones, and the ion transport matrix becomes

$$L_{jk}^{ii} = -\frac{f_t}{f_c} \frac{m_i n_i}{\tau_{ii}} \frac{I^2 T_i}{e^2 B^2} \begin{pmatrix} 0.53 + \alpha & -(0.625 + 1.5\alpha) \\ -(0.625 + 1.5\alpha) & 1.39 + 3.25\alpha \end{pmatrix}$$

to lowest order in Z^{-1}. Thus, the impurity transport is

$$\langle \Gamma_Z \cdot \nabla \psi \rangle = -Z^{-1} \langle \Gamma_i \cdot \nabla \psi \rangle$$

$$= \frac{f_t}{f_c} \frac{n_i T_i I^2}{m_i Z \Omega_i^2 \tau_{ii}} \left[(0.53 + \alpha) \frac{d \ln n_i}{d\psi} - (0.09 + 0.5\alpha) \frac{d \ln T_i}{d\psi} \right].$$

As usual, the flux of highly charged impurities is determined by the gradients in the bulk ion distribution rather than by the impurities themselves. The flux of impurities is normally inward, as we have earlier found in the Pfirsch–Schlüter regime. However, note that in the banana regime impurities move outward if the temperature gradient is much larger than the density gradient. This is referred to as the *temperature screening* effect, and requires the ions to be collisionless.

13

Advanced topics

In this chapter we collect some results of a more advanced or specialized nature than considered earlier in this book. Our presentation of these topics is necessarily incomplete and is intended to serve as an introduction to the research literature.

13.1 Poloidal rotation

Neoclassical theory has recently acquired renewed importance by a surprising turn of logic. Even if most of the heat transport in a tokamak is caused by plasma turbulence rather than Coulomb collisions, it is thought that neoclassical theory can still have important consequences for energy confinement. This is because 'zonal flows' (i.e., poloidal rotation) are believed to be instrumental in regulating the transport caused by plasma turbulence (Burrell, 1999). Plasma flows within a flux surface are likely to be governed by collisional processes even if the cross-field transport is anomalous.

Neoclassical theory predicts small poloidal rotation of order $V_\theta \sim \delta v_{Ti}$ in the steady state. Indeed, according to Eq. (12.20) the poloidal velocity of a pure plasma with circular cross section and large aspect ratio is equal to

$$V_\theta = \frac{\mu_{i2}}{\mu_{i1}} V_{2i} = -\frac{k}{m_i \Omega_i} \frac{dT_i}{dr},$$

with $k = -1.17$ in the banana regime, $k = 0.5$ in the plateau regime, and $k = 1.7$ in the collisional regime. If for some reason the poloidal rotation velocity happens to be different from this neoclassical prediction, collisions will tend to restore it to the neoclassical value.

Faster poloidal rotation can arise if the plasma is subject to a poloidal torque. For instance, plasma turbulence can produce local Reynolds stress, i.e., a non-vanishing average poloidal component of the inertial term

246

$\langle m_i n_i (\tilde{\mathbf{V}}_i \cdot \nabla) \tilde{\mathbf{V}}_i \rangle$ in the fluid momentum equation. This causes the plasma to rotate poloidally at a speed that is determined by the relative magnitude of the driving force and the neoclassical poloidal flow damping. In the banana regime, the latter is caused by friction between the circulating population, which carries the poloidal rotation, and trapped particles, which are locked in their magnetic well and cannot rotate.

We have established earlier that a divergence-free velocity field within a flux surface must be of the form

$$\mathbf{V}_a(\psi, \theta, t) = \omega_a(\psi, t)R\hat{\boldsymbol{\varphi}} + u_{a\theta}(\psi, t)\mathbf{B},\qquad(13.1)$$

see Eq. (8.24). In a low-collisionality plasma, the flow velocity of each species relaxes to this form on the bounce time scale. The density becomes uniform on the flux surface on the same time scale. After this initial transient, the toroidal and poloidal rotation evolve more slowly in response to collisions, radial transport, and any external forces that may be acting on the plasma. In this section, we derive a general expression for the evolution of the poloidal rotation under these circumstances, following Hirshman (1978) and Rosenbluth and Hinton (1996).

The parallel and toroidal velocities are

$$V_{a\parallel} = \omega_a I/B + u_{a\theta}B,$$
$$V_{a\varphi} = \omega_a R + u_{a\theta}I/R.$$

Hence

$$\langle B^2 \rangle u_{a\theta} = \langle BV_{a\parallel} \rangle - \omega_a I = \langle BV_{a\parallel} \rangle - \frac{I}{\langle R^2 \rangle}(\langle RV_{a\varphi} \rangle - Iu_{a\theta}),$$

so that

$$\left(\langle B^2 \rangle - \frac{I^2}{\langle R^2 \rangle} \right) u_{a\theta} = \langle BV_{a\parallel} \rangle - \frac{I}{\langle R^2 \rangle}\langle RV_{a\varphi} \rangle.\qquad(13.2)$$

If we note that $B^2 = I^2/R^2 + B_p^2$ and introduce the quantity

$$\hat{q}^2 = \frac{I^2}{2\langle B_p^2 \rangle}\left\langle \frac{1}{R^2} - \frac{1}{\langle R^2 \rangle} \right\rangle,$$

which coincides with the square of the safety factor q at large aspect ratio, then we can write (13.2) as

$$\langle B_p^2 \rangle(1 + 2\hat{q}^2)u_{a\theta} = \langle BV_{a\parallel} \rangle - \frac{I}{\langle R^2 \rangle}\langle RV_{a\varphi} \rangle.\qquad(13.3)$$

We have thus related the poloidal rotation to the parallel and toroidal flow velocities, and our next task is to write down equations for the evolution of these flows. In doing so, we shall restrict our attention to time scales shorter than the transport time (8.2) and the skin time (8.35), so that the density, temperature and magnetic field stay approximately constant.

Since $\langle B(\mathbf{V}_a \cdot \nabla)\mathbf{V}_a \rangle = 0$ the parallel dynamics is governed by

$$m_a n_a \frac{\partial \langle BV_{a\parallel} \rangle}{\partial t} = -\langle \mathbf{B} \cdot \nabla \cdot \boldsymbol{\pi}_a \rangle + \langle BF_{a\parallel} \rangle, \tag{13.4}$$

where $F_{a\parallel}$ denotes the parallel component of the friction from other species and any external force $\mathbf{F}_a$ that may be acting on the plasma component in question. (We have sometimes denoted this force by $R_{a\parallel}$, but avoid this notation here since the major radius R also appears in the equations.) Next, multiplying the toroidal component of the momentum equation by R and taking the flux-surface average gives a conservation equation for toroidal angular momentum,

$$m_a n_a \frac{\partial \langle RV_{a\varphi} \rangle}{\partial t} = \langle \mathbf{j}_a \cdot \nabla\psi \rangle - \langle R\hat{\boldsymbol{\varphi}} \cdot \nabla \cdot \boldsymbol{\pi}_a \rangle + \langle RF_{a\varphi} \rangle, \tag{13.5}$$

where we have related the magnetic force on species a to the radial component of its current $\mathbf{j}_a = n_a e_a \boldsymbol{\Gamma}_a$ by

$$\langle R\hat{\boldsymbol{\varphi}} \cdot (\mathbf{j}_a \times \mathbf{B}) \rangle = \langle R\mathbf{j}_a \cdot [(\nabla\varphi \times \nabla\psi) \times \hat{\boldsymbol{\varphi}}] \rangle = \langle \mathbf{j}_a \cdot \nabla\psi \rangle.$$

Finally, combining (13.3), (13.4) and (13.5) gives the following evolution equation for poloidal rotation,

$$m_a n_a \langle B_p^2 \rangle (1 + 2\hat{q}^2) \frac{\partial u_{a\theta}}{\partial t} = -\langle \mathbf{B} \cdot \nabla \cdot \boldsymbol{\pi}_a \rangle - \frac{I}{\langle R^2 \rangle} \langle \mathbf{j}_a \cdot \nabla\psi \rangle + \frac{I}{\langle R^2 \rangle} \langle R\hat{\boldsymbol{\varphi}} \cdot \nabla \cdot \boldsymbol{\pi}_a \rangle$$

$$+ \left\langle \mathbf{F}_a \cdot \left[\mathbf{B}_p + \left(1 - \frac{R^2}{\langle R^2 \rangle} \right) \mathbf{B}_\varphi \right] \right\rangle. \tag{13.6}$$

The first term on the right represents poloidal flow damping by parallel viscosity, and the second term is the torque arising from the radial current. The third term describes transport of angular momentum due to cross-field viscosity, and the last term represents the action of friction and external forces. The most remarkable feature of (13.6) is the way that the poloidal and toroidal dynamics are linked. A toroidal force can give rise to poloidal rotation, and a purely poloidal force does not produce poloidal spin-up in an entirely straightforward way. For instance, adding (13.6) for ions and electrons in a standard large-aspect-ratio equilibrium gives

$$\rho(1 + 2q^2)\frac{\partial V_\theta}{\partial t} = F_\theta - 2q\langle F_\varphi \cos\theta \rangle - j_r B_\varphi + \text{viscosity},$$

where ρ is the mass density and **F** the total force density. Note that the poloidal inertia is enhanced by a factor $(1 + 2q^2)$. The reason for this is the constraint (13.1), which relates the poloidal and toroidal rotation to each other. In general, it is not possible for the plasma to rotate poloidally without also rotating toroidally, at least at some parts of the flux surface. Therefore, if a poloidal torque is applied to the plasma, it is forced to rotate both poloidally and toroidally, which increases the effective inertia of the plasma.

In general, in order to determine the poloidal rotation in a plasma subject to a force that varies in time, one needs to solve time-dependent kinetic equations. It is not admissible to neglect the time derivative, as done when calculating cross-field transport, since poloidal flow damping occurs on the ion–ion collision time scale. This calculatation is therefore complicated, and in fact, correct calculations of poloidal flow damping in the banana regime were only published fairly recently (Morris, Haines and Hastie, 1996; Hinton and Rosenbluth, 1999). If the poloidal flow damping is treated as an initial-value problem, it turns out that most of the flow is damped on a time scale of $\epsilon \tau_i$, and that the damping is not exponential in time.

In a steady state, however, it is much easier to calculate the poloidal rotation since both sides of (13.5) must vanish. The poloidal rotation for each species can thus be found from (13.4),

$$\langle \mathbf{B} \cdot \nabla \cdot \pi_a \rangle = \langle BF_{a\parallel} \rangle,$$

where the parallel viscosity was calculated in Chapter 12. When combined with the corresponding equation for the 'heat friction' (12.38), this gives a system of equations analogous to (12.44) from which the poloidal rotation can be calculated as in Section 12.7, see Exercise 1 at the end of this chapter.

13.2 Toroidal rotation

The attentive reader may have noticed that the radial electric field only plays a minor role in neoclassical theory for axisymmetric plasmas. In the quasi-steady state described by the transport ordering (8.2), it does not affect cross-field transport or poloidal rotation. In fact, the only observable quantity where it shows up is the toroidal rotation frequency (8.22),

$$\omega_a(\psi) = -\frac{d\Phi_0}{d\psi} - \frac{1}{n_a e_a}\frac{dp_a}{d\psi},$$

where it appears in the same way for all species. Of course, there is a good reason for this. Suppose that we make a transformation from the laboratory frame to a frame rotating at the velocity

$$\mathbf{V} = -\hat{\varphi}R\frac{d\Phi_0}{d\psi}$$

of some flux surface ψ. The electric field measured in this frame vanishes on the flux surface in question since

$$\mathbf{E}' = \mathbf{E} + \mathbf{V} \times \mathbf{B} \simeq -\nabla\Phi_0 + \mathbf{V} \times \mathbf{B} = 0,$$

and the equation of motion for each species acquires new terms corresponding to the centrifugal force and the Coriolis force. These terms are, however, small if $\mathbf{V}$ is smaller than the thermal speed, in which case the only consequence of the coordinate transformation is to eliminate the radial electric field. Thus, as long as the flow velocity is small, the radial electric field cannot affect neoclassical transport.

In developing neoclassical theory, it is indeed normally assumed that the flow velocity of each species is smaller than the thermal velocity, $V_a \sim \delta v_{Ta}$. This is in contrast to Braginskii's equations, which assume sonic flow, $V \sim v_{Ti}$. However, it is possible to extend neoclassical theory to account for such large flows (Hinton and Wong, 1985; Catto, Bernstein and Tessarotto, 1987), and then the radial electric field does affect the transport. In this situation, the largest terms in the momentum equation are the electric and magnetic forces since the other terms, such as the pressure gradient, are smaller by the factor $\delta = \rho/L$,

$$\frac{\nabla p_a}{n_a e_a \mathbf{V}_a \times \mathbf{B}} \sim \frac{\rho_a}{L}.$$

Therefore the electromagnetic terms must approximately balance,

$$n_a e_a(-\nabla\Phi_{-1} + \mathbf{V}_{a0} \times \mathbf{B}) = 0, \qquad (13.7)$$

where Φ_{-1} and $\mathbf{V}_{a0}$ are the leading-order terms in the electrostatic potential and mean flow velocity. The reason for the subscript -1 on the potential is that it is larger by a factor δ^{-1} than otherwise assumed. For instance, this electrostatic potential varies more rapidly than the temperature, $e\nabla\Phi/\nabla T \sim \delta^{-1}$. It follows from (13.7) that $\mathbf{B} \cdot \nabla\Phi_{-1} = 0$, so that the potential is a flux function, $\Phi_{-1} = \Phi_{-1}(\psi)$. This implies that $\mathbf{V}_{a0}$ must be the same for all species and be equal to

$$\mathbf{V}_0 = -\hat{\varphi}R\frac{d\Phi_{-1}}{d\psi} = \hat{\varphi}\omega(\psi)R.$$

Thus, in lowest order the plasma on each flux surface rotates toroidally as a rigid body in such a way that the radial electric field vanishes in the local rest frame.

By invoking an H-theorem argument analogous to that in Section 8.3, it can be shown that the lowest-order distribution function must be Maxwellian in this frame,

$$f_{a0} = n_a \left(\frac{m_a}{2\pi T_a(\psi)} \right)^{3/2} e^{-m_a u^2 / 2 T_a(\psi)}, \qquad (13.8)$$

where $\mathbf{u} = \mathbf{v} - \mathbf{V}_0$, see Hinton and Wong (1985). The temperature is a flux function, but the density is not since parallel force balance requires

$$m_a n_a \mathbf{b} \cdot (\mathbf{V}_0 \cdot \nabla) \mathbf{V}_0 = -n_a e_a \nabla_\parallel \Phi_0 - T_a \nabla_\parallel n_a,$$

where the centrifugal force is $-m_a (\mathbf{V}_0 \cdot \nabla) \mathbf{V}_0 = m_a \omega^2 R \nabla R$, so that

$$\nabla_\parallel \left(\ln n_a + \frac{e_a \Phi_0}{T_a} - \frac{m_a \omega^2 R^2}{2 T_a} \right) = 0.$$

The density therefore varies poloidally in lowest order and has the form

$$n_a(\psi, \theta) = N_a(\psi) \exp \left(\frac{m_a \omega^2 R^2}{2 T_a} - \frac{e_a \Phi_0}{T_a} \right), \qquad (13.9)$$

where the integration constant $N_a(\psi)$ is a flux function, and we fix the arbitrary additive constant in the electrostatic potential so that Φ_0 vanishes on a flux-surface average. The poloidal variation of Φ_0 is determined by quasineutrality. For instance, in a pure plasma, $n_i = n_e$, which implies

$$\frac{e \Phi_0}{T_i} = \frac{m_i \omega^2 R^2}{2(T_e + T_i)}.$$

The ions accumulate on the outboard side of each flux surface by the action of the centrifugal force, thus creating a poloidal electric field that pulls the electrons with them.

The drift kinetic equation derived in Chapter 6 is not valid in such rapidly spinning plasmas since the guiding-centre drift is modified by the centrifugal and Coriolis forces. One must therefore resort to gyroaveraging the Fokker–Planck equation, which has the general form

$$\frac{df_a}{dt} = \frac{\partial f_a}{\partial t} + \dot{z}_k \frac{\partial f_a}{\partial z_k} = C_a(f_a - f_{a0}), \qquad (13.10)$$

where z_k are arbitrary phase-space coordinates. Hazeltine and Ware (1978) have derived a drift kinetic equation for plasmas with arbitrary large flows, but a simpler derivation is possible in the present context since we know that the lowest-order flow velocity is purely toroidal. The first step is to notice that the kinetic energy in the rotating frame is equal to

$$\frac{m_a u^2}{2} = \frac{m_a(\mathbf{v} - \mathbf{V}_0)^2}{2} = \frac{m_a(v^2 + \omega^2 R^2)}{2} - m_a \omega R v_\varphi$$

$$= \frac{m_a(v^2 + \omega^2 R^2)}{2} + m_a e_a \frac{d\Phi_{-1}}{d\psi}(\psi - \psi_*),$$

where $\psi_* = \psi - m_a R v_\varphi / e_a = -p_\varphi / e_a$ is a constant of motion because of axisymmetry, just as in the non-rotating case. Inserting this result for $m_a u^2 / 2$ in the Maxwellian (13.8) and using (13.9) for the density, gives the lowest-order distribution function

$$f_{a0} = N_a(\psi) \left(\frac{m_a}{2\pi T_a(\psi)} \right)^{3/2} e^{-H/T_a(\psi)},$$

where

$$H \equiv \frac{m_a v^2}{2} + e_a \frac{d\Phi_{-1}}{d\psi}(\psi - \psi_*) + e_a \Phi_0.$$

The function H thus defined is almost, but not quite, a constant of motion to the requisite accuracy. A true constant of motion is

$$H_* = \frac{m_a v^2}{2} + e_a \left[\Phi(\psi) - \Phi_{-1}(\psi_*) \right] \simeq H + e_a \frac{\partial^2 \Phi_{-1}}{\partial \psi^2} \frac{(\psi - \psi_*)^2}{2}.$$

As we did at the end of Section 11.1, we now introduce a function which is almost equal to f_{a0} but only depends on constants of motion,

$$f_{a*} = N_a(\psi_*) \left(\frac{m_a}{2\pi T_a(\psi_*)} \right)^{3/2} e^{-H_*/T_a(\psi_*)},$$

and write the distribution function as $f_a = f_{a*} + g_a$, with $g_a \ll f_{a*}$, so that the kinetic equation (13.10) becomes

$$\frac{dg_a}{dt} = C_a(f_{a*} - f_{a0} + g_a),$$

in analogy with (11.13). The drift kinetic equation follows upon gyro-averaging,

$$v_\parallel \nabla_\parallel \bar{g}_a = \overline{C_a(f_{a*} - f_{a0} + g_a)} = C_a(\overline{f_{a*} - f_{a0} + g_a}), \tag{13.11}$$

where we have neglected the cross-field drift on the left since g_a is small, and we have used the rotational invariance of the collision operator.

In addition, the time derivative has been disregarded by invoking the transport ordering (8.2). The difference between f_{a*} and f_{a0} is

$$f_{a*} - f_{a0} \simeq (\psi_* - \psi) \left[\frac{d \ln N_a}{d\psi} + \left(\frac{H}{T_a} - \frac{3}{2} \right) \frac{d \ln T_a}{d\psi} \right] f_{a0} - \frac{H_* - H}{T_a} f_{a0}$$

$$= -\frac{m_a R v_\varphi}{e_a} \left[\frac{d \ln N_a}{d\psi} + \left(\frac{H}{T_a} - \frac{3}{2} \right) \frac{d \ln T_a}{d\psi} + \frac{m_a R v_\varphi}{2 T_a} \frac{d\omega}{d\psi} \right] f_{a0}.$$

Recalling that $\mathbf{v} = \omega R \hat{\varphi} + \mathbf{u}$ and evaluating the gyroaverages

$$\overline{v_\varphi} = \frac{I u_\parallel}{RB} + \omega R,$$

$$\overline{v_\varphi^2} = \overline{(\mathbf{v}_\parallel \cdot \hat{\varphi})^2 + (\mathbf{v}_\perp \cdot \hat{\varphi})^2} = \left(\frac{I u_\parallel}{RB} + \omega R \right)^2 + \frac{v_\perp^2}{2} \frac{B_p^2}{B^2},$$

we can finally write the drift kinetic equation (13.11) as

$$\boxed{ u_\parallel \nabla f_{a1} - C_a(f_{a1}) = -u_\parallel \bar{f}_{a0} \sum_{j=1}^{3} A_j \nabla_\parallel \alpha_j, } \qquad (13.12)$$

where $f_{a1} = \bar{f}_a - \bar{f}_{a0} = \bar{f}_{a*} + \bar{g}_a - \bar{f}_{a0}$, the thermodynamic forces are

$$A_1 = \frac{d \ln N_a}{d\psi} + \frac{d \ln T_a}{d\psi},$$

$$A_2 = \frac{d \ln T_a}{d\psi},$$

$$A_3 = \frac{d \ln \omega}{d\psi},$$

and we have written

$$\alpha_1 = \frac{m_a R \overline{v_\varphi}}{e_a} = \frac{m_a}{e_a} \left(\omega R^2 + \frac{I u_\parallel}{B} \right),$$

$$\alpha_2 = \left(\frac{H}{T_a} - \frac{5}{2} \right) \alpha_1,$$

$$\alpha_3 = \frac{m_a^2 \omega R^2 \overline{v_\varphi^2}}{2 e_a T_a} = \frac{m_a^2 \omega}{2 e_a T_a} \left[\left(\omega R^2 + \frac{I u_\parallel}{B} \right)^2 + \frac{\mu R^2 B_p^2}{m_a B} \right].$$

Thus, the rotation shear $A_3 = d \ln \omega / d\psi$ enters naturally as a third driving force for transport in addition to the pressure and temperature gradients. Furthermore, even in the absence of rotation shear (i.e., if $d\omega/d\psi = 0$

locally), the rotation itself modifies the first-order distribution function f_{a1} and thus the transport since the radial electric field $\omega = -d\Phi_{-1}/d\psi$ enters in $\alpha_{1,2,3}$.

The solution of the drift kinetic equation (13.12) for a toroidally rotating plasma is beyond the scope of the present book. This equation has been solved in the banana and Pfirsch–Schlüter regimes by Hinton and Wong (1985), and Catto, Bernstein and Tessarotto (1987), who show that sonic rotation, $\omega R \sim v_{Ti}$, increases neoclassical transport. When the rotation is weak, which is the case in most experiments, the corrections to ordinary neoclassical transport are of order $(\omega R/v_{Ti})^2$. More importantly, the solution of (13.12) also reveals the cross-field transport of toroidal angular momentum. In fact, the toroidal rotation frequency ω obeys a radial transport equation just like the density and temperature. Its transport properties, i.e., the cross-field toroidal viscosity, control the toroidal rotation if the sources of angular momentum, such as neutral-beam injection, are given. In the case where the flow velocity is near sonic, the transport has been calculated in the papers just cited. In the weaker ordering $V = O(\delta v_{Ti})$, the calculation is more complicated. It has been carried out by Rosenbluth *et al.* (1971) in the banana regime, and by Hazeltine (1974) in the Pfirsch–Schlüter regime. Interestingly, there is no neoclassical enhancement of toroidal angular momentum transport in the Pfirsch–Schlüter regime, which is therefore essentially classical in a collisional plasma. In the banana regime, the cross-field toroidal viscosity scales as $\mu_i \sim q^2\rho_i^2/\tau_i$, and thus resembles the heat conductivity in the Pfirsch–Schlüter (rather than banana) regime.

13.3 Nonlinear transport in steep plasma profiles

Throughout this book we have treated what might be called 'conventional' plasma transport theory, in which the basic expansion parameter

$$\delta = \frac{\rho}{L_\perp}$$

is very small. (For the moment, we consider arbitrary aspect ratio and therefore make no distinction between the gyroradius ρ and the poloidal gyroradius ρ_p.) This assumption, upon which the entire theory rests, is a necessary one. Without it, there is no reason to expect that the plasma should be in local thermodynamic equilibrium. Analytical transport theory is then impossible as there is no Maxwellian to expand around. Yet, in experiments, especially in the tokamak edge where a steep 'pedestal' sometimes forms, the poloidal gyroradius can approach the radial scale length $L_\perp$, thus rendering conventional transport theory invalid.

There is, however, room to extend the theory to allow for steeper plasma profiles than are usually assumed while still retaining the assumption $\delta \ll 1$. The point is that not only is δ assumed to be small in the conventional theory, but it is effectively taken as infinitesimally small. As we saw in Chapter 8, this implies that to lowest order all densities and temperatures are constant on flux surfaces (unless there is strong toroidal rotation). In fact, the only two-dimensional feature of the plasma that survives in this ordering is the magnetic field inhomogeneity, which is what gives rise to the neoclassical enhancement of transport over the classical level. When δ is made larger, it turns out that poloidal asymmetries become possible. Physically interesting and analytically tractable extensions of neoclassical transport theory can be obtained by allowing the product of δ and some suitable combination of parameters (such as (9.7)) to be of order unity, while still requiring δ to be small. Such theories have been published by, e.g., Engelmann and Nocentini (1977), Hsu and Sigmar (1990), Rogister (1994), Helander (1998), and Fülöp and Helander (1999). Characteristic of these works is that they find that poloidal asymmetries appear spontaneously in the plasma, so that the density is no longer a lowest-order flux function.

Typically the first plasma parameter to develop a poloidal variation is the density, n_Z, of highly charged impurity ions, whose poloidal modulation is of the order

$$\frac{\tilde{n}_Z}{n_Z} \sim \delta_Z = \delta \hat{v}_{ii} Z^2, \tag{13.13}$$

where Z is the impurity charge number, $\hat{v}_{ii} \equiv L_\parallel / \lambda_{ii}$ is the collisionality, with λ_{ii} the mean-free path for bulk ions, and $L_\parallel \sim qR$ the connection length. Note that if the collisionality is not very small and if Z is reasonably large, δ_Z can easily be of order unity while δ remains small.

It is this window of opportunity to improve upon conventional transport theory that we shall explore in this section, by considering transport in a hydrogen plasma with a single species of highly charged impurities, such that

$$\delta \ll 1,$$

but

$$\delta_Z = O(1),$$

so that the impurity density has a significant poloidal variation. To keep the discussion as simple as possible, we take the hydrogen ions and the electrons to be in the banana regime, $\hat{v}_{ii} \ll 1$, and the impurities to be collisional, as is fairly typical of a plasma somewhat inside the last closed flux surface. Our orderings then imply that the impurity ions are highly charged, $Z \gg 1$, but that the total charge they carry is low, $n_Z Z \ll 1$. As a result, the electrostatic potential is approximately constant on flux surfaces, $\Phi \simeq \Phi_0(\psi)$, which can be verified *a posteriori*. Moreover, the perpendicular velocity (4.64) of the impurities is dominated by the $\mathbf{E} \times \mathbf{B}$ velocity,

$$\mathbf{V}_{Z\perp} \simeq \frac{\mathbf{b} \times \nabla\psi}{B} \frac{d\Phi_0}{d\psi} = \left(\frac{I\mathbf{b}}{B} - R\hat{\varphi}\right) \frac{d\Phi_0}{d\psi}.$$

As in Section 8.5, it then follows from the continuity equation, $\nabla \cdot (n_Z \mathbf{V}_Z) = 0$, that there must be a parallel impurity return flow equal to

$$V_{Z\parallel} = -\frac{I}{B} \frac{d\Phi_0}{d\psi} + \frac{K_Z(\psi)B}{n_Z},$$

where $K_Z(\psi)$ is an integration constant, which is determined by parallel momentum balance,

$$m_Z n_Z \mathbf{b} \cdot (\mathbf{V}_Z \cdot \nabla \mathbf{V}_Z) = -n_Z Z e \nabla_\parallel \Phi_1 - \nabla_\parallel p_Z - \mathbf{b} \cdot \nabla \cdot \boldsymbol{\pi}_Z + R_{Z\parallel},$$

where Φ_1 denotes the poloidally varying part of the electrostatic potential. If the radial electric field is of the same order as the temperature gradient, $e\Phi' \sim T'$, the flow velocities of both ion species are of order $V_\parallel \sim \delta v_{Ti}$. The ratio between the inertial term and the friction is then

$$\frac{m_Z n_Z \mathbf{b} \cdot (\mathbf{V}_Z \cdot \nabla \mathbf{V}_Z)}{R_{Z\parallel}} \sim \frac{m_z n_z V_\parallel^2 / L_\parallel}{m_i n_i V_\parallel / \tau_{iz}} \sim \frac{\delta}{Z \hat{v}_{ii}}.$$

For simplicity, we shall assume that this parameter is small, which is realistic in edge plasmas where the bulk ions are not too far into the banana regime. One can show that two additional simplifications arise from this assumption: the impurity viscosity becomes negligible, and the ion and impurity temperatures equilibrate. The reader is referred to Helander (1998) for details. Since the bulk ions are in the banana regime, their temperature is constant on flux surfaces, and we obtain

$$n_Z Z e \nabla_\parallel \Phi_1 + T_i \nabla_\parallel n_Z = R_{Z\parallel}. \tag{13.14}$$

To calculate the friction force, we need the perturbed ion distribution function, which in the banana regime is equal to

$$f_{i1} = -\frac{I v_\parallel}{\Omega_i} \frac{\partial f_{i0}}{\partial \psi} - \frac{e\Phi_1}{T_i} f_{i0} + g_i(v, \mu, \psi, \sigma),$$

where g_i vanishes in the trapped domain. The electron distribution is of a similar form (but with $m_i \to m_e$ and $e \to -e$), and the impurity density can be related to Φ_1 by quasineutrality,

$$n_Z = \frac{n_e - n_i}{Z} = \frac{2n_0}{T_0} \frac{e\Phi_1}{Z}, \tag{13.15}$$

where $2n_0/T_0 \equiv n_{e0}/T_e + n_{i0}/T_i$, and n_{i0}, T_{i0}, n_{e0} and T_{e0} are the densities and temperatures associated with the lowest-order, Maxwellian, distribution functions f_{i0} and f_{e0}. Using the collision operator (3.32) to calculate the parallel friction force between bulk ions and impurities gives

$$R_{Z\parallel} = -\int m_i v_\parallel C_{iZ}(f_{i1}) d^3v$$

$$= -\frac{p_i I}{\Omega_i \tau_{iZ}}\left(\frac{d\ln p_i}{d\psi} - \frac{3}{2}\frac{d\ln T_i}{d\psi}\right) + \frac{m_i n_i}{\tau_{iZ}}\left(u - \frac{K_Z}{n_Z}\right)B, \quad (13.16)$$

where

$$u = \frac{\tau_{iZ}}{n_i B}\int v_\parallel v_{iZ} g_i d^3v \quad (13.17)$$

is a flux function since $d^3v \propto B\,dv\,d\mu/v_\parallel$.

The poloidal flow K_Z appearing in (13.16) can be found from the solubility constraint $\langle BR_{Z\parallel} = 0\rangle$ that follows from (13.14). Inserting the resulting friction force in the parallel momentum equation (13.14) then gives

$$\left(T_i + \frac{n_Z Z^2}{2n_0}T_0\right)\nabla_\parallel n_Z = -\frac{p_i I}{\Omega_i \tau_{iZ}}\left(\frac{d\ln p_i}{d\psi} - \frac{3}{2}\frac{d\ln T_i}{d\psi}\right)\left(1 - \frac{\langle n_Z\rangle}{n_Z}\frac{B^2}{\langle B^2\rangle}\right)$$

$$+ \frac{m_i n_i u}{\tau_{iZ} n_Z}\left(n_Z - \frac{\langle n_Z B^2\rangle}{B^2}\right)B,$$

where we have used (13.15) to eliminate the electric field, and noted that $n_Z\tau_{iZ}$ is a flux function. This equation can be written in dimensionless form by introducing

$$n = n_Z/\langle n_Z\rangle,$$
$$b = B/\langle B^2\rangle^{1/2},$$
$$\alpha = \langle n_Z\rangle Z^2 T_0/2n_0 T_i,$$

$$\gamma = -\frac{eu}{I T_i}\langle B^2\rangle\left(\frac{d\ln p_i}{d\psi} - \frac{3}{2}\frac{d\ln T_i}{d\psi}\right)^{-1}, \quad (13.18)$$

and a modified poloidal angle coordinate ϑ defined by

$$\frac{d\vartheta}{d\theta} \equiv \frac{\langle \mathbf{B}\cdot\nabla\theta\rangle}{\mathbf{B}\cdot\nabla\theta},$$

so that the flux-surface average is equivalent to an average over ϑ. The parallel momentum equation then becomes

$$(1 + \alpha n)\frac{\partial n}{\partial\vartheta} = G\left[n - b^2 + \gamma(n - \langle nb^2\rangle)b^2\right], \quad (13.19)$$

where the parameter

$$G = -Z^2 \frac{IB}{\Omega_i \tau_{ii} \langle \mathbf{B} \cdot \nabla\theta \rangle} \left(\frac{d \ln p_i}{d\psi} - \frac{3}{2} \frac{d \ln T_i}{d\psi} \right)$$

measures the steepness of the plasma profile. It is straightforward to verify that $G = O(\delta_Z)$, so that the poloidal variation of the impurity density indeed satisfies the estimate (13.13). Physically, the parameter G measures the magnitude of the parallel friction force relative to the parallel pressure gradient.

Equation (13.19) governs the poloidal distribution of impurities on each flux surface. In conventional neoclassical theory, the pressure and temperature gradients are assumed to be so small that $G \ll 1$. It then follows from (13.19) that the impurities are evenly distributed, $n \simeq 1$. In the opposite limit of very steep gradients, $G \gg 1$, the friction force exceeds the pressure gradient and causes a rearrangement of impurities within each flux surface. If we expand n in powers of G^{-1},

$$n = n_0 + n_1 + O(G^{-2}) \qquad (G \gg 1), \qquad (13.20)$$

the lowest-order solution is in–out asymmetric (following $b^2(\vartheta)$),

$$n_0 = \frac{\gamma}{1 - \langle (1 + \gamma b^2)^{-1} \rangle} \frac{b^2}{1 + \gamma b^2}, \qquad (13.21)$$

while the first-order term n_1 contains up–down asymmetry and is determined by

$$(1 + \alpha n_0)\frac{\partial n_0}{\partial \vartheta} = G[n_1 + \gamma(n_1 - \langle n_1 b^2 \rangle)b^2]. \qquad (13.22)$$

In a torus with small inverse aspect ratio, $\epsilon = r/R \ll 1$, then poloidal variation of impurity density is small, $n - 1 = O(\epsilon)$. In the opposite limit of tight aspect ratio, there are very few circulating particles, so that $u \to 0$ and thus $\gamma \to 0$. Then $n_0 \to b^2$, implying a much larger impurity density on the inboard of the flux surface than on the outboard side.

This rearrangement of impurities has dramatic implications for collisional particle transport since it affects the ion–impurity friction force. The radial neoclassical flux of hydrogen ions is related to this force by Eq. (8.29), and thus becomes

$$\langle \Gamma_i^{\mathrm{neo}} \cdot \nabla\psi \rangle = \left\langle \frac{I R_{z\parallel}}{eB} \right\rangle = \langle \mathbf{B} \cdot \nabla\theta \rangle \frac{\langle pz \rangle I}{e \langle B^2 \rangle} \left[\left\langle \frac{n}{b^2} \right\rangle - 1 + \gamma(1 - \langle nb^2 \rangle) \right] G.$$

$$(13.23)$$

It increases linearly with the gradients when G is small, and is then proportional to $\langle b^{-2} - 1 \rangle G$, as is characteristic of Pfirsch–Schlüter transport. Note that the particle flux scales like the Pfirsch–Schlüter value although the bulk ions are in the banana regime. However, when the profiles become so steep that $G \gg 1$ then Eqs. (13.20)–(13.23) show that

$$\langle \boldsymbol{\Gamma}_i^{\text{neo}} \cdot \nabla \psi \rangle \propto G^{-1}.$$

The contributions to the flux from both n_0 and n_1 vanish, and as a result the neoclassical particle flux *decreases* with increasing gradients when the latter become sufficiently steep!

As the neoclassical channel is suppressed, classical transport becomes relatively more important. The classical particle flux (8.28) is

$$\langle \boldsymbol{\Gamma}_i^{\text{cl}} \cdot \nabla \psi \rangle = \left\langle \frac{R\hat{\boldsymbol{\varphi}} \cdot \mathbf{R}_{Z\perp}}{e} \right\rangle,$$

where $\mathbf{R}_{Z\perp}$ is the perpendicular friction force,

$$\mathbf{R}_{Z\perp} = \frac{m_i n_i}{\tau_{iZ}} \left(\mathbf{V}_{i\perp} - \mathbf{V}_{Z\perp} - \frac{3}{2m_i\Omega_i} \mathbf{b} \times \nabla T_i \right).$$

Since the difference in diamagnetic velocities is $\mathbf{V}_{i\perp} - \mathbf{V}_{Z\perp} \simeq \mathbf{b} \times \nabla p_i / n_i eB$, and $\nabla\varphi \cdot (\mathbf{b} \times \nabla\psi) = -B_\theta^2/B$, the classical flux becomes

$$\langle \boldsymbol{\Gamma}_i^{\text{cl}} \cdot \nabla \psi \rangle = -\frac{p_i B^2}{m_i \Omega_i^2 \tau_{iZ} n_Z} \left(\frac{d \ln p_i}{d\psi} - \frac{3}{2} \frac{d \ln T_i}{d\psi} \right) \left\langle \frac{R^2 B_\theta^2}{B^2} n_Z \right\rangle. \tag{13.24}$$

In a torus with large aspect ratio and circular cross section, $b^2(\vartheta) = 1 - 2\epsilon \cos\vartheta + O(\epsilon^2)$, and we can expand the impurity density as

$$n(\vartheta) = 1 + n_c \cos\vartheta + n_s \sin\vartheta + O(\epsilon^2).$$

The solution to (13.19) is then

$$n_s = \frac{2\epsilon(1+\alpha)G}{(1+\alpha)^2 + (1+\gamma)^2 G^2},$$

$$n_c = \frac{2\epsilon(1+\gamma)G^2}{(1+\alpha)^2 + (1+\gamma)^2 G^2}.$$

Thus, the in–out asymmetry increases monotonically with increasing gradient, while the up–down asymmetry has a maximum at $G = (1+\alpha)/(1+\gamma)$. It is now straightforward to evaluate the fluxes (13.23) and (13.24) to obtain

$$\langle \boldsymbol{\Gamma}_i^{\text{cl}} \cdot \nabla \psi \rangle + \langle \boldsymbol{\Gamma}_i^{\text{neo}} \cdot \nabla \psi \rangle = \frac{\epsilon^2 p z}{q^3 e} \left(1 + \frac{2q^2}{1 + \left(\frac{1+\gamma}{1+\alpha}\right)^2 G^2} \right) G.$$

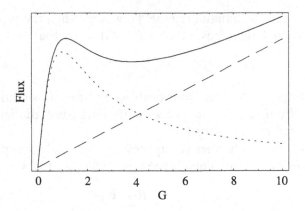

Fig. 13.1.   Neoclassical ion particle flux (dotted curve), classical flux (dashed line), and total flux (solid curve) versus gradient in a toroidal plasma with circular cross section, large aspect ratio, and safety factor $q = 3$. Conventional transport theory is only valid in the lower left corner, where the fluxes are linearly proportional to the gradients.

The second term in this expression represents the neoclassical contribution and exceeds the first, classical term by the Pfirsch–Schlüter factor $2q^2$, see Eq. (9.21), when the gradients are weak, $G \ll 1$. However, if the profiles are steep ($G \gg 1$) the neoclassical flux is suppressed and classical transport dominates. As the latter is not much affected by the weak $[O(\epsilon)]$ impurity redistribution, the flux then increases linearly with G. The total flux, shown in Fig. 13.1, is non-monotonic if $q > 2$.

Thus, when the density or temperature profile becomes sufficiently steep, transport becomes complicated. The parallel dynamics is no longer trivial, especially for highly charged impurities, which undergo a spontaneous rearrangement on each flux surface. This makes the neoclassical particle transport nonlinear. Energy transport is less affected since this is caused both by ion–ion collisions and ion–impurity collisions, but only the latter are affected by the redistribution.

13.4 Orbit squeezing

In the conventional transport ordering (8.1) one normally assumes that not only density and temperature vary slowly on the length scale of the gyroradius, but that the electric field $\mathbf{E} = -\nabla\Phi$ varies equally slowly. However, it is sometimes possible to develop a tractable transport theory when this is not the case. As we shall see, the electric field is then able to distort collisionless ion orbits, a phenomenon referred to as 'orbit squeezing' in the literature (Hazeltine, 1989; Shaing et al., 1998). This situation may occur in a transport barrier, where turbulence is suppressed as a result of a strongly sheared electric field, or at the plasma edge, where a sharp variation in electrostatic potential may arise as a consequence of boundary conditions.

To illustrate what happens to transport when orbit squeezing occurs, we consider classical heat ion transport in a plasma embedded in a straight, constant magnetic field in the presence of a transverse electric field,

$$\mathbf{B} = -\hat{\mathbf{z}}B,$$
$$\mathbf{E} = -\nabla\Phi(x) = -\hat{\mathbf{x}}\Phi'(x),$$

where (x, y, z) are Cartesian coordinates. We begin by analyzing collisionless particle dynamics to find how the gyro-orbits are distorted by the electric field. The magnetic field, $\mathbf{B} = \nabla \times \mathbf{A}$, can be derived from the vector potential $\mathbf{A} = -\hat{\mathbf{y}}xB$, and the Lagrangian (6.1) thus becomes

$$L = \frac{m}{2}(\dot{x}^2 + \dot{y}^2 + \dot{z}^2) - eBx\dot{y} - e\Phi(x).$$

The canonical momenta (6.7) are thus

$$p_x = m\dot{x},$$
$$p_y = m\dot{y} - eBx,$$
$$p_z = m\dot{z},$$

where p_y and p_z are constants of motion since the Lagrangian is independent of y and z. The quantity

$$\bar{x} = -p_y/eB = x - v_y/\Omega,$$

where $v_y = \dot{y}$ and $\Omega = eB/m$, is thus constant along the orbit. In the absence of an electric field, this corresponds to the x-coordinate of the guiding centre. Furthermore, since L is independent of time, the energy

$$H = \frac{mv^2}{2} + e\Phi(x) = \frac{m}{2}(\dot{x}^2 + \dot{z}^2) + V(x),$$

$$V(x) = \frac{m\Omega^2}{2}(x - \bar{x})^2 + e\Phi(x)$$

is also conserved. Here $V(x)$ plays the role of a potential well that confines the particle in the x-direction. Indeed, the Hamiltonian governing the motion in this direction is

$$E(x, p_x) = \frac{p_x^2}{2m} + V(x),$$

and is also a constant of motion.

Now suppose that it is sufficient to keep terms up to the second derivative in the Taylor expansion of $\Phi(x)$,

$$\Phi(x) \simeq \Phi_0 + x\Phi_0' + \frac{x^2}{2}\Phi_0''.$$

This means that although the electric field $E_x = -d\Phi/dx$ is allowed to
be large and to vary rapidly, we assume that its shear $E_x' = -d^2\Phi/dx^2$ is
nearly constant over the region of interest. The potential well $V(x)$ is then
parabolic, and its bottom is located at $x = X$ defined by $V'(X) = 0$,

$$V(x) = \frac{m\Omega^2 S}{2}(x - X)^2 + \text{constant},$$

$$X = \frac{x - (v_y - v_E)/\Omega}{S}, \tag{13.25}$$

where $v_E = -\Phi_0'/B$ is the $\mathbf{E} \times \mathbf{B}$ velocity at $x = 0$, and

$$S = 1 + \frac{m\Phi_0''}{eB^2}$$

is the so-called 'squeezing factor'. The particle thus executes a harmonic
oscillation, $x = \rho_x \sin\omega t$, with angular frequency $\omega = S^{1/2}\Omega$. The amplitude ρ_x depends on the energy of the particle. We do not consider the
case when $\Phi'' < 0$ and the electric field shear is so strong that $S < 0$.
Then $V(x)$ describes a potential hill rather than a well, and the particle
is not confined. In other words, the electric field is then strong enough to
accelerate the particle to infinity and effectively de-magnetize the ions.

Since gyro-orbits are distorted by the electric field when $S \neq 1$, we
expect this to affect cross-field transport. We thus consider a plasma
with significant orbit squeezing and endeavour to calculate the ion heat
transport that arises when the density and temperature vary in the x-
direction. As usual, to obtain a tractable problem we must assume that
this variation is slow on the length scale of the gyro-orbit, so that the
macroscopic length scale $L_\perp$ exceeds the squeezed gyroradius,

$$\rho_x \ll L_\perp.$$

This will allow us to neglect the time derivative in the kinetic equation
as being of second order in $\rho_x/L_\perp$ because of the diffusive nature of the
transport, in analogy with the discussion in Section 8.1. Additionally, we
assume that the gyrofrequency exceeds the collision frequency, $\omega\tau_{ii} \gg 1$,
so that the plasma ions are magnetized.

The calculation is greatly simplified by choosing appropriate independent variables in the kinetic equation. For this purpose, we note that

the position of a particle along a collisionless squeezed gyro-orbit can be expressed conveniently in terms of a phase, $0 \le \theta < 2\pi$, that evolves linearly with time according to

$$\frac{d\theta}{dt} = \omega.$$

Writing the ion kinetic equation in the phase-space coordinates $(X, y, z, H, \theta, p_z, t)$, then gives

$$\omega \frac{\partial f}{\partial \theta} = C(f), \qquad (13.26)$$

if we assume that the plasma is homogeneous in the y- and z-directions, so that $\partial f / \partial y = \partial f / \partial z = 0$. On the right-hand side, we need only include ion–ion collisions, so that $C(f) = C_{ii}(f)$.

The next step is to expand the distribution function in the smallness of the collision frequency, $f = f_0 + f_1 + \cdots$. In lowest order we then have $\partial f_0 / \partial \theta = 0$, implying that f_0 only depends on constants of motion, $f_0 = f_0(H, X, p_z)$. This function is deterimined by the next-order equation,

$$\omega \frac{\partial f_1}{\partial \theta} = C(f_0),$$

whose average over the angle θ provides a solubility constraint for f_0,

$$\oint C(f_0) d\theta = 0. \qquad (13.27)$$

This constraint, which requires that the effect of collisions on f_0 should vanish when averaged over an orbit, forces f_0 to be close to a Maxwellian. At the same time, f_0 cannot be exactly Maxwellian at each value of x since this would contradict the requirement that f_0 only depend on constants of motion. The solution of Eq. (13.27) is

$$f_0 = N(X) \left(\frac{m}{2\pi T(X)} \right)^{3/2} e^{-H/T(X)},$$

which differs from a true Maxwellian, $f_M = N(x)[m/2\pi T(x)]^{3/2} e^{-H/T(x)}$, since X depends on v_y, see Eq. (13.25). Indeed, f_0 is approximately equal to

$$f_0 \simeq f_M(x) + (X - x) \left(\frac{\partial f_M}{\partial x} \right)_H$$

$$= f_M(x) + (X - x) \left[\frac{N'}{N} + \left(\frac{H}{T} - \frac{3}{2} \right) \frac{T'}{T} \right] f_M(x), \qquad (13.28)$$

where $' = d/dx$ and we can truncate after only one term since $\rho_x / L_\perp$ has been assumed to be small. Note that the constraint (13.27) is satisfied because $(X - x)$ vanishes when averaged over the orbit. Also, note that the density in the Maxwellian is not $N(x)$ but $n(x) = N(x)\exp(-e\Phi/T)$ because of the potential energy entering in H.

We now turn to the problem of calculating the ion energy flux,

$$Q_x = \int H v_x f \, d^3 v,$$

which is the same as the heat flux, $q_x = Q_x - 5nTV_x/2$, since the particle flux $\Gamma_x = nV_x$, produced from ion–ion collisions vanishes. (As we shall see, this is true also when the gyro-orbits are squeezed.) For this calculation, we use a technique of wide applicability in transport theory. Rather than calculating the local energy flux, we seek its average over some region of width L, which is larger than the gyroradius but smaller than the macroscopic scale length,

$$\rho_x \ll L \ll L_\perp.$$

This does not interfere with our goal, which is to understand transport on the macroscopic length scale $L_\perp$. Thus we seek to evaluate

$$\bar{Q}_x = \frac{1}{L} \int_x^{x+L} H v_x f \, dx \, d^3 v = \frac{1}{L} \int \frac{H v_x f}{m\omega} \, d\theta \, dH \, dv_y \, dv_z, \qquad (13.29)$$

where in the second equality we have used

$$dx \, dv_x = d\theta \, dH / m\omega. \qquad (13.30)$$

For readers familiar with action-angle variables, this follows at once from the relations

$$dx \, dp_x = dJ \, d\theta,$$

$$\frac{dH}{dJ} = \omega,$$

where (J, θ) are action-angle variables for motion in the x-direction.

For readers less familiar with analytical mechanics, we digress briefly to prove Eq. (13.30). The issue is to prove that if we use (θ, H) as phase-space coordinates instead of (x, p_x), then the Jacobian is equal to $1/\omega$. To this end, consider two neighbouring orbits in phase space, corresponding to energies H and $H + \delta H$, respectively, and consider two curves of constant phase, θ and $\theta + d\theta$, see Fig. 13.2. The area between these four curves is equal to

$$dS = dx \, \delta p_x - \delta x \, dp_x,$$

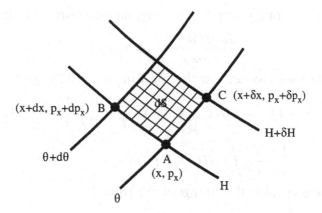

Fig. 13.2. The area of the shaded region is equal to $dS = dx\,\delta p_x - \delta x\,dp_x = d\theta\,dH/\omega$.

since this is the cross product of the vectors $\vec{AB}$ and $\vec{AC}$ corresponding to the increments $(\delta x, \delta p_x)$ and (dx, dp_x) along the two sets of curves. Since

$$dx = \dot{x}\,dt = \frac{\partial H}{\partial p_x}\frac{d\theta}{\dot{\theta}},$$

$$dp_x = \dot{p}_x\,dt = -\frac{\partial H}{\partial x}\frac{d\theta}{\dot{\theta}},$$

are the distances in phase-space by which the particle moves in time $dt = d\theta/\dot{\theta} = d\theta/\omega$, it follows that the area is

$$dS = \left(\frac{\partial H}{\partial p_x}\delta p_x + \frac{\partial H}{\partial x}\delta x\right)\frac{d\theta}{\omega} = \frac{d\theta\,dH}{\omega},$$

which implies that the Jacobian is indeed

$$\frac{\partial(x, p_x)}{\partial(H, \theta)} = \omega^{-1},$$

as asserted in Eq. (13.30).

Returning to Eq. (13.29), it is now a simple matter to calculate the heat flux, by introducing the function $G = (x - X)H$, so that

$$\frac{\partial G}{\partial \theta} = \frac{1}{\omega}\frac{dG}{dt} = \frac{Hv_x}{\omega},$$

and then integrating Eq. (13.29) by parts in θ,

$$\bar{Q}_x = -\frac{1}{L}\int\frac{G}{m}\frac{\partial f}{\partial \theta}d\theta\,dH dv_y\,dv_z = -\frac{1}{L}\int GC(f)\frac{d\theta\,dH\,dv_y\,dv_z}{m\omega},$$

where we have used the kinetic equation (13.26).

Finally, transforming back to our original phase-space variables,

$$\frac{d\theta\, dH\, dv_y\, dv_z}{m\omega} = dx\, d^3v,$$

and using the lowest-order distribution function (13.28) gives

$$\bar{Q}_x = -\int (x - X)HC\left[(X - x)\left(\frac{H}{T} - \frac{3}{2}\right)\frac{T'}{T}f_M\right](x)d^3v$$

$$= \frac{1}{S^2\Omega^2}\frac{T'}{T^2}\int v_y HC(v_y Hf_M)d^3v,$$

where we have recalled that terms of the form

$$\int v_y C(v_y Hf_M)d^3v = \int v_y HC(v_y f_M)d^3v = 0$$

vanish because of self-adjointness and momentum conservation in ion–ion collisions. This is also the reason why the ion particle flux vanishes. The remainder is essentially a matrix element of the ion–ion collision operator, see Eqs. (4.30),

$$\bar{Q}_x = \frac{nTT'}{mS^2\Omega^2\tau_{ii}}(M_{ii}^{11} + N_{ii}^{11}) = \frac{-2nTT'}{mS^2\Omega^2\tau_i} = -\frac{\kappa_\perp^i}{S^2}\frac{dT}{dx},$$

where $\kappa_\perp^i$ is Braginskii's perpendicular heat conductivity (4.48). In this expression the only signature of the sheared electric field is the appearance of the factor S^2 in the denominator. We conclude that the transport is enhanced if $\Phi'' < 0$ and suppressed if $\Phi'' > 0$. In the limit $S \to 0^+$, where the plasma is demagnetized, the heat conductivity becomes infinite.

Orbit squeezing also occurs in a tokamak. A sheared radial electric field distorts banana orbits if the banana squeezing parameter

$$S_b = 1 + \frac{m}{eB_p^2}\frac{d^2\Phi}{dr^2}$$

differs from unity. Note that it is the *poloidal* magnetic field that enters in this expression. Since this is usually an order of magnitude smaller than the toroidal field, banana orbits are more sensitive to radial electric fields than are gyro-orbits. The transport theory is, however, more complicated due to the fact that the sheared electric field not only distorts banana orbits but can also pull some guiding-centre orbits out of the plasma. In the case of a straight magnetic field and Larmor orbits, we saw that this only occurs if $S < 0$, but in a toroidal field the situation is different. A sheared electric field can de-confine particle orbits even if the squeezing factor S_b is moderate, if this electric field extends radially over more than one squeezed banana orbit. This implies that local transport theory cannot be applied in this case (Krasheninnikov and Yushmanov, 1994).

13.5 Neoclassical transport in stellarators

In this section we briefly outline some important features of neoclassical stellarator transport. This is a highly developed and complex branch of plasma physics, which cannot be done justice in this short section. Beyond the introductory discussion provided here, the interested reader is encouraged to consult the references given at the end of the chapter.

In a tokamak, the invariance of the toroidal canonical momentum p_φ implies that plasma particles are confined in the radial direction (Tamm's theorem, see Section 7.3). Since the stellarator lacks this toroidal symmetry, p_φ is not conserved and there are generally only two constants of motion: magnetic moment and energy. There is then no robust reason why particles should be confined by the magnetic field, and indeed, the collisionless guiding-centre trajectories tend to be very wide, or even intersect the wall, unless the magnetic field is optimized to prevent this. A similar situation occurs in a tokamak with large magnetic ripple caused by the discreteness of the toroidal field coils. The magnetic field strength may typically vary along a field line approximately like

$$B \simeq B_0[1 - \epsilon_1 \cos\theta - \epsilon_2 \cos(l\theta - n\varphi)],$$

where l and n are integers. The term $\epsilon_1 \cos\theta$ is caused by the toroidicity of the system, while the term $\epsilon_2 \cos(l\theta - n\varphi)$ has to do with non-axisymmetry. In the case of a tokamak with toroidal ripple, $l = 0$ and n is the number of toroidal field coils, while in a stellarator (l, n) represent the dominant mode numbers of the helical variation in magnetic field strength. As can be seen in Fig. 13.3, this leads to the appearance of local magnetic wells in which particles can be trapped, bouncing between local maxima of B. A particle confined in such a well tends to drift radially by a considerable distance, resulting in a very fat 'superbanana' orbit – or even a chaotic orbit – which may intersect the chamber wall.

In this situation neoclassical transport can become very large. As in the case of a tokamak, the character of the transport is sensitive to the collision frequency. If the collisionality is very low, the transport is proportional to the effective scattering frequency ν_{eff} of particles into and out of local magnetic wells multiplied by the square of the superbanana step size Δr,

$$D \sim \nu_{\text{eff}} \Delta r^2. \tag{13.31}$$

This is similar to, but more serious than, banana transport in an axisymmetric tokamak since ν_{eff} and Δr can be very large. However, the requirement of low collisionality – that superbanana orbits remain collisionless – is more stringent if the local magnetic wells are small. If the orbits are interrupted by collisions, a new collisionality regime appears that is peculiar to non-axisymmetric systems. In this regime, the time $1/\nu_{\text{eff}}$

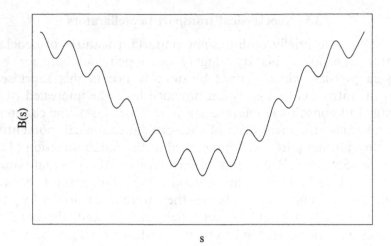

B(s)

s

Fig. 13.3. Variation of magnetic field strength along a field line in a stellarator or rippled tokamak.

after which a trapped particle is scattered out of a local magnetic well is shorter than the time required to complete a full superbanana orbit. Its radial excursion is thus of the order v_d/v_{eff}, where v_d is the drift velocity, and the diffusion coefficient thus becomes

$$D \sim v_{\text{eff}} \left(\frac{v_d}{v_{\text{eff}}} \right)^2 = \frac{v_d^2}{v_{\text{eff}}}.$$

Note that this D is inversely proportional to the collision frequency. It thus becomes very large in a hot plasma where the collision frequency is low, but saturates at the level given by the superbanana diffusion rate (13.31).

Modern stellarators avoid such large neoclassical transport by a clever optimization of the magnetic field structure. There are a number of ways to achieve this. In a *quasihelical* or *quasisymmetric* device, the magnetic field coils and the plasma currents are arranged in such a way that the magnetic field strength depends on the poloidal and toroidal angles, in Boozer coordinates, only through the linear combination

$$\alpha = n\varphi - l\theta,$$

where l and n are integers. In a toroidal system this cannot be achieved exactly, but if the aspect ratio is not too small it can be done to a fair approximation – hence the term *quasi*helical. Since then $B = B(\psi_p, \alpha)$ it is useful to express the guiding-centre Lagrangian (7.40) in coordinates (ψ_p, θ, α), replacing the toroidal angle by $\varphi = (\alpha + m\theta)/n$. The Lagrangian then becomes

$$\bar{L} = \frac{m}{2B^2} \left[\left(B_\theta + \frac{lB_\varphi}{n} \right) \dot\theta + \frac{B_\varphi}{n} \dot\alpha \right]^2$$

$$+ Ze \left[\left(\psi_t - \frac{l\psi_p}{n} \right) \dot\theta - \frac{\psi_p}{n} \dot\alpha \right] - \mu B(\psi_p, \alpha) - Ze\Phi(\psi_p),$$

and since it does not depend on θ, the corresponding canonical momentum,

$$p_\theta = \frac{\partial \bar{L}}{\partial \dot\alpha} = \frac{mv_\parallel}{B} \left(B_\theta + \frac{lB_\varphi}{n} \right) + Ze \left(\psi_t - \frac{l\psi_p}{n} \right), \qquad (13.32)$$

is conserved. Thus, in a quasihelical stellarator this replaces the toroidal canonical momentum p_φ. For similar reasons as in Tamm's theorem the constancy of p_θ implies radial confinement: the second term in (13.32) is essentially a radial coordinate, and since this term is larger than the first one in a Larmor radius expansion the particle must stay close to a given flux surface. The orbits are thus narrow and neoclassical transport is small. In a quasihelical stellarator with $n = 0$, the magnetic field strength depends only on the poloidal Boozer angle θ. Such a device is called *quasiaxisymmetric* as it is toroidally symmetric in 'Boozer space'. For similar reasons, a quasihelical system with $l = 0$ is said to have *quasipoloidal* symmetry.

Another way to suppress neoclassical transport in a stellarator is to make the magnetic field *quasiomnigenous*. This means that the net radial magnetic drift of a particle nearly vanishes on a bounce average. From Chapter 6, we know that drift motion conserves the longitudinal adiabatic invariant,

$$J(\varphi - q\theta, \psi_p, H, \mu,) = \oint mv_\parallel ds,$$

which is a function of the magnetic field line $(\varphi - q\theta, \psi_p)$ on which the bounce orbit is located, as well as energy and magnetic moment. Thus, if the drift motion across the flux surface vanishes on a bounce average, surfaces of constant J must coincide with flux surfaces. The condition for omnigeneity is that $J = J(\psi_p, H, \mu)$. In reality this can only be achieved approximately, in so-called quasiomnigenous systems. It is not difficult to realize that in such a device the minimum magnetic field strength, $B_{min}(\varphi - q\theta, \psi_p)$, along each field line must be the same for each field line on a given magnetic surface. The reason for this can be understood by considering a deeply trapped particle. Such a particle has very small parallel velocity and thus $J \simeq 0$. Since J is conserved, it stays close to zero as the particle drifts across the magnetic field. This implies that $v_\parallel$ remains small. Since the energy, $H = \mu B_{min}$, is conserved, B_{min} must stay the same over the drift orbit.

We conclude with a final definition: a magnetic equilibrium in which the instantaneous guiding-centre drift velocity – as opposed to the bounce-averaged one – lies within the flux surface is called *isodynamic*. Such a system is completely free from neoclassical transport. This is clearly a more stringent requirement than omnigeneity, and is therefore more difficult to achieve in practice.

Further reading

The topics discussed in this chapter represent active fields of current research. The most complete discussion of poloidal flow damping in the banana regime is given by Hinton and Rosenbluth (1999), whose paper contains references to earlier work. Toroidal rotation is treated in the papers cited in Section 13.2, and a representative overview of neoclassical theories for plasmas with large gradients or distorted orbits can be found in the papers mentioned in Sections 13.3 and 13.4. Analytical mechanics of stellarator particle orbits is discussed very transparently by Cary and Shasharina (1997). There are two major reviews of neoclassical transport in stellarators: a review paper by Kovrizhnykh (1984) and the book by Wakatani (1998), which contains a wealth of material on most aspects of stellarators. The recent review papers by Boozer (1998) and Wobig (1999) give useful, shorter overviews of stellarator physics in general and also contain references to the most important literature in this field, including neoclassical theory.

Exercise

1. Calculate the poloidal rotation in a low-collisionality plasma caused by a constant force acting on the ions.

   Solution: Adding an external force to the right of Eq. (12.45) gives rise to the system of equations

   $$\hat{\mu}_{i1} u_{i\theta} + \hat{\mu}_{i2} \frac{2q_{i\theta}}{5p_i} = \frac{\langle BR_{i\parallel} \rangle}{\langle B^2 \rangle},$$

   $$\hat{\mu}_{i2} u_{i\theta} + \hat{\mu}_{i3} \frac{2q_{i\theta}}{5p_i} = l_{22}^{ii} \left(\frac{V_{2i}B}{\langle B^2 \rangle} + \frac{2q_{i\theta}}{5p_i} \right),$$

   which has the solution

   $$u_{i\theta} = \frac{1}{\langle B^2 \rangle} \frac{(\hat{\mu}_{i3} - l_{22}^{ii})\langle BR_{i\parallel} \rangle + l_{22}^{ii}\hat{\mu}_{i2}\langle V_{2i}B \rangle}{\hat{\mu}_{i1}(\hat{\mu}_{i3} - l_{22}^{ii}) - \hat{\mu}_{i2}^2}.$$

   Inserting the viscosity coefficients (12.48) appropriate for the banana regime gives

   $$u_{i\theta} = \frac{1.88(1 + 0.98y)\tau_{ii}\langle BR_{i\parallel} \rangle/m_i n_i - 1.17y\langle V_{2i}B \rangle}{y(1 + 0.462y)\langle B^2 \rangle},$$

with $y = f_t/f_c$. The second term in the numerator is the ordinary poloidal rotation caused by the ion temperature gradient (11.58), while the first term represents rotation produced by friction. This result has been used to calculate the poloidal rotation resulting from friction between thermal ions and fast ions produced by fusion reactions or neutral-beam injection (Hinton and Kim, 1993; Rosenbluth and Hinton, 1996).

14

Experimental evidence for neoclassical transport

In this brief concluding chapter we outline the experimental evidence for neoclassical transport that has accumulated from tokamak experiments around the world.

Galeev and Sagdeev (1968) published the first theoretical paper on 'banana diffusion' in the near collisionless regime. While there were a few isolated experimentally consistent observations (e.g., Sigmar et al., 1974) the bulk of well-diagnosed experiments on the major tokamaks started to appear in the late 1980s. By now, there exists a broad and solid record which we briefly outline here by quoting some of the experimental highlights. In addition to our own selection, we would like to refer the interested reader to the excellent Introduction and Discussion sections in the paper by Houlberg et al., (1997). All experiments mentioned below are tokamaks.

• The electrical conductivity in tokamaks follows the Spitzer prediction, and trapped-electron corrections are essential: without the neoclassical correction to parallel resistivity, poloidal flux evolution and sawtooth response location at $q = 1$ cannot be explained in TFTR (Zarnstorff et al., 1990) and in JET (Ward, 1994).

• The bootstrap current was theoretically discussed by Bickerton, Connor and Taylor (1971), and by Kadomtsev and Shafranov (1972). Following a historic early discovery of the bootstrap current in the Wisconsin Octopole by Zarnstorff and Prager (1984, 1986), the bootstrap current was demonstrated to exist in TFTR by carefully zeroing out two opposing neutral beam induced currents (Zarnstorff et al., 1988), and similarly in JET (Challis et al., 1989), DIII-D (Forest et al., 1994) and JT-60 (Kikuchi et al., 1990). In all of these experiments agreement with theory also requires employing neoclassical resistivity corrections. To get agreement in TEX-TOR, it was also necessary to invoke the theoretical expression for the

272

neoclassical parallel ion flow velocity (Hothker *et al.*, 1994). Schwarzschild wrote an illuminating bootstrap current article in *Physics Today* (1986).

• Energy confinement in tokamaks is usually worse than the neoclassical prediction, particularly in the electron channel. However, ion heat transport can be neoclassical when plasma turbulence is reduced. Such improved (near neoclassical) values for ion thermal conductivity and particle diffusivity are observed in TFTR plasmas with reversed magnetic shear equilibria, which are instrumental in suppressing otherwise dominating turbulent transport. The collisional ion thermal conductivity even drops to levels below the standard neoclassical value, which should be the irreducible lower limit. This suggests that the neoclassical theory used here was lacking some relevant non-standard features, e.g., by assuming that the ion Larmor radius in the poloidal magnetic field was uniformly much smaller than the ion temperature gradient scale length and by ignoring potato-orbit effects (Levinton *et al.*, 1995). In regions with internal transport barriers, neoclassical ion transport has also been reported in other large tokamaks: JET (JET Team, 1999), ASDEX-U (Wolf, 2000), and JT-60U (Sakamoto *et al.*, 2000). In quiescent DIII-D plasmas, the ion thermal conductivity in a deuterium plasma is at the neoclassical level over most of the plasma profile (Strait *et al.*, 1996). Consequently, the inferred fusion yield (had this been a D-T plasma) reaches a Q-value of 0.32 in this modest-size experiment. The overall ion energy confinement is usually close to neoclassical in the spherical tokamak START (Roach *et al.*, 2000), see Fig. 14.1.

• While neoclassical theory of poloidal flows cannot by itself explain observed enhanced confinement, their pivotal role in observed suppression of turbulence was treated theoretically by Kim, Diamond and Groebner (1991); Novakovskii, Liu *et al.* (1997); Ernst, Coppi *et al.* (1998); and by Zhu, Horton and Svgama (1999), and is shown to play an important role in the observed suppression of turbulence in TFTR (Meade and the TFTR Group, 1991); (JET Team, 1991) and DIII-D (Burrell *et al.*, 1994). For an encompassing review see Burrell (1999).

• In quiescent plasmas particle transport experiments encompassing both impurity ions and electrons (Levinton *et al.*, 1995) have made contact with detailed neoclassical predictions, including carbon peaking on axis in H-mode and PEP-mode (pellet enhanced performance) plasmas (Lauro-Taroni *et al.*, 1994), as well as neoclassical temperature screening of impurity transport (JET Team, 1991; Giannella *et al.*, 1989) and poloidal asymmetries of impurity ion densities. Central impurity peaking as a consequence of friction between bulk ions and impurities was predicted early by Braginskii (1958) and Taylor (1961), and neoclassically by Connor (1973), Rutherford (1974) and many others since then. It has been observed

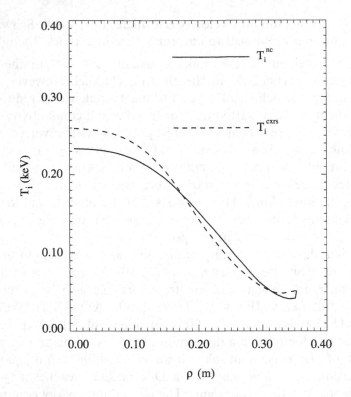

Fig. 14.1.   The measured (T_i^{cxrs}) and neoclassically predicted (T_i^{nc}) ion temperature profiles in the START tokamak.

in almost every tokamak, e.g., T-4 (Vershkov and Mirnov, 1974), Alcator
(Petrasso et al., 1986), ASDEX (Kaufmann et al., 1988; and Fussmann et
al., 1989), TEXT (Synakowski et al., 1989), PBX (Ida et al., 1989), and
TFR (TFR Group, 1982). An illustrative detailed analysis of theory versus
experiment can be found in Wenzel and Sigmar (1990).

• Closely related to impurity peaking are the neoclassical predictions for
poloidal asymmetries of the main ion and impurity densities. Up–down
asymmetries arise from parallel friction and ensuing variations of the
ion pressure within the flux surface combined with the drive by radial
gradients and plasma rotation (Rutherford, 1974; Hsu and Sigmar, 1990;
Helander, 1998; Fülöp and Helander, 1999). The last three authors extend
Rutherford's linear theory into the nonlinear neoclassical regime, giving
rise to bifurcated solutions, as mentioned in Section 13.3. These effects
have been observed in numerous tokamak experiments on Alcator A
(Terry et al., 1977), on DIII-D (Burrell and Wong, 1979), on PDX (Brau
et al., 1983), and more recently, again on DIII-D (Groebner et al., 1990)
and Alcator C-MOD (Rice et al., 1997). The latter experiment seems to

have impurity density asymmetries that are larger than the neoclassical prediction.

In addition to these observations in tokamaks, neoclassical transport predictions have also been verified in stellarators (Wobig, 1999), and in straight magnetic fields classical transport has been observed both parallel and perpendicular to the field (Burke, Maggs and Morales, 1998).

Appendix
Useful formulas

Throughout this book Standard International (SI) units are adopted, but in the research literature cgs units are still widely in use. To convert any expression from SI to cgs units, make the replacements

$$\mathbf{B} \rightarrow \frac{\mathbf{B}}{c},$$

$$\epsilon_0 \rightarrow \frac{1}{4\pi},$$

$$\mu_0 \rightarrow \frac{4\pi}{c^2}.$$

The inverse transformation is more complicated, and is described in Jackson (1975). The only departure from SI units we make is in the definition of temperature, where we suppress Boltzmann's constant, writing T instead kT. Thus, temperature is measured in units of energy (joule) instead of kelvin.

Basic definitions and natural constants

$$e = 1.602 \cdot 10^{-19} \text{ C}$$
$$\epsilon_0 = 8.854 \cdot 10^{-12} \text{ F/m}$$
$$\mu_0 = 4\pi \cdot 10^{-7} \text{ H/m}$$
$$c = \frac{1}{\sqrt{\mu_0 \epsilon_0}} = 2.998 \cdot 10^8 \text{ m/s}$$
$$m_e = 9.109 \cdot 10^{-31} \text{ kg}$$
$$m_i = 1.673 \cdot 10^{-27} \text{ kg (hydrogen)}$$
$$v_{Ta} = \sqrt{2T_a/m_a}$$

$$x_a = v/v_{Ta}$$
$$\Omega_a = e_a B/m_a$$

Collision times and frequencies

$$\hat{v}_{ab} = \frac{n_b e_a^2 e_b^2 \ln \Lambda}{4\pi \epsilon_0^2 m_a^2 v_{Ta}^3}$$

$$\tau_{ab} = \frac{3\pi^{1/2}}{4\hat{v}_{ab}}$$

$$\tau_{ei} = \frac{12\pi^{3/2}}{2^{1/2}} \frac{m_e^{1/2} T_e^{3/2} \epsilon_0^2}{n_i Z^2 e^4 \ln \Lambda}$$

$$\tau_i = 2^{1/2}\tau_{ii}$$

$$\tau_e = \tau_{ee}$$

$$v_D^{ab}(v) = \hat{v}_{ab} \frac{\phi(x_b) - G(x_b)}{x_a^3}$$

$$v_s^{ab}(v) = \hat{v}_{ab} \frac{2T_a}{T_b}\left(1 + \frac{m_b}{m_a}\right)\frac{G(x_b)}{x_a}$$

$$v_\parallel^{ab}(v) = 2\hat{v}_{ab}\frac{G(x_b)}{x_a^3}$$

$$v_{ei}(v) = \frac{\hat{v}_{ei}}{x_e^3}$$

Collision operators

$$\mathcal{L} = \frac{1}{2}\frac{\partial}{\partial \xi}(1 - \xi^2)\frac{\partial}{\partial \xi} = \frac{2hv_\parallel}{v^2}\frac{\partial}{\partial \lambda}\lambda v_\parallel \frac{\partial}{\partial \lambda}$$

$$C_{ei}(f_e) = v_{ei}(v)\left(\mathcal{L}(f_e) + \frac{m_e \mathbf{v}\cdot \mathbf{V}_i}{T_e}f_{Me}\right)$$

$$C_{ie}(f_i) = \frac{\mathbf{R}_{ei}}{m_i n_i}\cdot \frac{\partial f_i}{\partial \mathbf{v}} + \frac{m_e n_e}{m_i n_i \tau_{ei}}\frac{\partial}{\partial \mathbf{v}}\cdot \left[(\mathbf{v} - \mathbf{V}_i)f_i + \frac{T_e}{m_i}\frac{\partial f_i}{\partial \mathbf{v}}\right]$$

$$C_\alpha(f_\alpha) = \frac{1}{v^2 \tau_s}\frac{\partial}{\partial v}[(v^3 + v_c^3)f_\alpha(v)] + \frac{v_b^3}{v^3 \tau_s}\mathcal{L}f_\alpha$$

$$C_{aa}^m(f_a) = v_{aa}(v)\left(\mathcal{L}f_a + \frac{m_a \mathbf{v}\cdot \mathbf{u}}{T_a}f_{Ma}\right) \qquad \text{(model operator)}$$

$$\mathbf{u} = \int \mathbf{v}v_{aa}(v)f_a d^3v \Big/ \int v_{aa}(v)\frac{m_a v^2}{3T_a}f_{Ma}d^3v$$

Appendix

Matrix elements

$$M_{ab}^{jk} = \frac{\tau_{ab}}{n_a} \int v_{\parallel} L_j^{(3/2)}(x_a^2) C_{ab} \left[\frac{m_a v_{\parallel}}{T_a} L_k^{(3/2)}(x_a^2) f_{a0}, f_{b0} \right] d^3 v$$

$$N_{ab}^{jk} = \frac{\tau_{ab}}{n_a} \int v_{\parallel} L_j^{(3/2)}(x_a^2) C_{ab} \left[f_{a0}, \frac{m_b v_{\parallel}}{T_b} L_k^{(3/2)}(x_b^2) f_{b0} \right] d^3 v$$

$$M_{ab}^{jk} = M_{ab}^{kj}$$

$$\frac{N_{ab}^{jk}}{T_a v_{Ta}} = \frac{N_{ba}^{kj}}{T_b v_{Tb}}$$

$$M_{ab}^{j0} = -N_{ab}^{j0}$$

$$M_{ab}^{00} = -\frac{1 + y_{ab}}{(1 + x_{ab}^2)^{3/2}} = -N_{ab}^{00}$$

$$M_{ab}^{01} = -\frac{3}{2} \frac{1 + y_{ab}}{(1 + x_{ab}^2)^{5/2}} = -N_{ab}^{10}$$

$$M_{ab}^{11} = -\frac{13/4 + 4x_{ab}^2 + 15x_{ab}^4/2}{(1 + x_{ab}^2)^{5/2}}$$

$$N_{ab}^{11} = \frac{27}{4} \frac{T_a}{T_b} \frac{x_{ab}^2}{(1 + x_{ab}^2)^{5/2}}$$

$$M_{ab}^{02} = -\frac{15}{8} \frac{1 + y_{ab}}{(1 + x_{ab}^2)^{7/2}} = -x_{ba} N_{ba}^{02}$$

$$M_{ab}^{12} = -\frac{69/16 + 6x_{ab}^2 + 63x_{ab}^4/4}{(1 + x_{ab}^2)^{7/2}}$$

$$N_{ab}^{12} = \frac{225}{16} \frac{T_a}{T_b} \frac{x_{ab}^4}{(1 + x_{ab}^2)^{7/2}}$$

$$M_{ab}^{22} = -\frac{433/64 + 17x_{ab}^2 + 459x_{ab}^4/8 + 28x_{ab}^6 + 175x_{ab}^8/8}{(1 + x_{ab}^2)^{9/2}}$$

$$N_{ab}^{22} = \frac{2625}{64} \frac{T_a}{T_b} \frac{x_{ab}^4}{(1 + x_{ab}^2)^{9/2}}$$

Here $x_{ab} = v_{Tb}/v_{Ta}$ and $y_{ab} = m_a/m_b$.

Velocity-space averages

$$\{F(v)\} = \int F \frac{mv_\parallel^2}{nT} f_M d^3v = \frac{8}{3\sqrt{\pi}} \int_0^\infty F(x) e^{-x^2} x^4 dx$$

$$\{v_D^{ab}\}\tau_{ab} = \sqrt{1 + x_{ab}^2} + x_{ab}^2 \ln \frac{x_{ab}}{1 + \sqrt{1 + x_{ab}^2}}$$

$$\{x_{ab}^2 v_D^{ab}\}\tau_{ab} = \frac{1}{\sqrt{1 + x_{ab}^2}}$$

$$\{x_{ab}^4 v_D^{ab}\}\tau_{ab} = 2 \frac{1 + 5x_{ab}^2/4}{(1 + x_{ab}^2)^{3/2}}$$

In a torus with $h = B_0/B$, $B_0^2 = \langle B^2 \rangle$,

$$f_t = 1 - \frac{3}{4} \int_0^{\lambda_c} \frac{\lambda d\lambda}{\langle \sqrt{1 - \lambda/h} \rangle} \simeq 1.46 \sqrt{\epsilon}$$

$$\left\langle \int A(v) \frac{m_a v_\parallel (hv_\parallel - HV_\parallel)}{hT_a} f_{a0}\, d^3v \right\rangle = f_t n_a \{A\}$$

Fluid equations

$$\left. \frac{dn_a}{dt} \right|_a + n_a \nabla \cdot \mathbf{V}_a = 0$$

$$\left. m_a n_a \frac{d\mathbf{V}_a}{dt} \right|_a = -\nabla p_a - \nabla \cdot \pi_a + e_a n_a (\mathbf{E} + \mathbf{V}_a \times \mathbf{B}) + \mathbf{R}_a$$

$$\frac{3}{2} n_a \left. \frac{dT_a}{dt} \right|_a + p\nabla \cdot \mathbf{V}_a = -\nabla \cdot \mathbf{q}_a - \pi_a : \nabla \mathbf{V}_a + Q_a$$

Drift velocity

$$\mathbf{v}_d = \dot{\mathbf{R}}_\perp = \frac{\mathbf{E} \times \mathbf{B}}{B^2} + \frac{v_\perp^2}{2\Omega} \mathbf{b} \times \nabla \ln B + \frac{v_\parallel^2}{\Omega} \mathbf{b} \times \kappa$$

$$\kappa = (\mathbf{b} \cdot \nabla)\mathbf{b}$$

$$\mathbf{b} = \mathbf{B}/B$$

Tokamak magnetic field

$$\mathbf{B} = I(\psi)\nabla\varphi + \nabla\varphi \times \nabla\psi$$

$$\mathbf{b} \times \nabla\psi = I\mathbf{b} - RB\hat{\boldsymbol{\varphi}}$$

Neoclassical parallel viscosity coefficients

$$\langle \mathbf{B} \cdot \nabla \cdot \mathbf{\Pi}_a \rangle = 3\langle (\nabla_\parallel B)^2 \rangle \left(\mu_{a1} u_{a\theta} + \mu_{a2} \frac{2q_{a\theta}}{5p_a} \right)$$

$$\langle \mathbf{B} \cdot \nabla \cdot \mathbf{\Theta}_a \rangle = 3\langle (\nabla_\parallel B)^2 \rangle \left(\mu_{a2} u_{a\theta} + \mu_{a3} \frac{2q_{a\theta}}{5p_a} \right)$$

$$\mu_{a1} = K_{11}^a,$$

$$\mu_{a2} = K_{12}^a - \frac{5}{2} K_{11}^a,$$

$$\mu_{a3} = K_{22}^a - 5 K_{12}^a + \frac{25}{4} K_{11}^a$$

$$K_{jk}^a \simeq \frac{2p_a}{5} \{ x^{2(j+k-1)} / v_T^a(x) \} \qquad \text{(Pfirsch–Schlüter regime)}$$

$$K_{jk}^a = \frac{\pi^{1/2} p_a q R}{3 v_{Ta}} (j+k)! \qquad \text{(Plateau regime)}$$

$$K_{jk}^a = \frac{m_a n_a \langle B \rangle^2}{3 \langle (\nabla_\parallel B)^2 \rangle} \frac{f_t}{1 - f_t} \{ v_D^a x^{2(j+k-2)} \} \qquad \text{(Banana regime)}$$

Bibliography

H. Alfvén, Arkiv för matematik, astronomi och fysik **27A** (no. 22), 1 (1940).

C. Angioni and O. Sauter, Phys. Plasmas **7**, 1224 (2000).

L.A. Arzimovich, *Elementary Plasma Physics* (Blaidsell Publishing Company, New York, 1965).

R. Balescu, *Transport Processes in Plasmas* (North-Holland, Amsterdam, 1988), Vols. I and II.

P.L. Bhatnagar, E.P. Gross and M. Krook, Phys. Rev. **94**, 511 (1954).

R. Bickerton, J. Connor and J. Taylor, Nature Phys. Sci. **229**, 110 (1971).

D. Biskamp, *Nonlinear Magnetohydrodynamics* (Cambridge University Press, Cambridge, 1993).

A.H. Boozer, Phys. Fluids **24**, 1999 (1981).

A.H. Boozer, Phys. Plasmas **5**, 1647 (1998).

S.I. Braginskii, Sov. Phys. JETP **58**, 358 (1958).

S.I. Braginskii, in *Reviews of Plasma Physics*, edited by M.A. Leontovich (Consultants Bureau, New York, 1965), Vol. 1, p. 205.

K. Brau, S. Suckewer and S.K. Wong, Nucl. Fusion **23**, 1657 (1983).

A.T. Burke, J.E. Maggs and G.J. Morales, Phys. Rev. Lett. **81**, 3659 (1998).

K. Burrell, Phys. Plasmas **6**, 4418 (1999).

K.H. Burrell and S.K. Wong, Nucl. Fusion **19**, 1571 (1979).

K.H. Burrell *et al.*, Phys. Plasmas **1**, 1536 (1994).

J.D. Callen and R.A. Dory, Phys. Fluids **15**, 1523 (1972).

J.R. Cary and S.G. Shasharina, Phys. Plasmas **4**, 3323 (1997).

P.J. Catto, I.B. Bernstein and M. Tessarotto, Phys. Fluids **30**, 2784 (1987).

P.J. Catto, R.J. Hastie, I.H. Hutchinson and P. Helander, Phys. Plasmas **8**, 3334 (2001).

C.D. Challis *et al.*, Nucl. Fusion **29**, 563 (1989).

C.S. Chang and F.L. Hinton, Phys. Fluids 29, 3314 (1986).

S. Chapman, Phil. Trans. R. Soc. London Ser. A **98**, 1 (1916).

S. Chapman and T.G. Cowling, *The Mathematical Theory of Non-Uniform Gases*, 3rd ed. (Cambridge University Press, Cambridge, 1970).

H.A. Claassen, H. Gerhauser, A. Rogister and C. Yarim, Phys. Plasmas **7**, 3699 (2000).

J.W. Connor, Plasma Phys. **5**, 765 (1973).

J.W. Connor and R.J. Hastie, Nucl. Fusion **15**, 415 (1975).

J.W. Connor *et al.*, Nucl. Fusion **13**, 211 (1973).

M. Coronado and H. Wobig, Phys. Fluids **29**, 527 (1986).

S.C. Cowley, P.K. Kaw, R.S. Kelly and R.M. Kulsrud, Phys. Fluids B3, 2066 (1991).

R.O. Dendy, *Plasma Dynamics* (Oxford University Press, Oxford, 1990).

F. Engelmann and A. Nocentini, Nucl. Fusion **17**, 995 (1977).

D. Enskog, *Kinetische Theorie der Vorgänge in mässig verdünnten Gasen* (Almqvist & Wiksell, Uppsala, 1917), Inaugural dissertation, Uppsala University; Arkiv för matematik, astronomi och fysik **16**, 16, 1 (1922).

E.M. Epperlein and M.G. Haines, Phys. Fluids **29**, 1029 (1986).

D. Ernst *et al.*, Phys. Rev. Lett. **81**, 2454 (1998).

C.B. Forest *et al.*, Phys. Rev. Lett. **73**, 2444 (1994).

J.P. Freidberg, *Ideal Magnetohydrodynamics* (Plenum Press, New York and London, 1987).

T. Fülöp and P. Helander, Phys. Plasmas **6**, 3066 (1999).

G. Fussmann, *Particle Transport in Magnetically Confined Plasmas* (Institute of Physics Publishing, Bristol and Philadelphia, 2001).

G. Fussmann *et al.*, J. Nucl. Mater. **162**, 14 (1989).

A.A. Galeev and R.Z. Sagdeev, Sov. Phys. JETP **26**, 233 (1968).

A.A. Galeev and R.Z. Sagdeev, in *Reviews of Plasma Physics* Vol. 7. (Consultants Bureau, New York, 1979).

R. Giannella *et al.*, *Proceedings of the 16th European Conference on Controlled Fusion and Plasma Physics* (European Physical Society, Petit-Lancy, Switzerland, 1989), Vol. I, p. 209.

H. Goldstein, *Mechanics*, 2nd ed. (Addison-Wesley, Reading, Massachusetts, 1980).

R.J. Goldston and P.H. Rutherford, *Introduction to Plasma Physics* (Institute of Physics Publishing, London, 1995).

H. Grad, Phys. Fluids **10**, 137 (1967).

R. Groebner *et al.*, Phys. Rev. Lett. **64**, 3015 (1990).

A.B. Hassam and R.M. Kulsrud, Phys. Fluids **21**, 2271 (1978).

R.D. Hazeltine, Plasma Phys. **15**, 77 (1973).

R.D. Hazeltine, Phys. Fluids **17**, 961 (1974).

R.D. Hazeltine, Phys. Fluids B**1**, 2031 (1989).

R.D. Hazeltine and J.D. Meiss, *Plasma Confinement* (Addison-Wesley, Redwood City, 1992).

R.D. Hazeltine and F.L. Waelbroeck, *The Framework of Plasma Physics* (Perseus Books, Reading, Massachusetts, 1998).

R.D. Hazeltine and A. Ware, Plasma Phys. **20**, 673 (1978).

R.D. Hazeltine, F.L. Hinton and M.N. Rosenbluth, Phys. Fluids **16**, 1645 (1973).

P. Helander, Phys. Plasmas **5**, 3999 (1998).

P. Helander, Phys. Plasmas **7**, 2878 (2000).

F.L. Hinton, in *Handbook of Plasma Physics*, edited by M.N. Rosenbluth and R.Z. Sagdeev (North-Holland, Amsterdam, 1983), Vol. 1, p. 147.

F.L. Hinton and R.D. Hazeltine, Rev. Mod. Phys **48**, 239 (1976).

F.L. Hinton and Y.-B. Kim, Phys. Fluids B**5**, 3012 (1993).

F.L. Hinton and M.N. Rosenbluth, Plasma Phys. Control. Fusion **41**, A653 (1999).

F.L. Hinton and S.K. Wong, Phys. Fluids **28**, 3082 (1985).

S.P. Hirshman, Phys. Fluids **20**, 589 (1977).

S.P. Hirshman, Phys. Fluids **21**, 224 (1978a).

S.P. Hirshman, Phys. Fluids **21**, 1295 (1978b).

S.P. Hirshman, Nucl. Fusion **18**, 917 (1978c).

S.P. Hirshman, Phys. Fluids **23**, 1238 (1980).

S.P. Hirshman, Phys. Fluids **31**, 3150 (1988).

S.P. Hirshman and D.J. Sigmar, Phys. Fluids **19**, 1535 (1976).

S.P. Hirshman and D.J. Sigmar, Nucl. Fusion **21**, 1079 (1981).

K. Hothker *et al.*, Nucl. Fusion **34**, 1461 (1994).

W. Houlberg, S. Shaing, S. Hirshman and M. Zarnstorff, Phys. Plasmas **4**, 3230 (1997).

C.T. Hsu and D.J. Sigmar, Plasma Phys. Control Fusion **32**, 499 (1990).

C.T. Hsu, P.J. Catto and D.J. Sigmar, Phys. Fluids B**2**, 280 (1990).

K. Ida, R.J. Fonck, S. Sesnic, R.A. Hulse, B. LeBlanc and S.F. Paul, Nucl. Fusion **29**, 231 (1989).

J.D. Jackson, *Classical Electrodynamics* (Wiley, New York, 1975).

JET Team, *Plasma Physics and Controlled Nuclear Fusion Research 1990* (International Atomic Energy Agency, Vienna, 1991), Vol. I, p. 27.

JET Team, Nucl. Fusion **39**, 1743 (1999).

B.B. Kadomtsev and V.D. Shafranov, Nucl. Fusion Suppl. (1972) p. 209.

A.N. Kaufman, Phys. Fluids **3**, 630 (1960).

M. Kaufmann *et al.*, Nucl. Fusion **28**, 827 (1988).

M. Kikuchi *et al.*, Nucl. Fusion **30**, 343 (1990).

Y.B. Kim, P.H. Diamond and R.J. Groebner, Phys. Fluids **B3**, 2050 (1991).

L.M. Kovrizhnykh, Nucl. Fusion **24**, 851 (1984).

S.I. Krasheninnikov and P. Yushmanov, Phys. Plasmas **1**, 1186 (1994).

L.D. Landau, Physikalische Zeitschrift der Sowjetunion **10**, 154 (1936).

L.D. Landau and E.M. Lifshitz, *The Classical Theory of Fields*, Course of Theoretical Physics, Vol. 2 (Pergamon Press, Oxford, 1975).

L.D. Landau and E.M. Lifshitz, *Mechanics*, Course of Theoretical Physics, Vol. 1 (Pergamon Press, Oxford, 1976).

L. Lauro-Taroni *et al.*, *Proceedings of the 21st European Conference on Controlled Fusion and Plasma Physics* (European Physical Society, Petit-Lancy, Switzerland, 1994), Vol. I, p. 102.

F.M. Levinton *et al.*, Phys. Rev. Lett. **75**, 4417 (1995); E. Mazzucato *et al.*, Phys. Rev. Lett. **77**, 3145 (1995).

E.M. Lifshitz and L.P. Pitaevskii, *Kinetic theory*, Course of Theoretical Physics, Vol. 10 (Pergamon Press, Oxford, 1981).

Y.R. Lin-Liu and R.L. Miller, Phys. Plasmas **2**, 1666 (1995).

R.G. Littlejohn, J. Plasma Phys. **29**, 111 (1983).

E.K. Maschke and H. Perrin, Plasma Phys. **22**, 579 (1980).

P.J. McCarthy, Phys. Plasmas **6**, 3554 (1999).

D. Meade and the TFTR Group, *Plasma Physics and Controlled Nuclear Fusion Research, 1990* (International Atomic Energy Agency, Vienna, 1991), Vol. I, p. 9.

R.C. Morris, H.G. Haines and R.J. Hastie, Phys. Plasmas **3**, 4513 (1996).

S. Novakovskii *et al.*, Phys. Plasmas **4**, 4272 (1997).

R.D. Petrasso *et al.*, Phys. Rev. Lett. **57**, 707 (1986).

J. Rice, J. Terry, E. Marmar and F. Bombarda, Nucl. Fusion **37**, 241 (1997).

C.M. Roach *et al.*, Nucl. Fusion **41**, 11 (2001).

B.B. Robinson and I.B. Bernstein, Ann. Phys. (N.Y.) **18**, 110 (1962).

A. Rogister, Phys. Plasmas **1**, 619 (1994).

M.N. Rosenbluth and F.L. Hinton, Nucl. Fusion **36**, 55 (1996).

M.N. Rosenbluth, R.D. Hazeltine and F.L. Hinton, Phys Fluids **15**, 116 (1972).

M.N. Rosenbluth, W. McDonald and D. Judd, Phys. Rev. **107**, 1 (1957).

M.N. Rosenbluth, P.H. Rutherford, J.B. Taylor, E.A. Frieman and L.M. Kovrizhnikh, *Plasma Physics and Controlled Nuclear Fusion Research, Madison 1971* (International Atomic Energy Agency, Vienna, 1971), Vol. I, p. 495.

R. Roussel-Dupré and A.V. Gurevich, J. Geophys. Res. **101** A2, 2297 (1996).

P.H. Rutherford, Phys. Fluids **13**, 482 (1970).

P.H. Rutherford, Phys. Fluids **17**, 1782 (1974).

Y. Sakamoto *et al.*, in *Fusion Energy* (Proc. 18th International Conference, Sorrento, International Atomic Energy Agency, Vienna, 2000), paper IAEA-CN-77/EX6/04.

B. Schwarzschild, in *Physics Today*, Nov. 22 (1986).

V.D. Shafranov, in *Reviews of Plasma Physics*, edited by M.A. Leontovich (Consultants Bureau, New York, 1966), Vol. 2, p. 103.

K.C. Shaing and J.D. Callen, Phys. Fluids **26**, 1526 (1983).

K.C. Shaing, A.Y. Aydemir and R.D. Hazeltine, Phys. Plasmas **5**, 3680 (1998).

K.C. Shaing, R.D. Hazeltine and M.C. Zarnstorff, Phys. Plasmas **4** 1371 (1997); K.C. Shaing, R.D. Hazeltine, and M.C. Zarnstorff, Phys. Plasmas **4** 1375 (1997).

K.C. Shaing, M. Yokohama, M. Wakatani and C.T. Hsu, Phys. Plasmas **3**, 965 (1996).

I. Shkarofsky, T. Johnston and M. Bachynski, *The Particle Kinetics of Plasmas* (Addison-Wesley, Reading, MA, 1966).

D. Sigmar, J. Clarke, R. Neidigh and K. Vandersluis, Phys. Rev. Lett. **33**, 1976 (1974).

D.V. Sivukhin, in *Reviews of Plasma Physics*, edited by M.A. Leontovich (Consultants Bureau, New York, 1966), Vol. 4, p. 93.

E.R. Solano and R.D. Hazeltine, Phys. Plasmas **1**, 548 (1994).

L. Spitzer, *Physics of Fully Ionized Gases* (Interscience, New York, 1955).

L. Spitzer and R. Härm, Phys. Rev. **89**, 977 (1953).

E.J. Strait *et al.*, Phys. Rev. Lett. **77**, 2714 (1996).

H. Sugama and W. Horton, Phys. Plasmas **3**, 304 (1996).

E.J. Synakowski, R.D. Bengtson, A. Ouroua, A.J. Wootton and S.K. Kim, Nucl. Fusion **29**, 311 (1989).

M. Taguchi, Plasma Phys. Control Fusion **30**, 1897 (1988).

I.E. Tamm and A.D. Sakharov, in *Plasma Physics and the Problem of Controlled Thermonuclear Reactions*, edited by M.A. Leontovich (Pergamon Press, Oxford, 1961).

E. Tandberg-Hansen and A.G. Emslie, *The Physics of Solar Flares* (Cambridge University Press, 1988).

J.B. Taylor, Phys. Fluids **4**, 1142 (1961).

J.B. Taylor, Phys. Fluids **7**, 767 (1964).

J.L. Terry *et al.*, Phys. Rev. Lett **39**, 1615 (1977).

TFR Group, Nucl. Fusion **22**, 1173 (1982).

B.A. Trubnikov, in *Reviews of Plasma Physics*, edited by M.A. Leontovich (Consultants Bureau, New York, 1965), Vol. 1, p. 105.

V.A. Vershkov and S.V. Mirnov, Nucl. Fusion **14**, 383 (1974).

M. Wakatani, *Stellarator and Heliotron Devices* (Oxford University Press, 1998).

J.P. Wang and J.D. Callen, Phys. Fluids B**4**, 1139 (1992).

D.J. Ward, Plasma Phys. Control. Fusion **36**, 673 (1994).

R. Weening, Phys. Plasmas **4**, 3254 (1997).

P. Welander, Arkiv för fysik **7**, 507 (1954).

K. Wenzel and D. Sigmar, Nucl. Fusion **30**, 1117 (1990).

J. Wesson, *Tokamaks*, 2nd ed. (Oxford University Press, Oxford, 1997).

R.B. White, *Theory of Tokamak Plasmas* (North-Holland, Amsterdam, 1989).

G.B. Whitham, *Linear and Nonlinear Waves* (Wiley, New York, 1974).

H. Wobig, Plasma Phys. Control. Fusion **41**, A159 (1999).

R.C. Wolf *et al.*, in *Fusion Energy* (Proc. 18th International Conference, Sorrento, International Atomic Energy Agency, Vienna, 2000), paper IAEA-CN-77/EX4/04.

M.C. Zarnstorff and S. Prager, Phys. Rev. Lett. **53**, 454 (1984); Phys. Fluids **29**, 298 (1986).

M.C. Zarnstorff *et al.*, Phys. Rev. Lett. **60**, 1306 (1988).

M.C. Zarnstorff *et al.*, Phys. Fluids B**2**, 1852 (1990).

S.B. Zheng, A.J. Wooton and E.R. Solano, Phys. Plasmas **3**, 1176 (1996).

P. Zhu, W. Horton and H. Sugama., Phys. Plasmas **6**, 2503 (1999).

Index